THE
HIDDEN
IVIES

THE
HIDDEN
IVIES

[FIFTY TOP COLLEGES
FROM AMHERST TO WILLIAMS
THAT RIVAL THE IVY LEAGUE
2nd ed.]

Howard R. Greene, M.A., M.Ed.,
and Matthew W. Greene, Ph.D.

COLLINS
REFERENCE

SECOND EDITION

Designed by Jackie McKee

The Library of Congress has catalogued the previous edition as follows:

Library of Congress Cataloging-in-Publication Data
Greene, Howard 1937-
   The hidden ivies : thirty colleges of excellence / Howard R. Greene and Matthew W. Greene.
      p. cm.
   Includes index.
   ISBN 0-06-095362-4
      1. College choice–United States–Handbooks, manuals, etc. 2. Universities and colleges–United States–Entrance requirements–Handbooks, manuals, etc. 3. Universities and colleges–United States–Admission–Handbooks, manuals, etc. I. Greene, Matthew W., 1968- II. Title.

LB2350.5 G74 2000
378.1'61–dc21   00-043157

ISBN 978-0-06-172672-9

14 15 16 ❖/RRD 10 9 8 7

A university should be a place of light, of liberty, and of learning.

—Benjamin Disraeli

Training is everything.
The peach was once a bitter almond;
cauliflower is nothing but cabbage
with a college education.

—Mark Twain, *Pudd'nhead Wilson*

CONTENTS

ACKNOWLEDGMENTS]

One of the many benefits of having been grounded in a traditional liberal arts education is the exposure we received to many of the great social and literary thinkers like Samuel Johnson, the eighteenth-century writer.

Johnson said, "Knowledge is of two kinds. We know a subject ourselves, or we know where we can find information upon it." As counselors and writers in the educational field, we possess a fair amount of knowledge about higher education, individual colleges, and the planning process for the transition from high school to college. But there is so much that we cannot know by ourselves. Fortunately, we know where to turn for further information and insight. This book could not have been produced without the support and cooperation of the many individuals who have had much to teach us: the administrative leaders of the colleges represented in this book and the students who are the beneficiaries of the superb education provided by these institutions. In all cases, their enthusiasm for their respective colleges and the honest portrayal of their strengths and special qualities make this book meaningful. To them all we tip our faded mortarboards and thank them for giving so freely of their time and thought.

To our former editor and publisher, Diane Reverand, we can only declare our gratitude for her vision of the importance of higher education and the college admissions process, and her confidence in us to cast a brighter light on a complex and sometimes bewildering subject. We would also like to thank Bruce Nichols and Stephanie Meyers for their support in launching new editions of the Greenes' Guides at this key moment in time. And Toni Sciarra, though you handed the ball off on the eighty-yard line, you did help us march down the field during the past several years, and we thank you for your efforts and contributions.

We are delighted, as well, to thank our associate at Howard Greene and Associates, Doris Forest, who knows instinctively when it is necessary to convert into an action plan concepts and ideas that have a way of floating in space. Her ability to help us move forward in an orderly fashion and to connect with the sources of information in a timely manner has been essential to the completion of this task. We also appreciate the work of Marki Grimsley, our summer intern, whose own liberal arts education and communication and organizational skills were extremely helpful in beginning our work on this second edition.

Our mission in writing this book for students and parents is to create greater awareness of the small, distinctive cluster of colleges and universities of excellence that are available to gifted college-bound students. We would be surprised, indeed, to encounter a student or parent throughout the country or outside the United States who has not heard of the exclusive band of elite institutions known as the Ivy League. They have become, in our modern economic universe of marketing, the ultimate brand name for quality education and prestige. It is a challenge not to become intoxicated with the thought of graduating from one of these famous schools. Each year these eight institutions— Brown, Columbia, Cornell, Dartmouth, Harvard, Princeton, the University of Pennsylvania, and Yale—are flooded with applications from thousands of fully qualified individuals who, despite the fact that they have

done everything one could ask of them as students, will not be invited to enroll in any of these schools. Other universities such as Cal Tech, Duke, Georgetown, MIT, Stanford, Virginia, and the group historically known as the "Little Ivies" (including Amherst, Bowdoin, Middlebury, Swarthmore, Wesleyan, and Williams) have scaled the heights of prestige and selectivity and also turn away thousands of our best and brightest young men and women. This is the present situation, and it is one that will be with us for a number of years down the road.

Parents can prepare for their role of supporter and guide to their children as early as junior high school, since numerous surveys have revealed that they are the number-one influence in their children's choice of college. The more you are armed with information on the breadth of top college choices to aim for and what is required for admission, the better able you will be to serve your son or daughter. Students can profit by reading—in the early years of high school—about the factors that make for an outstanding college and which ones might match up best for them. If they then follow the essential steps we outline, they will better their odds of being accepted to the college of their choice.

Let's go directly to the heart of the matter: There are two dynamics at work that have combined to intensify the competition for admission to the best colleges. The number of high school graduates of exceptional ability and motivation who have set their sights on gaining the very best college education they can have is rising dramatically each year, as is the cost of receiving such an opportunity. Admissions deans at the selective colleges are shaking their heads in astonishment at the academic and other achievements of applicants to their schools. Never before have they seen so many young men and women who have attained such outstanding records in their high schools; performed at such high levels on the SAT and ACT; and garnered so many awards for their artistic, athletic, and intellectual capabilities. They worry, just as parents do, about the constant rise in tuition, fees, and room-and-board expenses. It has become an ongoing contest of increasing financial aid in order to stay ahead of the annual increases in the cost of a selective college education, which, at the top end, is now in the $40,000-to-$50,000-per-year range.

In 2008, some 1.6 million bachelor's degrees were conferred, a number expected to break 1.8 million by 2016. A great many of the best students are seeking guidance on identifying the best colleges and how to get into them. We have written this book to guide you on both

counts. Parents and students are in a continual frenzy over the preparation for and admission to a top college. The cost factor, and the perception that the prestige factor will determine the quality of life their children will achieve, have turned them into their own research-and-guidance teams who seek information from every source. *The Hidden Ivies* fills a large gap in the college admissions literature by providing the thoughtful researcher with descriptive, responsible, and insightful discussions of an exciting group of colleges and universities toward which many families are increasingly turning.

Both students and parents are greatly concerned about fulfilling the dream of attending a top college or university. As the number of outstanding candidates to the most selective schools continues to escalate, the competition for admission becomes more intense since the number of available places has not increased over time. To make matters worse, in the past several years virtually all of the Ivies and their counterparts have experienced a higher than anticipated yield—the percentage of admitted students who choose to enroll—of accepted candidates, which leads to overenrollment and drives down the number of applicants who can be accepted in subsequent years! This is, in fact, one of the virtues of the top colleges: their determination to retain their historic enrollment in order to deliver outstanding teaching and a sense of community that has much to do with the learning experience.

Every day in the offices of Howard Greene and Associates, we meet with families who are embarking on the search for the best college for their children. One of the very first questions we are asked by many parents concerns the odds of their son or daughter getting into one of the Ivies. The next question, invariably, concerns what other top colleges to consider if this is either an unrealistic or uncertain goal. As the cost of a college degree continues its relentless climb, the name recognition of an institution becomes extremely important to the educational consumer, both to certify that the benefits of such a program are worth the high sticker price and to assure acceptance to graduate school and a good job at a later date. A commonly heard refrain of concerned parents is that they have so little real knowledge of, let alone insight into, the top colleges beyond the prestigious ones. Here is the point at which thousands of ambitious students need direction to these other great institutions, as well as greater understanding of their particular styles of teaching, the areas of academic and faculty strengths, and the predominant social and learning environments.

We are not surprised by the public's lack of awareness of the many other colleges and universities that stand for excellence and may, in fact, be the most appropriate educational choice for many students. With such a vast array of public and private colleges available, the consumer can easily be overwhelmed in sorting the wheat from the chaff. Highly sophisticated marketing and image building by the majority of colleges make for quite a challenge in differentiating the great from the mediocre. What you will find here is a window into those colleges and universities that we consider to be the Hidden Ivies, those institutions that meet all the criteria we believe set the standard for excellence. Some, like Duke and Amherst, are certainly well-known in their own right. Others you may be hearing about for the first time. Any ambitious and talented high school student would do well to enjoy four years at any one of these colleges. We encourage you to launch your college search with an open mind, a spirit of adventure, and the confidence that

you will ultimately enjoy your collegiate years and be prepared to move on to a fulfilling personal and professional life. One of our students, when asked in our survey what we should have asked, said, "If this book is called *The Hidden Ivies*, you should have asked why students at these schools chose not to attend the real Ivies." We think we have answered this question through our discussion of what makes an Ivy an Ivy, the criteria for selecting a school of excellence, the value of the liberal arts education at the Hidden Ivies, and the individual college descriptions.

What makes *The Hidden Ivies* unique in the library of college guidebooks is that we bring to the reader the voices and experiences of the educational leaders as well as those of students at the institutions under review. We are gratified by the spirit of generosity on the part of both groups. They have shared with us through lengthy interviews and written responses their vision of their institution and its essential qualities. Too many guidebooks attempt to capture the definitive picture of a college through a few interviews with students or a brief visit to the campus. We believe strongly that the true nature of any educational enterprise is determined by its leaders, who are responsible for setting its goals and policies and then implementing them. Present and past students are in a position to articulate whether these goals are met by what they gained from their time on campus. We have followed into the colleges reviewed here many of the students we counseled so that we have a sense of their personal growth and satisfaction over time. This confirms our sense of what type of student and personality will thrive on a particular campus.

We believe that our professional experience in counseling college-bound students and researching colleges for more than forty years enables us to determine the essential components that create an outstanding educational experience. By identifying a cohort of fifty colleges of exceptional merit and an additional group of wonderful institutions that merit recognition, we want to help ambitious students select the best available opportunity that fits their interests, personality, learning style, and goals for their future.

The degree of enthusiasm for their respective colleges expressed both by students and the institutional leaders serves as confirmation of our selection of the colleges included in *The Hidden Ivies*. We are deeply indebted to the individuals who took precious time from their heavy schedules to cooperate in this review. The opportunities for intellectual and personal development they provide for present and future generations of gifted young men and women stands as the ultimate reward for their labors.

In our book *Inside the Top Colleges*, we describe the fundamental reasons that underlie the intense drive to enroll in the eight Ivy League and twelve of their peer colleges today and the impact on the student who actually gets to experience life on one of these campuses. To make the point, of some 700,000 freshmen entering private, not-for-profit colleges and universities in 2007, under 50,000 were enrolled in the fifty Hidden Ivies (about 7 percent). Of some 1.5 million students taking the SAT, only about 70,000 scored above 700 on the critical reading section (under 5 percent).

Despite the fact that not every talented student can gain admission to so small a segment of our vast higher educational system, the dream continues in the minds and hearts of a great many parents from the moment their children are born. Eventually it becomes the overriding ambition of tens of thousands of students as well. Success for both the parent and the child is measured often by the prestige of the college to which one is admitted. The perceived reality in our intensely competitive society is that a first-rate education from a prestigious institution confirms the quality of the diploma, and thus drives more and more people to set their sights on making it into one of the most elite institutions.

In interviewing more than 4,000 enrolled students in the top colleges and universities for *Inside the Top Colleges*, we received confirmation that one of their primary reasons for choosing to enroll was the prestige and name recognition attached to their school. They assumed that the quality of education they would receive and their prospects for their future positively correlated with the prestige factor. While we brought to light some of the negative factors that are to be found in these great universities, we did find a high overall level of student satisfaction with and appreciation for the education they received. The major benefits were the expertise of the faculty, the exceptional resources available for learning in and out of the classroom and, of equal or greater importance, the talents and energies of their fellow students. We will review some of the less desirable factors that influenced their experience, such as access to faculty and large classes, since the cluster of select colleges in this book meets these issues successfully.

Let us discuss up front a question in the minds of most families: Is a college degree worth the huge financial investment and commitment of time? The present cost of a four-year education averages $14,333 in state and $25,200 out of state at the public universities and $34,132 at the private colleges each year. The most elite private colleges, however, now have an average cost above $40,000 per year, including room and board and student fees. But keep in mind that these schools have exceptional levels of scholarship support because of their commitment to enrolling gifted students of all backgrounds. They also have the endowments to meet this goal. The phrase repeated constantly by the admissions staff, put simply, is: Ignore the sticker price and concentrate on your ability to qualify for admission, since we have such generous financial assistance for those who demonstrate need.

To return to the question of whether a family can justify the high costs of a college degree, the answer is a resounding yes!

College graduates not only earn more over the course of their lifetimes but are significantly less at risk to be unemployed. These are average data, of course, and do not account for the personal and intellectual enhancements that can affect the quality of one's life. We argue here that the benefits of attending a Hidden Ivy or a college of similar character and quality are significant in terms of teacher and peer standards, campus resources, a definite sense of campus community, and intellectual challenge. All of the factors contribute to a student's lifelong learning experience. We also make the case for the continuing value of a liberal arts education, one of the defining strengths exhibited by the Hidden Ivies.

In this new century, we will continue to be confronted with a rising tide of students who will swell college campuses. Elementary and high school enrollment has surpassed the peak level reached in 1977, the summit of the post–World War II baby boom, increasing to some 53 million students. Enrollment in grades 9 through 12 will increase by 21 percent. High school graduates have grown to over 3.1 million, and the number of college students to over

15 million. Space availability in the most selective colleges and universities will remain static at the same time that more and more aspiring scholars and future professional leaders in all fields will vie for these limited places. In a society in which access to the best career opportunities is often linked to the name factor of the college and graduate diploma, the race for admission to the very best of our colleges will continue at a pace unimagined a generation ago. Admissions deans at the top-end colleges and universities have expressed both surprise and delight at the acceleration in the numbers of applications from outstanding high school seniors, but have noted the difficulties associated with not having enough places to offer all those talented applicants.

There is, however, another side to the condition of America's universities. A report from the Carnegie Commission on Higher Education describes in negative terms the conditions of teaching and learning in the top research universities, which underscores the importance for students to have accurate and relevant information in making the right choice of college. The Carnegie study confirmed what we as professional counselors and educators have known right along since we speak with undergraduates across the country: The great majority of universities are growing their enrollments at the same time that they are in desperate need of containing or reducing the costs of operating their institutions.

Economic efficiency is all too easily translated into larger classes to maximize the time and talent of the tenured faculty, the use of graduate students who are pursuing their own studies under great pressure and, in what is fast becoming a permanent fixture on campuses, part-time instructors who spend infrequent time with their students because they typically have to hold such positions at several colleges to eke out a living. Administrators are spending much of their time seeking cost-efficient methods of teaching with modern digital technology. Going down this path involves even greater separation of teachers from their students, who find that they have less and less interaction in the classroom with the experienced and well-known faculty and certainly little or no opportunity to seek their guidance outside the classroom. Many of the best of the public universities have recognized these issues and have instituted such programs as honors colleges, residential learning programs, and special honors or critical-thinking curricula in order to emulate their smaller liberal arts brethren. They have found these special programs to be attractive to large numbers of outstanding high school students who find the combination of an elite residential learning program at state school prices too appealing to turn down. This is a trend in the right direction. However, today only a relatively small number of colleges have the financial resources, a sense of historic mission, and the determination to limit growth so that they can provide their students with the essential ingredients of an excellent education. Further on, and in our book *The Public Ivies*, we discuss the best of the public universities and the advantages they have to offer an ambitious student. Here, however, we put our focus on the small and medium-size private colleges and universities. These are the Hidden Ivies.

Inside the Top Colleges was a groundbreaking study of twenty of the most elite colleges in America in that it surveyed in great detail the experience of the students who attend them.* The knowledge of the many strengths and some of the weaknesses of student life in these major colleges will serve as a background in describing the Hidden Ivies. As we found, every institution has its strengths and its weaknesses. We thought it important to college-bound students to uncover those factors that can result in a highly positive or a potentially negative college experience.

We begin this book with a discussion of those primary factors that give an Ivy League college its level of excellence and thus its reputation, as well as those elements that could be improved upon. We then identify fifty institutions that we believe merit similar consideration. We have included our top choices of those independent, residential liberal arts colleges and small universities that we believe match up with the Ivies. Note the emphasis given to the residential nature of all fifty institutions because of the profound and lasting impact of four years spent interacting continuously with a broad variety of students and faculty who are present to teach and advise. The further students get into their college years, the greater the value they place on the interaction with their peers and teachers. This, rather than location, enrollment, nature of the food and dormitories, and the availability of alcohol, as examples, will determine their affective and intellectual development and level of satisfaction.

The colleges and universities we have selected for inclusion all provide a shared community experience, a virtual living/learning laboratory for

* The twenty colleges and universities included in *Inside the Top Colleges* (formerly titled *The Select*) are: Brown; University of California, Berkeley; the University of Chicago; Columbia; Cornell; Dartmouth; Duke; Georgetown; Harvard; Johns Hopkins; MIT; the University of North Carolina, Chapel Hill; Northwestern; the University of Pennsylvania; Princeton; Stanford; Wesleyan; Williams; the University of Wisconsin, Madison; and Yale.

intelligent young men and women at a most impressionable stage of their lives. They are likely to acquire the following critical skills or instincts: cooperation, leadership, sharing, deferring appropriately to the needs and sensitivities of others, social and intellectual discourse, conflict resolution, collaboration, mentoring, self-reliance, coping with competition, independence, control of one's emotions, delayed gratification, organizing and planning, finding one's place in the immediate and the larger community, appreciating religious and cultural differences, making decisions that count, not procrastinating, not blaming others for one's actions, and learning the truth that intelligence without character, personal integrity, and a working set of values can be a dangerous thing.

We hasten to add that this is not an exhaustive list. There are many outstanding colleges and universities that could have been included. However, we have cause to place these particular institutions in the top group. We recognize in the appendices another cluster of outstanding colleges and universities for their benefit to students in search of a great educational experience. We know from our many years of experience as counselors how valuable this guidance can be for concerned students and their parents. Just as "Ivy bound" students must broaden their college search to include some of the Hidden Ivies and other schools of excellence, those applying with confidence to the Hidden Ivies must expand their view of potential colleges and universities that will meet their aspirations. We hope that our discussion of the Hidden Ivies and the criteria that make them top colleges will serve as a template for students and parents for evaluating the many other options of higher learning available to them.

In our first edition of *The Hidden Ivies*, we included thirty colleges and universities. For this updated, second edition, we sought to include additional colleges that might be less well-known to readers, as well as to make consistent the overall group by including the rest of the top colleges originally left out of the book. *Inside the Top Colleges* was written in the late 1990s to provide insights about a selection of public and private colleges and universities. When we came to write *The Hidden Ivies*'s first edition in 2000, we therefore left out of that book such schools as Williams and Duke that we had covered in *Inside the Top Colleges*. For this new edition of *The Hidden Ivies*, we wanted to bring to you a consistent group of the top liberal arts colleges and universities of excellence not part of the Ivy League. Thus, you will see some very recognizable names herein, which suggests that these institutions are by no means "hidden." Similarly, you will note that they and others in the book are as competitive to get into and as challenging to succeed in as the Ivies. Nevertheless, we believe that, as a group, these fifty colleges offer every good student a range of diverse opportunities to consider.

We hope also that this book will prove of particular value to the increasing number of college-bound students of color and from nontraditional backgrounds. Students of African American, Hispanic, Native American, and Asian heritage now represent more than a quarter of the four-year college population. That proportion will continue to grow during the next decade. These young men and women, like all ambitious students, deserve the opportunity to know which educational direction they can take to ensure a future of security and meaningful engagement in the affairs of our country. As affirmative-action policies are challenged and even dismantled in many of our great public universities, many of the private colleges we discuss in *The Hidden Ivies* continue to be proactive in offering disadvantaged students access to excellent educational opportunities. Their financial strength is a vital source of support for any talented, disadvantaged candidate regardless of racial or ethnic background. We heartily endorse their commitment and would like them to be identified through this guide.

We attempt in chapter 1 to educate college-bound students on the value of acquiring a liberal arts education. While it appears a contradiction in terms at first glance, exposure to the full range of intellectual disciplines—from the social sciences, humanities, and the arts to sciences and mathematics—is vital for a successful professional future when the prevailing notion is that a college education is solely a training ground to prepare graduates for employment. From the constant flow of questions from parents and students alike, we recognize the importance of explaining the value—personal, intellectual, and professional—of spending four critical undergraduate years learning how to think critically; read and analyze advanced ideas and concepts; think for oneself by developing the art of disciplined evaluation; and discover the pure joys of art, music, literature, science, and the many other fields one will encounter in a first-rate liberal arts college. The emphasis on grounding their education with an ethical and moral awareness is implicit in the teaching of a majority of the faculty. It is not a mere coincidence that all of the private colleges selected as Hidden Ivies offer solely or primarily a liberal arts education at the undergraduate level. This is true of all of the Ivy League institutions, to make our point.

In chapter 2 we detail the criteria that we believe make a college or university an outstanding institution. Careful attention is given to each college's mission statement because this is the vision or blueprint, if you will, by which the leaders decide on the programs, policies, and content of the curriculum and nature of teaching. You will learn how so many institutions of excellence can define their environments and learning approaches in differing ways. We describe common elements, such as the ratio of the size of the endowment to that of the student body, and what this means for educational excellence, the educational attainments of the faculty, the physical facilities for learning (library, science, visual and performing arts, and athletic programs and facilities) in relation to the student population, the resources for advising and learning support, graduate school admission in all of the major disciplines, and the response by the administration to student needs. We share with the reader input from present undergraduates on those elements that most determine the quality of experience in the individual college descriptions.

We then provide in chapter 3 a summary of the overall character of each of our cluster of Hidden Ivies as gleaned both from the input of the administrators and the students surveyed. We also take advantage of our many years of experience with students attending these colleges and an awareness of what kind of individual would function best in each case. Information on cost, requirements for admission, and special qualities or talents that are sought by the admissions committee are included. This is not intended to be still another standard guidebook or rankings list, filled with mere facts and information provided mainly by the institutions themselves. The descriptions address the unique personality of each college, primarily through the voices of the students, whom we believe speak for a majority of the student body, and the key administrators who generously shared their views and expertise with us. The colleges will be discussed in terms of the criteria we have determined are of paramount importance in judging an institution of higher education.

In gathering the data for *Inside the Top Colleges*, a detailed questionnaire was completed by students on all twenty of the elite campuses. This time, however, we surveyed the college leaders of each school as well, from the president to the deans of faculty and admissions, the provost, and the professional support staff. A sample questionnaire that students and administrators completed is included in the appendices. Many college counselors who have years of experience in guiding students to the top colleges also shared with us their recommendations for the Hidden Ivies list. As we reviewed and expanded our list of schools for this new edition, and updated the profiles of those colleges and universities we introduced in the first, we were gratified to see these institutions' continued growth and development. We were also taken aback by the incredible rise in their endowments, application numbers, and class profiles, accompanied by greater selectivity, higher applicant yields, and more unpredictability in admissions. As we discuss in *Making It into a Top College*, selective college admissions has become only more challenging these days, and this is particularly true at the Hidden Ivies. As you explore these colleges through this book and then on their Web sites and campus tours, we know you will be as impressed as we have been by what each of these institutions has to offer.

Few studies of colleges and universities have directly asked both the student, who is the primary consumer, and the educational leaders responsible for providing the education specific questions regarding the nature and quality of their experience other than superficial and anecdotal queries. We believe that we have garnered important information and insights that will prove of great value to college-bound students, their parents, counselors who are engaged in guiding their young charges, and the college leaders and teachers who run these institutions. There is no other book in the marketplace that focuses exclusively on those colleges, public and private, that are equal to or almost as impressive in all respects as the Ivy institutions.

In chapter 4 we conclude *The Hidden Ivies* with our list of key recommendations for successful admissions as well as advice for the inexperienced parent and college-bound student from those who have recently gone through the college search process. We detail tips on what they did to assist their children, what mistakes they would now avoid, what they have found makes for a right or wrong choice of school, and what misperceptions they had to which they would alert you. We encourage you to keep in mind that all of the colleges of excellence included here will lead you to where it is you think you want to go. It is the choice of routes and the nature and style of travel you prefer that will differ. This is what we will help you to discover. So let us begin our journey through the land of the Hidden Ivies.

ONE

Every session we have with parents and students on college planning inevitably leads to this question: "Why should my child go to a liberal arts college? How will it prepare him/her for a job or career? Why not jump right into a specialized major?" In this chapter, we answer that question. We have found that there is a great deal of misunderstanding and a general lack of information about the nature of a liberal arts education, its value in society, and its role in preparing students for graduate programs and careers. Many students are unaware of the differences between a "college" and a "university," between a graduate and an undergraduate education and degree, and between degree programs specializing in technical, business, arts, or other fields and those offering or demanding study across the arts and sciences disciplines.

We defend the value of a liberal arts education, building on the work of many prominent scholars who have argued that an education spanning multiple academic disciplines and re-

quiring that students learn core concepts, methods, and content builds unparalleled strengths in reasoning, understanding, and communication, preparing students for any academic or professional challenge they may choose. At the same time, we avoid for the most part the disagreements among "liberal" and "conservative" thinkers on necessary reforms in liberal arts education and on whether changes over the past several decades have been "good" or "bad." However, we must add that in our minds, liberal arts colleges have changed for the better and significant choice exists among these institutions to give students a great deal of leeway in determining which curriculum and environment best suits their needs and interests.

Goals of a Liberal Arts Education

As Nathan Glazer has stated, "Liberal education has meant many things, but at its core is the idea of the kind of education that a free citizen of a society needs to participate in it effectively."* In a complex, shifting world, it is essential to develop a high degree of intellectual literacy and critical-thinking skills; a sense of moral and ethical responsibility to one's community; and the ability to reason clearly, to think rationally, to analyze information intelligently, to respond to people in a compassionate and fair way, to continue learning new information and concepts over a lifetime, to appreciate and gain pleasure from the beauty of the arts and literature and to use these as an inspiration and a solace when needed, to revert to our historical past for lessons that will help shape the future intelligently and avoid unnecessary mistakes, and to create a sense of self-esteem that comes from personal accomplishments and challenges met with success.

- Think and problem-solve in a creative, risk-taking manner.

- Express ideas and feelings in organized, logical, coherent, descriptive, rich language, both orally and in writing.

- Analyze, organize, and use data for meaningful solutions.

- Develop the capability of setting goals with appropriate information and research and then achieve these goals with proper means.

- Help define a personal-value and ethical system that serves throughout life in making the challenging decisions one will face.

- Have the capacity and instinct to work in a cooperative, collaborative manner with others in one's professional and community life.

These are ambitious goals! How different colleges and universities achieve them reveals variations in educational philosophy, institutional personality and history, and particular social and academic strengths and missions. All the liberal arts colleges share a commitment to disciplinary and student diversity, intellectual and otherwise. To varying degrees, these colleges require students to pursue courses in key academic subject areas, some with more specificity, others with a great deal of freedom, in order to expose students to multiple areas of knowledge, diverging perspectives on the world, and different paths to scientific, ethical, social, and humanistic understanding.

Content is important, but so are process and style. Liberal arts colleges may expect students to master a core body of knowledge, including Western and, increasingly, non-Western masterworks in fields ranging from physics to music to government, comparative literature, history, and language. Students will build on their secondary school education by majoring in one or more specific areas of knowledge (academic "disciplines" or "fields"), but will pursue areas of interest within key general academic areas: the physical sciences; mathematics; the humanities (history, English and other languages, visual and performing arts, and so forth); and the social sciences (political science, sociology, etc.). So-called cross-disciplinary courses of study are offered in such areas as women's studies, African Ameri-

*Glazer, Nathan. "After the Culture Wars," *New York Times Review of Books*, 7/26/98:6.

can studies, environmental studies, and social psychology. Students will be exposed to a wide range of subjects that they may not have encountered previously: anthropology, genetics, philosophy, criminology, economics, engineering sciences, religion, education. But all of this will be in the context of a broad-based approach to learning. One cannot graduate from a liberal arts college without having experienced course work in a multitude of subject areas. The goal: an intelligent and "well educated" student who can converse knowledgeably about a wide range of topics and who has learned how to learn about anything under the sun.

Thus, process and style undergird a liberal arts education. Students learn how to think, approach problems, write, present information intelligibly, and make coherent arguments in their field of choice and others they may encounter. A liberal arts education challenges students' conceptions and pushes them to ask difficult questions, question established answers, and develop their own arguments through logical reasoning and the discovery of new understandings. A liberal arts education helps a student specialize in at least one particular area, but also to see and make connections among multiple fields of inquiry. As Ernest Boyer, a former president of the Carnegie Foundation for the Advancement of Teaching, has argued, traditional research designed to promote the advancement of knowledge should be complemented by the "scholarship of integration," which makes connections across disciplines; the "scholarship of application,"

which concentrates on the interrelationship between theory and practice; and the "scholarship of teaching," which both educates students and attracts them to the academic world. Such a view of scholarship clearly relates to the goals of a liberal arts education.

Learning How to Learn: The Luxury of Time

Alan Ryan writes, "At its best, liberal education opens a conversation between ourselves and the immortal dead, gives us voices at our shoulders asking us to think again and try harder."* How many of us, academics or not, would not relish the notion of taking four years of our lives to keep open that conversation, enjoying what we call the luxury of time, to think, to make connections, to question, and to learn? Howard Bowen, a highly respected teacher/researcher of higher education, writes in an early important study:

> As compared with others, college-educated people on the average are more open-minded toward new ideas, more curious, more adventurous in confronting new questions and problems, and more open to experience. They are likely to be more rational in their approach to issues. They are more aware of diversity of opinions and outlooks, of the legitimacy of disagreement. . . . They are less authoritarian, less prejudiced, and less dogmatic. At the same time, they are more independent and autonomous in their views, more self-confident and more ready to disagree. They are more cosmopolitan.[†]

---

* Ryan, Alan. *Liberal Anxieties and Liberal Education* (New York: Hill & Wang, 1998). Quoted in Glazer, Nathan. "After the Culture Wars," *New York Times Review of Books*, 7/26/98:6.

[†] Bowen, Howard R. *Investment in Learning: The Individual and Social Value of American Higher Education* (Johns Hopkins University Press, reprint edition, January 1977).

Bowen's point is that attempts to measure the value-added benefit of a college education often consider the wrong issue. The major benefit of a liberal arts education is that it will produce the kinds of educated leaders that will benefit our economic, political, social, and family lives.

John Wooden, the great basketball coach at UCLA, once quipped, "It's what you know after you learn everything that counts." A liberal arts education, particularly one obtained in a residential college setting, seeks to provide that learning experience and that sense of knowing a lot, but also knowing what you do not know. "We go to college," the poet and teacher Robert Frost said, "to be given one more chance to learn to read in case we haven't learned in high school. Once we have learned to read, the rest can be trusted to add itself to us." As with the majority of his observations and commentary, Frost is full of irony in his view of the purpose of the liberal arts experience. His message, as true today as it was many years ago (when Frost was at Dartmouth and Amherst), is that reading intelligently and analytically, with a critical mental eye, will enable one to carry on his/her education for the remainder of a lifetime. In a sense, once one has "learned to read" in its broadest meaning, he/she is prepared to go out into the world. It is the definition of reading that counts.

Ernest Martin Hopkins, a former president of Dartmouth, characterized in his 1929 convocation address to the entering class the view of the essential qualities of the liberal arts college: "The liberal arts college is interested in the wholeness of life and in all human activity. . . . It is characterized as liberal because it recognizes no master to its limit to seek knowledge and no boundaries beyond which it has not the right to search. Its primary concern is not with what men and women shall do but with what they shall be." The liberal arts education is the means by which outstanding young men and women will develop those skills and qualities of mind and spirit that will enable them to lead productive and valuable lives. This means not only for their own well-being, but also for the good of their families, communities, and the larger society. The colleges we have selected are among the leaders in higher education in preparing young adults to take their places as responsible and enlightened leaders in the world. To think critically and with a conscience, to be resilient in an accelerating world of technological, intellectual, cultural, and social change, are critical skills for the individual and society's maintenance.

AN EXAMPLE OF THE LIBERAL ARTS IN ACTION

Pomona College's faculty has a ten-point list of what skills it teaches to help students succeed in later life through exposure to the arts and sciences curriculum. The college educates students to:

1. Read literature critically.
2. Use and understand the scientific method.
3. Use and understand formal reasoning.
4. Understand and analyze data.
5. Analyze creative art critically.
6. Perform or produce creative art.
7. Explore and understand human behavior.
8. Explore and understand a historical culture.
9. Compare and contrast contemporary cultures.
10. Think critically about values and rationality.

Trends in the Liberal Arts over Time

The liberal arts colleges in America are dynamic institutions, constantly if slowly evolving in reaction to their environment; the demands of students, parents, graduates, policy makers, and others; and the leadership of administrators and faculty. Important changes have taken place in the liberal arts and in postsecondary education over the past several decades, including diversification of staff, student body, and curriculum; provision of financial aid; accentuation of the continuing struggle between teaching and research; expansion of educational access; and pursuit of graduate degrees. We discuss these trends and others on the following pages.

Inclusion

Perhaps one of the most powerful changes in liberal arts colleges since the 1950s has been the expansion of educational opportunities for nonwhite, non-Protestant, non-wealthy, and non-male students. The policy of exclusion at the elite institutions based on racial, ethnic, religious, and socioeconomic factors is clearly a thing of the past. These colleges have redefined their missions and have worked assiduously to attract, retain, and assist a student body that better reflects American society.

In addition to doing away with formal, public, or hidden policies of exclusion, selective colleges have actively tried both to change their image and broaden their appeal and impact. They recruit students from nontraditional backgrounds by visiting their schools and writing to them. They hire special admissions officers of color, put together targeted informational material, work with the College Board to identify talented test-takers, and form close associa-

tions with organizations like A Better Chance, Questbridge, and Prep for Prep, which help students of color succeed and go on to college. Colleges have boosted financial aid resources and expanded efforts of alumni to identify appropriate candidates. They have hired multicultural advisors and staffed centers on campus to encourage tolerance and diversity and to support each student's needs. Of course, this is a continuing process, and colleges, not to mention American society, are learning about what it takes to promote and maintain diversity in a way that is successful for everyone. But a major shift in thinking has occurred so that not only do colleges prohibit and discourage exclusion, they see inclusion and diversity as essential elements of their educational missions.

Liberal arts colleges have also continued to diversify their faculty, who serve as role models and offer differing voices to expose students to multiple perspectives. In the past, class snobbery extended to the college faculty as well as to the students. Very few teachers were not white, male, Protestant, and educated from the same small band of institutions. Today, liberal arts colleges certainly cross-fertilize each other's faculties, but representation of multiple groups and viewpoints on the faculty has dramatically increased.

Not only are the faculty more diverse at the liberal arts colleges, but the courses they teach have broadened and fragmented. At most liberal arts colleges today, students can access courses in new areas of scholarship, including African American studies, women's studies (gender studies), environmental studies, non-Western literature, ethics and science, and so forth. The "core curriculum" of the great works of Western literature, history, philosophy, mathematics, sciences, and the arts has splintered and branched. Few colleges have completely dismantled the "core." Most have found a middle path combining "traditional" scholarship with exposure to new and alternative ("critical," "postmodern," "non-Western") fields of study.

Finally, we should mention the prominent role of athletics as a means of recruitment among disadvantaged and nontraditional groups over time. This was one of the first vehicles by which colleges identified talented students in nontraditional environments. While the Ivies still do not offer athletic scholarships, their need-based financial aid has allowed them to attract and enroll "scholar athletes" regardless of ability to pay. The fact that at many selective liberal arts colleges some 50 percent of students are receiving some sort of financial aid indicates their willingness to promote socioeconomic diversity. Many potential students and families may still react negatively to the cost of private liberal arts colleges and the image they may still carry of being snobbish, elitist places. However, while these colleges do not have the natural diversity that many state universities likely attract, they have pursued diversity through active recruitment of various student populations. And the public universities have launched similarly active programs to attract students of color and other underrepresented groups.

Liberal Arts Colleges: A Small but Important Sector

Only 4 percent of today's collegiate student body is enrolled in residential liberal arts colleges, yet these institutions remain a vital force in setting high standards of teaching and intellectual rigor. Many larger institutions have modeled their undergraduate liberal arts programs on the traditional residential college model. What also appears to be true is that the small band of such colleges carries a dispropor-

tionate influence in higher education. A major Carnegie report on teaching, as well as many prominent educators, urge the large universities to follow the residential colleges' emphasis on intellectual standards and teaching and interaction with students. The growth in honors programs and residential colleges within larger universities testifies to this fact. (See collegiateway.org for more on residential colleges.)

What drives the degree of excellence of education at the liberal arts colleges is the caliber and commitment of the faculty. Endowment building and alumni support are the keys to creating and sustaining a first-rate faculty. This is the essence of the commitment of these colleges. One does not find an inexperienced graduate teaching assistant or an adjunct instructor, or a large lecture hall outfitted with a large video screen. An education in the liberal arts fits hand in glove with a teaching style that emphasizes faculty-student interaction; discussion-based learning; and opportunities to practice writing, speaking, and becoming involved.

There is a natural tension between the more career- and vocationally directed curriculum offered at the majority of colleges today, particularly in the public colleges and universities, and the more general liberal arts–based programs that emphasize a coherent intellectual experience for students. The latter group aims to provide exposure to the full range of disciplines—the humanities, the sciences, the arts, the social sciences, and languages—that have influenced our development as a democratic nation and people. We will discuss below the important skills for success in virtually all endeavors of work and living. It is noteworthy that the best of the career-oriented programs model the liberal arts curriculum to the extent of requiring students to take courses in the humanistic tradition.

The Success of the Liberal Arts: Why Does Such an Education Work?

What is it about a liberal arts education that makes it successful, and what is the definition of success? What are the advantages of the liberal arts "experience"? Here, we will discuss the other factors that make a liberal arts education work.

Positive Effects of the Residential College Experience

It is difficult to extricate liberal arts education from its traditional American home base, the residential college. We are speaking here, in general, of the classic small to midsize colleges (1,500 to 5,000 undergraduate students) that offer a liberal arts curriculum through a four-year program leading to the bachelor of arts (B.A.) degree. And we are talking more specifically about the Hidden Ivies, the select group of colleges we are recommending. They have in common with the Ivy League colleges (see chapter 2) the advantages of large financial endowments relative to their enrollment, very high admissions selectivity, a topflight faculty, a large administrative/student personnel support team, excellent facilities for academic and residential life, a strong ethos of community and one's place within it, and very high rates of graduation within four or five years. For graduates, there is a high percentage of placement in graduate schools, fellowships won, eventual completion of masters and doctoral degrees, and support from alumni. In researching the most selective colleges, we have found that particular schools like Dartmouth, Williams, and Princeton had extremely positive ratings by students of faculty, sense of

community, school spirit, support of adult mentors and staff, peer interaction and influence, rate of graduation, and graduate school admission.* The positive effects of all of these interacting factors contribute to student happiness, learning, and success.

Factors that Make a Liberal Arts Education Work

These are some of the most important factors that lead to student success during and after a liberal arts education:

- **Faculty commitment to teaching and interacting with students.** There is a clear emphasis on the value of the undergraduate, the student pursuing the B.A. degree, as opposed to the graduate, the student pursuing the M.A., Ph.D., or other advanced degree. In part, this is because graduate students are found mainly on the campuses of universities rather than colleges. Faculty at the freestanding liberal arts colleges and smaller universities with strongly developed "colleges of liberal arts" (or "arts and sciences") tend to self-select. In other words, they choose to serve at these institutions because they are committed to teaching, mentoring, and interacting with their undergraduate students.

- **Opportunities to participate in research projects.** Since there are no or few graduate students present, undergraduate students often have the chance to work with their full-time professors on challenging research, writing, and internship experiences.

Inside the Top Colleges.

- **Intimacy of and opportunity for peer interaction.** Residential living and easy participation in campus organizations—athletic, social, political, journalistic, musical, academic—foster community connections and student-to-student learning and engagement. Since most students live on campus in dormitories, particularly during the early college years, they forge bonds with one another that last a lifetime, and they continue to learn from one another informally during "downtime" and even during formal, residentially based mentoring and faculty seminars.

- **Opportunities for leadership.** Due to the smaller size of the colleges and their emphasis on involvement, students can develop their leadership skills in community service, class leadership, sports captaining, club presidencies, and so forth. They can lead a *balanced* life on campus, combining academics, extracurricular involvement, and social life. They can identify themselves in their college community by taking on the mantle of leadership in one area or another.

- **A broader exposure to activities and experiences.** Due to the soft boundaries in small to medium-size institutions, whether in academic disciplines or extracurricular life, students encounter multiple learning opportunities, pushing them to expand their perspective and to take on new roles.

- **Accessible support services.** Students find it easier to get academic, financial, personal, and career counseling in a more intimate setting.

- **An historic and articulated sense of the college.** This is conveyed through the classroom, the dormitory, and the activities available on campus. Students gain an understanding of the mission of the school, whether through the emphasis on the honor code at Haverford College or a commitment to a core liberal arts curriculum at Reed College. A sense of identity, individually defined by each student but broadly shared by the college community, permeates college life and stays with students when they leave.

- **A strong commitment to developing essential skills.** An emphasis on promoting rigorous critical thinking, analytical abilities, and communication skills, especially writing, inspires the entire curriculum.

- **Synthesis of teaching and research.** Faculty are required to combine teaching with scholarship in order to be up-to-date in their fields and serve as models for students. The teaching and research processes inform one another as faculty learn from students and combine historical knowledge of their field with current findings and theories.

- **Small classes.** Fewer students in each class and a lower student–full-time-faculty ratio, on the whole, provide more opportunity for discussion, questions, and collaborative student projects. There is a difference between listening to a lecture twice weekly by a professor in a class of 500 students, reinforced by one class a week of forty students taught by a graduate teaching assistant (the model at the larger public universities for many of the introductory and even second-level courses), and learning in a class of fifteen to twenty-five, or even forty to seventy-five, taught by a full-time professor in a more discussion-oriented or Socratic format (the model at the smaller liberal arts colleges). In fact, the larger universities have instituted "critical thinking" and "honors" programs, offering more challenging courses taught by professors in a seminar style, in order to emulate the format of the smaller colleges. Educators recognize the value of addressing students' multiple learning styles in collaborative, more personalized, and more intellectually stimulating ways.

- **Less emphasis on multiple-choice exams.** With the classroom format and teaching style discussed above, the liberal arts colleges can more easily conduct the examination of student knowledge in ways other than the multiple-choice formats so necessary to test large numbers of students. And when professors give essay examinations or multiple writing assignments, they themselves do the grading rather than assigning the task to a graduate-student assistant. Undergraduates thus receive more informed feedback from professors and are able to express themselves more fully and subtly in their answers. Learning the writing and thinking skills necessary to make coherent arguments is a key facet of the examination process in the liberal arts college.

- **Engagement in the classroom and on campus.** Students have opportunities for engagement on many fronts. The peer

influence that raises everyone's personal expectations results in a very high rate of graduation, which varies from 85 to 97 percent of each class. This is extraordinary when you consider that the national rate of graduation from college over a six-year period now averages 54 percent! Even at the most selective large universities, the rate of graduation in a normal time frame is only 65 to 75 percent.

Successful Graduate School Admission

As the B.A. or B.S. degree has become more common in society, the M.A., M.B.A., Ph.D., J.D., M.S.W., M.S., M.D., or other graduate degree has become even more important in distinguishing individual accomplishments and abilities. There is impressive data to reinforce the advantage of choosing the right college on an individual basis, looking for the college that will provide the best foundational and formative educational experience where students can pursue the right courses in their areas of interest. Most families are naive in thinking that one has to attend an Ivy college in order to qualify for the best graduate schools. The liberal arts colleges we review here and many others have equally distinguished records in training and preparing their undergraduates for professional and academic graduate school programs. Let us offer the following as an example in the disciplines of science and engineering for students who attain a Ph.D.: The following Hidden Ivies—Carleton, Grinnell, Haverford, Oberlin, Pomona, Swarthmore, and Reed—on the basis of the number of students per 100 enrolled, send more of their undergraduates on to Ph.D.'s than all of the large and distinguished research universities, with the exception of MIT, and at an equal or better ratio than Columbia, Yale, or Harvard. Of course, the latter group's total numbers are much higher. Other liberal arts schools like Bryn Mawr, Amherst, and Williams also have a remarkable record of preparing students for doctoral work in science and technology. Some of the midsize universities profiled here, such as Vanderbilt, Johns Hopkins, and Washington University, very much rival the Ivies in their preparation of students for graduate work and undergraduate research opportunities on campus.

We often talk about attending a top liberal arts college as "not closing any doors." In fact, attending such a college can open many previously unknown doors. Even in an area as competitive and well-known as medicine, liberal arts colleges do an outstanding job of sending their graduates on to become M.D.'s. These students often do not major in a hard science. The Association of American Medical Colleges (AAMC) reports that of all students in 2007 who entered medical school, some 5,000 of a total of almost 18,000 first-year students reported majoring in a nonscience subject. For example, 725 humanities majors entered medical school. That compares to almost 10,000 biological science majors. And 1,931 social science majors were enrolled. The AAMC's guide explains some admission goals this way:

Medicine demands superior personal attributes of its students and practitioners. Integrity and responsibility assume major importance in the classroom and research laboratory as well as in relationships with patients and colleagues. Medical schools also look for *evidence of other traits such as leadership, social maturity, purpose, motivation, initiative, curiosity, common sense, perseverance, and breadth of interests.* Anyone considering a career as a physician must be able to relate to people effectively. The increasing emphasis on a team approach to medical care adds another dimension to the need for this skill. Because of the demanding nature of both the training for and the practice of medicine, motivation is perhaps the most salient nonintellectual trait sought by most admission committees.*

STRONG MEDICAL COLLEGE PREPARATION

These Hidden Ivies were among the top producers of applicants to U.S. medical schools in 2008. Their company on the list includes the large public research universities and the Ivies. The number of applicants is listed in parentheses.

Duke (372)
Johns Hopkins (352)
Northwestern (339)
Notre Dame (338)
Washington (308)
Stanford (304)
Emory (300)
Southern California (244)
Vanderbilt (215)
Tulane (188)
Rice (175)
Boston College (164)
Georgetown (149)
Chicago (146)

*Association of American Medical Colleges. "Medical School Admission Requirements: United States and Canada," 49th edition. 1998 (Washington, D.C.: AAMC): 22–23. Italics added.

Tufts (145)

Rochester (140)

Sound familiar? Some of these traits sought by medical school admission committees are exactly those fostered by the top liberal arts colleges, where students can pursue sciences course work and research complemented by broad study in areas of the humanities and social sciences.

Some other reasons for liberal arts colleges' successful preparation for graduate admission are: the close-at-hand model of faculty as teachers, researchers, and mentors with a passion for their disciplines; the full use of all campus facilities by undergraduates, since they are not second-class citizens vis-à-vis graduate and postdoctoral students; the greater opportunity for cross-disciplinary studies, given the flexibility of faculty and curricula; more funding for internships and fellowships at the undergraduate level; and stronger guidance from teachers concerning graduate schools since they know their students well. We can sum up all of this by noting that the undergraduate is the chief reason for the existence of the college.

Writing in 1987, Marian Salzman and Nancy Better noted the shift in thinking by corporate America toward the liberal arts graduate's skills in adapting, relating, thinking, and "learning how to learn." They offer a wonderful quote from Roger B. Smith, speaking in 1984 as chairman of General Motors:

> There is almost no phase of business life that can be successfully conducted without the benefit of humanistic values and insights. . . . [S]tudying the humanities gives you a sense of perspective. In discussions of corporate management nowadays, we often find the word "stewardship"—and with good reason. Certainly, a corporation must serve the interest of its customers, employees, and shareholders. But that alone is not enough. In the cities and towns where its facilities are located, it must also act with a sense of responsibility for the natural environment, the economic health of the country, and ultimately, the welfare of future generations. People who have studied history and philosophy find it easy to maintain such a broad perspective. They know that what a corporation does can have moral implications that may reach far beyond the making of goods and the earning of profits. . . . [To] thrive—even to survive—you have to be able to envision new things, as well as new ways of doing old things. You have to extrapolate on the basis of what worked in the past. You have to be able to organize and reorganize your operations to achieve economy and eliminate redundancy. And you need the power to imagine how the course of events might be changed, and by what kinds of intervention.*

In the information- and service-driven economy of the new century, the skills of the creative learner in utilizing new sources of information in new ways seem even more valuable to employers, public and nonprofit agencies, and graduate programs.

Multiple Approaches to Multiple Intelligences

Emotional intelligence, emotional competencies, and multiple intelligences are some of

---

*Salzman, Marian L., and Nancy Marx Better. *Wanted: Liberal Arts Graduates* (New York: Anchor Press, 1987): 9–10.

the concepts now being discussed as alternative visions of human intelligence. Psychologist Howard Gardner has identified seven distinct types of intelligence, from musical to interpersonal, and continues to refine his conceptual approach to the ways in which we all learn.

HOWARD GARDNER'S SEVEN TYPES OF
INTELLIGENCE WITH RESPECT TO
GIFTED/TALENTED CHILDREN

- Linguistic—Children with this kind of intelligence enjoy writing, reading, telling stories, or doing crossword puzzles.

- Logical-Mathematical—Children with lots of logical intelligence are interested in patterns, categories, and relationships. They are drawn to arithmetic problems, strategy games, and experiments.

- Bodily-Kinesthetic—These kids process knowledge through bodily sensations. They are often athletic, dancers, or good at crafts such as sewing or woodworking.

- Spatial—These children think in images and pictures. They may be fascinated with mazes or jigsaw puzzles or spend free time drawing, building with Legos, or daydreaming.

- Musical—Musical children are always singing or drumming to themselves. They are usually quite aware of sounds others may miss. These kids are often discriminating listeners.

- Interpersonal—Children who are leaders among their peers, who are good at communicating, and who seem to understand others' feelings and motives possess interpersonal intelligence.

- Intrapersonal—These children may be shy. They are very aware of their own feelings and are self-motivated.

(Source: Southwest Opportunities Program)

Daniel Goleman's study of the particular "soft" skills and talents that make for star performers in the broad spectrum of careers and jobs are, in our opinion, directly related to the experience of a first-rate liberal arts education—in and out of the classroom. Studies indicate that the higher the level of position and responsibility, the less important technical and cognitive skills are and the more important competence in emotional intelligence becomes.

Goleman defines emotional intelligence as:

The capacity for recognizing our own feelings and those of others, for motivating ourselves, and for managing emotions well in ourselves and in our relationships. It describes abilities distinct from, but complementary to, academic intelligence, the purely cognitive capacities measured by IQ.*

Goleman's outline of critical competency skills clearly reflects the training in and out of the classroom that a student receives in a challenging, stimulating, diverse liberal arts college.

* Goleman, Daniel. *Working with Emotional Intelligence* (New York: Bantam Books, 1998): 317.

It takes a high IQ and subsequent grade achievement and test scores to make it into and through college and graduate school today in order to gain entry into the top career tracks. IQs there are generally in a range well above average (115) to superior (125+). What makes the greater differential is emotional intelligence.

Goleman found that employers looking for signs of a person's capability to learn on the job seek the following talents, as opposed to technical skills, in hiring young employees:

- Listening and oral communication skills.

- Adaptability and creative responses to setbacks and obstacles.

- Personal management, confidence, motivation to work toward goals, a sense of wanting to develop one's career and take pride in accomplishments.

- Group and interpersonal effectiveness, cooperativeness and teamwork, skills at negotiating disagreements.

- Effectiveness in the organization, wanting to make a contribution, leadership potential.

As Goleman says, "Of the seven desired traits, just one was academic: competence in reading, writing, and math." In a study of the traits corporations seek in the M.B.A. candidates they wish to hire, the same key skills were identified, the three most sought-after capabilities being communication skills, interpersonal skills, and initiative. Graduate schools of business, in kind, look for these characteristics in candidates for admission. Risk taking, perseverance, independent and creative thinking, and communication skills in all respects are essential traits to develop. Rigid specialization is a road to a dead end in the present and future world.

THE EMOTIONAL COMPETENCE FRAMEWORK

Personal Competence

These competencies determine how we manage ourselves:

- **Self-awareness:** Knowing one's internal states, preferences, resources, and intuitions. These include emotional awareness, accurate self-assessment, and self-confidence.

- **Self-regulation:** Managing one's internal states, impulses, and resources. These include self-control, trustworthiness, conscientiousness, adaptability, and innovation.

- **Motivation:** Emotional tendencies that guide or facilitate reaching goals. These include achievement drive, commitment, initiative, and optimism.

Social Competence

These competencies determine how we handle relationships:

- **Empathy:** Awareness of others' feelings, needs, and concerns. These include understanding others, developing others, service orientation, leveraging diversity, and political awareness.

- **Social Skills:** Adeptness at inducing desirable results in others. These include influence, communication, conflict management, leadership, change catalyst, building bonds, collaboration and cooperation, and team capabilities.

(Source: Daniel Goleman. *Working with Emotional Intelligence* (New York: Bantam Books, 1998)

In the next chapter, we review those criteria that we feel are essential in evaluating not only the Hidden Ivies, but any college or university, especially those emphasizing a liberal arts education. Once we have laid out the guidelines for considering and critiquing liberal arts colleges, we will discuss those colleges and universities that we believe do an outstanding job of delivering a liberal education today.

TWO

What Makes an Ivy an Ivy?
(And a Hidden Ivy?)

When we talk about the "Ivy League," we are usually grasping at something more than just a name. The designation of being an "Ivy," one of the top colleges in America, carries connotations of prestige, the highest quality of education attainable, intellectual rigor and challenge, and the ultimate in sophistication and selectivity. As we discussed in *Inside the Top Colleges*, it is important for parents and students to look beyond the halo effect that surrounds these colleges and often masks particular academic and social issues that may be of concern. Every college and university deserves similar critical evaluation, but our research and counseling experience has made clear that the Ivies and other most selective colleges share certain qualities that make them truly impressive educational institutions.

Our goal is to discern those characteristics that make an Ivy an Ivy and then to draw together a list of criteria that we have considered in presenting the Hidden Ivies. These

are the key factors that families can apply in evaluating any college of potential interest. These criteria serve to showcase the strengths of the Ivies and what make the Hidden Ivies equal to or almost as good as the Ivy schools. Some of the factors we have identified are quantitative in nature, while others capture more amorphous qualities. *The Hidden Ivies* is not meant to be a rankings guidebook or a quantitative evaluation of these colleges. We aim to provide a set of criteria that help to capture the essence of these excellent institutions and illuminate their particular qualities. We believe these guidelines will be helpful in evaluating any college or university that is striving to provide a topflight liberal arts education.

The Ivy League

What began as an elite athletic conference in the 1950s has grown over time into the collection of the eight educationally elite Ivy League colleges and universities that we know today: Brown, Columbia, Cornell, Dartmouth, Harvard, the University of Pennsylvania, Princeton, and Yale. What holds them together still is a shared tradition of academic excellence, a high degree of selectivity, and agreement regarding the award of scholarships solely on the basis of financial need. In other words, the Ivies do not offer athletic or merit-based scholarships. In the last few decades, additional colleges and universities have become as selective and as well known as the Ivies. These include Stanford, MIT, Duke, Northwestern, Johns Hopkins, and the University of Chicago.

Every day parents and students in our offices say, "We want the Ivies," or when asked about the colleges in which they are interested, they respond, "The Ivies plus . . ." Often, these families are not even sure which colleges constitute the Ivy League. Nor are they aware of the significant differences among each of the eight Ivy schools, the level of competition to get into them, and the level of work required to succeed once there. What they do know, however, is the reputation, the wealth of resources, and the academic and intellectual caliber that are available behind the Ivy walls.

What makes an Ivy an Ivy? Important factors in a few key categories are: the faculty, the resources, the educational and leadership team, important educational and other outcomes, the student body, and other special elements that contribute to stature and success. We encourage all families engaged in the college search to use these criteria to evaluate any institution of interest.

The Faculty

As Henry Adams said, "A teacher affects eternity; he can never tell where his influence stops." When examining a topflight academic institution, one is struck first and foremost by the quality of the faculty on campus. And when one talks to students at these colleges and universities, one hears again and again about the importance of the teaching faculty in providing for a good (or bad) educational experience. A number of reports from the Boyer Commission and the Carnegie Foundation for the Advancement of Teaching have leveled much criticism of the 125 or so research universities and the trend toward poor teaching and a lack of concern for undergraduates. It is at the expense of undergraduates that these universities seek excellence and prestige in their graduate programs, and it is the graduate students who do a great deal of teaching at the research universities, freeing up top faculty to do more research and writing. The fame of many of the top research universities is often based on the graduate schools, not on the undergraduate liberal arts colleges. We start with the faculty because they are the living endowment of a college, and it is they who over time serve most strongly to transmit the ideas and content that constitute a first-rate education. What are the important aspects of the faculty to consider?

Academic Preparation

What is the quality and nature of the degrees that the faculty hold? At the top colleges, almost all of the faculty hold the highest possible degree in their discipline. Usually, that means a doctorate (Ph.D.) in their area of academic expertise or, in the case of the arts, an advanced degree in their field of specialization. Additionally, these professors have continued to develop and share knowledge of their field as evidenced in their teaching and writing as well as ongoing training. Look for teaching awards, full-length book publications in academic presses, top journal publications, and the source of undergraduate and graduate degrees as evidence of strong faculty academic preparation.

Accessibility

Good faculty maintain their accessibility to students and a commitment to sharing their knowledge and passion with undergraduates. They inspire and cultivate those in their charge through seminars, after-class meetings, extensive office hours, communications via e-mail, and training in current teaching methodologies.

Work Opportunities with Undergraduates

Look for the opportunities that faculty provide for research projects, assistantships, and internships with themselves or other sources. In addition to classroom work, undergraduates may have the opportunity to pursue publication of joint research projects with their professors, act as research assistants in their areas of interest over the summer or during the semester, and find internships off campus with the assistance and sponsorship of teachers.

Low Faculty-Undergraduate Ratio

What is the ratio of faculty to undergraduates on campus? And what is the percent of time devoted to undergraduate teaching and advis-

ing? Low class-teaching loads and smaller class sizes allow faculty to devote more time to individual students; to get to know them by name and by their interests; and to advise students on appropriate research projects, courses, and choices of major. Bear in mind that educators self-select, just as students should, the type of academic environment they most value. Those who love to teach and interact with students are more likely to be found in a college that encourages and rewards excellence in teaching.

High Level of Full-Time Faculty

A disturbing trend in higher education has been the demise of full-time faculty members and their replacement by part-time, visiting, or adjunct professors and graduate teaching assistants. At a lower rate of pay and with less institutional support, these replacement faculty typically have less experience and less commitment to the college. While some may indeed be better teachers than some tenured faculty (those full-time professors who, in essence, after about six years at a university have been offered lifetime employment there), many will move on in a year or two and will not be available to offer advising or support over the long term. The Ivies and other top colleges have a very high level of full-time tenured faculty on campus.

Endowed Chairs for Teaching Faculty

An endowed chair is a high-paying, tenured teaching position in a specific discipline that is typically named for a generous donor or important college figure. The chair is funded in perpetuity through the college's endowment, a permanent fund for the support of the college. Thus a faculty member holding such a chair has the security of position, and freedom from additional pressures, to teach a small number of students with a high degree of attention and challenge. As one can imagine, such endowed chairs attract and retain topflight teacher/ scholars.

Faculty Diversity

What is the ratio of male to female teachers, and the percent of faculty of color or from non-traditional backgrounds? A diverse faculty contributes to a diverse intellectual environment in which students learn from a variety of perspectives and role models. It helps to foster student diversity, improving the wealth of voices on campus. We believe that such a climate is the most conducive to a balanced and mind-expanding education.

Teacher Training and Recognition Programs

Colleges that care about teaching tend to implement policies and programs that reflect their commitment. Teacher training programs mentor and educate faculty (and graduate students) on the best practices and develop teachers' skills in and out of the classroom. Universities can promote standards to increase the quality of instruction and they can make information available on goals for teaching, student assessments of faculty, and standards for review of teaching practice. In terms of evaluating faculty for promotion and tenure, teaching and advising can be recognized as an important aspect of rewarding professors for their commitment to the undergraduate student.

The Resources

The resources of a college or university are indicative of its ability to provide a top-quality educational experience for students. From faculty to salaries to computer networks and systems to athletic facilities to library holdings, a college's resource base, best represented by its endowment, makes the difference between an excellent and an adequate education.

A Large Endowment Relative to the Size of the Student Body

While there are differences among the individual institutions, all of the Ivies have large endowment funds in relation to the numbers of students (undergraduate and graduate) on campus. One of the best ways of determining a college's ability to provide for its students, faculty, and graduates is to examine the school's endowment in total numbers and as a ratio of the undergraduate population. The total endowment is the overall size of the college's holdings in funds and investments. Income from interest on the endowment allows the college to provide for scholarships and financial aid, teaching chairs, new buildings, maintenance, and residential life. Colleges and universities seem to be incessantly on a mission to raise funds for their endowment, and one can see why. Without these resources, tuition would skyrocket even more than it already has and the colleges would be unable to offer the kind of academic and social experience that students demand.

The commitment of the Ivies and the Hidden Ivies to undergraduate education is in part indicated by the wealth of the school in relation to the number of undergraduates on campus. The ability to provide for undergraduate education means that students entering these top colleges have access to a huge array of resources. Some of the smaller colleges have, relative to their undergraduate student population, a larger endowment than some of the premier, larger state universities. Some key qualities flow from such secure holdings at the Ivies and their counterparts.

COLLEGES WITH LARGEST ENDOWMENTS PER STUDENT, 2007

Among private liberal arts colleges and universities, Princeton is number one in the nation, and has been for some time now. Harvard, with the largest overall endowment among all universities, is number three. Here are the nontechnical or graduate-oriented colleges and universities in the top thirty:

Princeton University
Yale University
Harvard University
Stanford University
Pomona College
Grinnell College
Amherst College
Swarthmore College
Rice University
Williams College
Wellesley College
Berea College
Dartmouth College
University of Notre Dame
Bowdoin College

A number of the Hidden Ivies make the top-fifty list in this category, which is pertinent to the factors we include in our discussion of their strengths and priorities.

(Source: National Association of College and University Business Officers, *Chronicle of Higher Education*, 2008)

The Quality of the Library

The library (or libraries) is the core of any college. In all its aspects, from computer and information technology to book collections and magazine and journal subscriptions to trained staff and physical plant, the library facilitates a student's ability to learn, research, and study. Technological sophistication allows students to search the college's library, as well as other libraries, perhaps even from the student's own dorm room. Articles and reports in digital form are increasingly becoming essential components of student research papers and assignments. Ample book collections on campus, particularly in specialized areas, allow students to check out titles quickly and easily, while unique special collections and rare works give students the ability to research primary source material and pursue innovative directions. Prodigious magazine and journal subscriptions allow both faculty and students access to past and current research across multiple disciplines. Some top libraries are even designated holding sites for the Library of Congress.

Strong library resources include professionally trained staff who are able to help students negotiate extensive holdings, conducting sophisticated on-site and computer-based bibliographic and index searches. Such staff can assist faculty in promoting strong research and study skills among students.

They often teach seminars in their areas of expertise.

The best colleges have unparalleled modern library *systems* composed of multiple sites on campus, often separated into specific substantive areas such as the sciences, the fine arts, English literature, law, and so forth. These libraries have widely available study and reading spaces. They are open in at least one location around the clock, and they are located conveniently close to campus. Libraries are an essential learning space for students on campus; it is difficult to overemphasize the importance of their place in one's college career. Ivy-level library systems offer private spheres for student reflection and work, public spaces for the learning community, and extensive resources for the pursuit of knowledge.

Technological Sophistication

Access to the library system is only one important element of campuswide digital technology. Additional aspects of a technologically sophisticated college community include Internet access throughout academic and residential centers; training opportunities for faculty, staff, and students; a computer or information technology center with professional staff, wide availability of computers throughout campus, and the use of advanced technology for instruction and research. These resources are essential for the most up-to-date use of computer hardware and software, and they provide students with the opportunity to remain at the cutting edge of information usage and analysis. Faculty (and employers) have come to assume a level of knowledge of and facility with computers on the part of their students. They expect advanced research, daily Internet, IM, and e-mail communication, papers that are laser-printed and competently format-

ted, extensive bibliographies and citations, charts and graphs, and digital media. Students should have access to top-of-the-line technology and the support staff to assist them in applying it to their academic pursuits.

Modern Scientific Research Facilities

Up-to-date science and technology facilities offer students and faculty the means to engage in meaningful research. High-quality modern labs with specialized equipment, supported by the necessary research materials, attract and help to retain strong faculty, who in turn lure and work with exciting, high-caliber students.

Residential Life

The quality, nature, and availability of residential housing and facilities for social interaction on campus define an Ivy college. Social facilities can include pubs, dining halls, student centers, and space for student organizations. One must appreciate the magnitude of the impact of a high quality of residential life on the student experience. The informal, lasting, and meaningful interactions fostered by the residential college experience create social bonds and the kind of intellectual community that defines a top college. Colleges are experimenting with suite-style living, where a cluster of two or more individual student rooms are joined through a common living room, for example. Special-interest housing, described later, can play an essential social and intellectual role for many students. Some schools are connecting faculty, sometimes on a live-in basis, with particular student dorms of living clusters. And, of course, trained student residence hall advisors play an important role in counseling and mentoring students on an informal basis. Among the Hidden Ivies, almost all the colleges make on-campus residential housing available for 100 percent of the undergraduate student body.

Athletic Facilities

The ability to participate in athletics at many levels across multiple sports allows students additional opportunities to connect with their peers and the college. Topflight athletic facilities for both intercollegiate and intramural sports, the availability of these resources for all students, the quality of coaches and trainers, equal opportunities for participation for men and women, and a wide variety of athletic programs are emblematic of the Ivies and the Hidden Ivies.

The Arts

Arts facilities, from museums and art studios to performing arts spaces and radio, television, and film studios, are another creative outlet for students. Look for student opportunities to participate in theater, film studies, college radio, and museum research. College holdings may include special art and manuscript collections donated to the school, historic film libraries, specially constructed performance spaces, and individual practice rooms and studios.

Extracurricular Programs

The quality and diversity of extracurricular programs that require college funding goes back to the endowment. Schools with available resources can support a multitude of student clubs and activities, including outdoor clubs, debate teams, community service foundations, highly professional newspapers and journals,

internship placements, and political organizations. Extracurricular programs often require physical space, faculty advising, basic funding, and legal and insurance backing from the college. The availability of such programs can indicate an active, involved, and firmly supported student body.

Special Interest Groups

Ivy colleges support and sponsor a variety of special interest groups. In addition to receiving meeting space and financial support, these groups (Asian American societies, African American student organizations, international student alliances, foreign language studies, or college political groups) may even have access to so-called affinity housing. This is an opportunity for like-minded students to live in their own dorm or their own cluster within a dorm. Substance-free dorms, sometimes called "chem-free," coed and single-sex fraternity and sorority houses, and academic honor societies are other living spaces for special interest groups.

Foreign Study and Intercollegiate Consortia Programs

Ivy institutions support an array of special learning centers at home and abroad and programs for international study that are sponsored by the institution itself, often with faculty leadership from the college. These programs allow students to pursue language, cultural, and professional study that is underwritten and facilitated by the college. Additionally, consortia agreements with other institutions allow students to take courses and pursue exchange programs in particular areas of interest. Cooperative 3/2 or 3/3 degree programs allow students the opportunity to pursue the liberal arts at their home institution and then to gain a bachelor's or a master's degree in engineering or business, for example, in cooperation with another institution for an additional two or three years of study. Such opportunities dramatically expand the resources that the specific college can offer on its own.

Support Staff

It is important to note the training and number of staff in support services such as academic, personal, financial, and career advising. Professional resource support for students with learning disabilities and attentional disorders provides assistance in negotiating challenging courses and the entire college system. Career advising helps students to make the transition from school to work and to find internships in their area of interest. Personal, health, and mental health counseling helps students to make positive choices and handle the stresses of college life.

Financial Aid

Our final emphasis in the area of resources is on the financial aid budget of a college. In our opinion, this is one of the most important aspects of a college's resource base and one of the most important indicators of its ability to attract a strong and diverse student body without regard to financial need. This is how these institutions implement their philosophy of a student body based on a meritocracy of talent rather than an aristocracy of wealth and privilege. As families have begun to learn, the "sticker price" of a college's tuition plus room and board, like that of a new automobile, is often not the actual cost that the family will

bear in paying for their son's or daughter's education. Prodigious endowment funds for scholarships, loans, and work-study allow colleges to defray the financial burden. As college costs continue to rise and students' average levels of debt increase accordingly, this is important. The composition of the aid package offered to a student can vary and reveal major differences among colleges. Some use special talent awards, for example, merit scholars or varsity athletes, to attract and enroll particular students. However, the eight Ivy League institutions have agreed to offer only *need-based* financial aid. Thus they do not fund merit, athletic, or other awards. Yet at the Ivies, anywhere from 45 to 60 percent of students receive some sort of need-based financial aid, and these most selective colleges represent most of the schools in the small group of institutions that are truly *need-blind* in their admissions process, which means they do not take a family's or a student's ability to pay into account *at all* in making an admissions decision.

We believe that a college's ability to provide financial aid for its students and remove as much as possible the money factor in determining which students to accept or reject leads to a more socioeconomically diverse student body composed of bright, talented students from around the world. This diversity helps to make these college campuses interesting and challenging places to study. But even these wealthy colleges will increasingly become subject to the "barbell effect," whereby the neediest students are financially supported and the wealthiest students easily pay the full cost of tuition, but a middle group of students and families, let us say those with incomes between $50,000 and $150,000 (still a lot of money!), will become less and less able to pay annual costs of over $50,000 while not qualifying for any aid based on financial need. Thus expensive and competitive institutions could still develop two clusters of students: those with the most need and those with the least need, while those with some need turn to less expensive options like the public universities. Concern over these issues has led a number of the Ivies and other selective colleges to begin reforming their financial aid calculations to make it easier to grant aid to so-called middle-income families and to reduce or eliminate the loan component of aid packages. See our discussion of this trend in *Making It into a Top College*. However, such moves require ample college funding to be workable, and we are back to the focus on endowment. It remains to be seen whether the Ivies and the Hidden Ivies will agree to give some or more merit-based aid, to take a higher percentage of funds from endowments to significantly lower tuition costs, or to find more creative solutions to these issues.

The Educational and Leadership Team

At the helm of any college or university is a team of administrators, faculty, and staff who play an essential role in articulating the mission of the institution, developing its resources, and leading the school through changes in course. Often students and parents do not consider the leaders of a college when evaluating choices in higher education, but we would argue that strong and well-educated leaders foster healthy growth and progress in their institutions. Top colleges are characterized by leaders who often play the part of public intellectuals, representing their college at home and in the media and taking stands in important current debates on issues. They have the capability to attract to their campus outstanding teachers and administrative leaders, as well as to convince individuals of means to lend financial support to important programs.

Background of Chief Leaders

Focus on the education, training, and professional track record of the president, chancellor, dean of faculty, and dean of academic affairs. These individuals are likely to have advanced academic degrees in a particular discipline; training in administrative, legal, or management areas; and publications in journals, books, and respected magazines. They are members of foundation boards, directors of corporations, advisors to political leaders, and outspoken advocates in public debate. To what degree do these leaders articulate the mission of their institution and their respective ability to carry this out? What positions have they taken on issues of importance to maintaining or changing the core of the school? Look, for example, at President James Wright, completing his tenure at Dartmouth College. A former provost and dean of the faculty at the college with many years of experience, he has espoused a vision of the quality of campus life and equal access to its resources, and he has emphasized the goal of continuing to develop the combination of top teaching and scholarship among the faculty at this smaller Ivy League college. While in some ways flirting with controversy by taking a clear stand, with the backing of the board of trustees, against predominately single-sex fraternities and sororities, he laid out a long-range view of how residential life will develop at the school.

Effective leaders make statements, set priorities, and bring in talented support staff.

Support Staff

Who are the key people in the academic and student personnel roles and what is their experience? Are they full-time, trained in their areas of expertise? What backgrounds have they had at other institutions? The dean of residential life, dean of admissions, director of the career and graduate counseling center, and the dean of first-year students, for example, can all play significant roles in an undergraduate student's educational experience. These members of the support staff can develop programs, apply for grants, and advocate for students' interests in their areas of emphasis. They can represent and advance their college at professional meetings and through parents and students.

The Mission

What is the mission of the college and how well is it articulated to parents, students, alumni, faculty, donors, and peer academic institutions? How well is that mission implemented? The mission is more than a set of fund-raising goals. It speaks to the college's sense of itself—its history, values, and ideas. The mission is really the identity of a school, and top colleges are clear about who they are, where they are going, and how they are going to get there.

An Honor Code

One of our favorite examples of a distinguishing personality trait of a college is the presence or lack of an honor code. Such a code of academic and personal honesty and principles of good scholarship can be seen as a subtext of campus tone and spirit. It is important to look at the role of campus leaders in expressing the importance of an honor code and to look for examples of its successful implementation. Elements of an honor code—such as the right of students to take exams independently, without professorial monitoring, and the responsibility of students to report any cheating or plagiarism on the part of fellow students—can be reflected in admission application questions that focus on values and ethics. The college may conduct seminars and information sessions exploring the ramifications of the honor code and its place in campus life. Faculty may be required to discuss the honor code at the beginning of each course. Some honor codes are written, requiring students to sign once or regularly in agreeing to the code's requirements. Others are essentially well-understood and integrated principles that govern campus life. Find a well-articulated honor code and you will typically find a well-developed sense of mission. Here, for example, is a portion of Vanderbilt's statement regarding its honor code and its vital role in the life of the institution:

> The honor system is a time-honored tradition that began with the first classes at Vanderbilt in 1875. Students established the system and continue to manage it today. It rests on the presumption that all work submitted as part of course requirements is produced by the student, without help from any other source unless credit is given in a manner prescribed by the instructor. Cheating, plagiarizing, or otherwise falsifying results of study are specifically prohibited. . . . Responsibility for the preservation of the system falls on the individual student who, by registration, acknowl-

edges the authority of the Honor Council. Students are expected to demand of themselves and their fellow students complete respect for the honor code.

While honor codes are certainly about more than just cheating, addressing issues of ethics, social responsibility, and character, they address an all-too-prevalent issue for today's students. One study noted the prevalence of cheating in school, as well as the fact that a "significant minority" of students are interested in addressing the issue of cheating, says Don McCabe, a Rutgers University professor who has been studying the issue for a decade. McCabe is convinced that if those students are given the opportunity to lead the charge against cheating, most others would follow.*

The Student Body

The students at top colleges are impressive in many respects. They are scholars, athletes, volunteers, and musicians, with such highly developed résumés that admissions officers, alumni, and parents shake their heads in disbelief, muttering, "I couldn't get in today!" The intellectual environment created by the mixing of a diverse, intelligent, and motivated student body fosters an atmosphere of creative growth, debate, scholarship, and bonding. Look at who the students are on campus and you will gain an understanding of the impact of living and learning in their midst. One of the most important findings in our interviews with thousands of undergrads in the elite colleges was the powerful influence of their peers on the individual's learning experience. Those who expressed

their love for the college pointed to the influence of classmates as much as to the faculty.

Selectivity

What does it mean to be a "selective" college? Such a distinction translates into the ability of the school to actively choose among a group of talented students to craft a class comprised of students with unique abilities. The more selective a college, the greater its luxury to pick and choose among the best as it forms each class. Many of the Hidden Ivies have selectivity rates—the percentage of those who apply who are admitted—that approach or equal those in the Ivy League. The selectivity factor plays out on campus in such areas as intellectual preparedness, motivation, range of interests, and diversity of backgrounds, talents, and energy. Such a class fosters the ability of faculty to teach curious and bright students. This attracts dedicated professors to teach at a particular institution and leads to the higher level of intellectual and social stimulation. We know from our conversations with college students and from the research in *Inside the Top Colleges* the major significance of the peer group in defining the nature of the learning experience. Who students study with in the classroom and socialize with in the commons, the clubhouse, and the dormitory is as important to them as who is doing the teaching and the courses they are taking.

Diversity

When we speak of the diversity of the student body, we mean much more than racial or ethnic

*Marklein, Mary Beth, "Revealing the Answer to Cheating" (*USA Today*, 1/5/00): 9D.

diversity. This type of diversity is important, yet we add other components for consideration: socioeconomic, geographic, international, intellectual, artistic and athletic, political, religious, and sexual orientation. A diverse student body reflects a college's ability to attract, select, and retain students with individual interests and talents, dedications and values, backgrounds and perspectives. Such diversity fosters intellectual and social growth. It forces students to challenge their assumptions and to learn.

Attracting and Supporting the Students

One of the ways top colleges encourage and facilitate diversity is through generous financial aid that is grounded in a significant endowment. We mentioned before how endowment translates into need-blind, or nearly need-blind, admissions at the most selective colleges. These schools have a serious commitment to matching the financial needs of all students. This allows them to attract talented applicants from many backgrounds, to support these students once admitted, and to build a class based primarily on aptitude and interest. Other factors that help to entice and retain top students are a national reputation, high-quality academic and residential facilities, and exciting and well-trained faculty.

OTHER INTANGIBLES

What else is it that drives the engine of a top college? What other key factors seem to define these institutions, together and individually?

Personality and Culture

There is an impalpable dynamic of an overriding culture and ethos that defines a college or university. These elements leave a lasting imprint on a student's education, thinking, and behavior. What is it that creates a friendly or competitive environment, an intellectually curious and stimulating environment, or a hostile and "work for the grades" ethos? Is there a religious or ethical foundation that underlies campus and classroom life or an honor code that establishes the greatest sense of the place? What are the history and the collective traditions of the institution that form its identity and which so strongly affect students?

Single-Sex Institutions

While all of the Ivies are now coeducational learning environments, some of the Hidden Ivies are still single-sex colleges. There is, of course, a great deal of debate on the merits and drawbacks of student learning in such an environment. On the one hand, researchers argue and students maintain, young women excel in a single-sex academic environment, where they are supported to compete in traditionally male disciplines like science and engineering and where socially and intellectually they find an educational setting more attuned to their inclination to cooperate with classmates and learn through discussion. However, others point out that in these kinds of environments, women suffer from a lack of exposure to male cultures and points of view and that eventually women will have to learn to communicate and work with their male counterparts. We should also note that there are a number of all-male colleges that have their own arguments as to why these schools work well for male students. There is a wealth of pedagogic and psychological data in both directions, and what is clear is that the decision to attend a single-sex versus a coed college is largely an individual one. Some young women will do better in one or the other environment, often for very personal reasons. The single-sex women's colleges we have chosen to discuss here do maintain a single-sex setting, but they are linked in one way or another to peer coed environments. They excel academically and offer their students extensive support, understanding of the particular issues that young women face in higher education, sanctuary from the social pressures often found on a coed campus, and an opportunity to learn with and from other high-caliber female students. These single-sex colleges are not for everyone, but they can be important additional considerations for many students.

Prestige

In *Inside the Top Colleges*, we point out the dangers of "the halo effect," of focusing on the name of a college and ignoring the appropriateness of the institution for yourself and the real academic, social, and environmental aspects of campus life. While we would continue to argue against students focusing on names only, it is possible to talk about "the prestige factor" in terms of the institution's reputation and what this prestige confers on its students and graduates. A college's reputation is important, and following the lines of a self-fulfilling prophecy, there is some sort of intangible value to attending a school that is highly regarded. There are the alumni connections to be made, the famous faculty to encounter, the clubby tra-

ditions to join and be part of throughout one's life. But prestige is a two-sided coin. You can consider a college's reputation as a real factor in evaluating that school, but you must be careful not to confuse prestige with absolute quality or to choose the "better" school according to myth or rankings without understanding the appropriateness of fit for yourself.

Outcomes

In evaluating the "success" of a college, you must at some point begin to quantify outcomes. How well is the college accomplishing what it set out to do? How well is its mission being fulfilled and what measures can you apply to judge those criteria that the school itself sees as important? We have discussed a number of criteria that we feel are important in assessing the quality of any college or university in the liberal arts. Here are some specific criteria that you might use to dig under the surface at these institutions. The Hidden Ivies do very well by these measures:

- The *retention rate* of students at the end of the first and second years. How many students are returning to campus? If this number is low, say below 70 percent, something is wrong.

RETENTION AND GRADUATION DATA

ACT regularly reports national trends data on college retention and completion rates. Students and parents are often shocked when we comment on how appallingly low these rates are at many colleges and universities. First- to second-year retention rates track the percentage of freshmen who return to college for their sophomore year. Graduation rates are divided into four-, five-, and six-year categories, and you should note that many colleges regularly report six-year graduation rates, rather than four-year statistics. It is clear from ACT's data that retention and graduation rates rise with the level of selectivity of the college, whether public or private. Not surprisingly, then, the Ivies and Hidden Ivies regularly retain and graduate eighty to over 90 percent of their entering students, largely within the traditional four-year time frame.

While the national data and rates represent the full spectrum of undergraduate types, from traditional college-bound eighteen- to twenty-two-year-olds to more mature, returning, or part-time adult learners, the statistics do emphasize the extraordinary rates of retention and graduation of undergraduates at the select colleges we have included in the Hidden Ivies.

We have identified retention as an essential factor in our selection of high-quality institutions because, as we know from our professional experience, the more selective the institution is in its admissions policy, the greater the intellectual motivation and preparation or readiness of the student body, and the greater the sense of belonging to the community. The higher the quality of the faculty teaching and student contact and engagement, the greater will be the retention rate and eventual graduation rate. Feeling connected socially and academically is a key element in retention and graduation, as the National Survey of Student Engagement (NSSE at nsse.iub.edu) demonstrates.

Here are some data points from ACT (act.org) on first-year retention and graduation rates in 2008 for four-year private institutions, which you can compare to individual Hidden Ivies and other colleges you are considering:

Retention (Percent Returning for Their Second Year)

Highest Degree Offered by College

	All	Bachelor's	Master's	Doctoral
Selectivity of College				
Highly Selective	91.4	93.5	87.9	91.4
Selective	80.8	78.7	80.1	83.7
Traditional	70.0	67.1	70.9	73.1
Liberal	63.3	63.2	62.5	70.3
Open	66.3	63.6	67.2	74.3
All	72.9	69.6	72.3	80.4

Graduation (Percent Graduating in Four, Five, or Six Years)
Highest Degree Offered by College

	Bachelor's			Master's			Doctoral		
Selectivity of College	4 year	5 year	6 year	4 year	5 year	6 year	4 year	5 year	6 year
Highly Selective	82.6	86.7	86.9	74.9	80.4	82.9	73.0	81.9	85.1
Selective	62.6	69.8	70.9	57.2	66.7	68.2	54.2	66.0	67.8
Traditional	35.8	44.7	45.2	39.9	51.0	53.7	38.2	52.1	54.6
Liberal	29.4	43.2	36.6	28.7	43.9	37.5	35.2	40.1	46.0
Open	39.6	49.4	49.4	37.4	49.4	53.1	46.7	54.1	61.2
All	48.4	56.1	57.2	44.9	55.4	57.5	52.9	63.4	67.0

- Percent who *graduate* in a four- or five- or six-year period. Statistics are now often reported as a five- or six-year graduation rate, which tells you something about expectations of a college career these days. Again, if after five years only 40 percent of students have graduated . . . there is a problem.

- Number who enroll in *graduate schools*: range, variety, and quality of the graduate programs. Today, one's graduate degree, be it in medicine, law, engineering, business, or social work, has become ever more important for job opportunity and security. One of the strengths of the liberal arts colleges is placement in competitive graduate programs.

SUCCESSFUL GRADUATE SCHOOL ADMISSION

There is impressive data to reinforce the importance and advantage of choosing the right college on an individual basis. Most families are naive in their thinking that one has to attend or has an advantage by attending an Ivy college in order to qualify for the best graduate schools.

The colleges we review here, and many others, have equally distinguished records in training and readying their undergraduates for professional and academic graduate school programs. The reasons for this are those we articulate throughout our discussion of the unique qualities these schools possess.

- Level of activity and quality of *recruitment* for seniors, and the overall picture of job placement. If students are not going on to graduate school directly, where are they headed? What kind of corporate and other recruitment is taking place on a college's campus?

- Kinds of *careers* recent graduates have chosen. One would hope to see some diversity in choice here, from finance and banking, law and medicine to social action, teaching, research, and religious careers. However, the kinds of career choices made by graduates can help to indicate the general tone and personality of the campus.

- Number of *fellowships, awards,* and *honors* received by graduates. Such prestigious awards as the Rhodes and Fulbright scholarships indicate strength of academic preparation.

- *Alumni involvement.* Involvement in annual giving and in such volunteer duties as interviewing, advising on careers, internship mentoring, and job placement are a measure of satisfaction and commitment. Alumni giving represents vast potential in terms of program sponsorship and financial aid and in terms of the operating budget. Alumni networking and communication are a lifelong benefit for the graduate.

Some Landmarks to Consider in Evaluating a College or University

Statistical Data

Selectivity for admissions, as it represents a quality student body.

Retention of students and graduation rate as a sign of student motivation, loyalty, commitment, and satisfaction, as well as of college resources and programs.

Percentage of students attending graduate school and where they attend, as a sign of motivation and reputation of the college and strength of academic program.

Recipients of graduate and other national/international fellowships.

Job placement and campus visits by companies and other organizations.

Faculty reputation, degrees, accessibility, salary level, etc.

Campus resources for quality of life and educational delivery.

Financial-aid programs (percent of students on aid, "need-blind" admissions, etc.), as a sign of ability to attract outstanding students and a diverse student body.

Endowment relative to enrollment.

Student to full-time faculty ratio for undergraduates.

Average undergraduate class size, median class size, percentage of classes below twenty-five students.

Number and breadth of majors and courses offered.

Foreign study and/or internship opportunities and participation.

Participation in sports, intramural, and campus activities.

Overall diversity (racial/ethnic, international, socioeconomic, religious, geographic).

Special Qualities of Campus Life

Campus spirit, morale, sense of community, pride in the institution.

Satisfaction factor—would students attend again?

Satisfaction with fellow students and faculty.

A purposeful environment and a balanced social/academic/activity formula.

Feeling of academic challenge, academic concern, support.

Applying the Criteria

We have laid out what we intend to be a road map for exploring not only the Ivies and the Hidden Ivies, but also any college or university under consideration. Our hope is that students and families will use the criteria and model of assessment we have presented here to explore any institution of interest. Most data are easily obtainable from guidebooks, Web sites, news magazines, and the colleges themselves. The factors we discuss represent important elements of a college education for any student, but particularly those pursuing the liberal arts. No institution can be perfect in every way. However, the Hidden Ivies have shown across the board that, according to the criteria that make an Ivy an Ivy, they are some of the best colleges and universities in the country.

A note about the statistical data: Every effort was made to include the most accurate and up-to-date information about each college. Unfortunately, despite recent efforts to improve and standardize data collection and reporting, there remain inconsistencies and gaps among the colleges. Sometimes this is due to variances in the ways in which the schools report their data; sometimes the colleges do make it difficult to find the information one is looking for. We encourage readers to view with a critical and careful eye data about institutions. There are a number of good data sources available to you, and we consulted several of them for the majority of the information contained in our college profiles. They include:

- The Common Data Set initiative (common dataset.org), Peterson's (petersons.com), *U.S. News & World Report* (usnews.com), the College Board (collegeboard.com).

- The College Navigator (nces.ed.gov/ collegenavigator).

- The National Association of College and University Business Officers (NACUBO, nacubo.org).

- The University and College Accountability Network (ucan-network.org) and colleges' own Web sites.

The kinds of statistics you will find include:

- Number of undergraduates: The number of degree-seeking undergraduates, typically as reported for fall 2008.

- Total number of students: The number of degree-seeking undergraduates plus graduate students. Most of the schools listed here have few or no graduate students, so the number of undergraduates will be the same as or close to the total number of students.

- Ratio of male to female: The percentage of first-year students of each sex. You'll see some surprising ratios here that will be fairly consistent from year to year at a particular college, and which could impact the overall climate on campus and your odds of admission.

- Tuition and fees for 2008–2009: The sticker price for the core aspects of your educational cost. When you add in room and board and other expenses, the total cost of attendance at the Hidden Ivies can exceed $50,000.

- Percentage of students receiving need-based financial aid: Unless otherwise noted, this is

The Hidden Ivies: Admissions Data

	Applied	Admitted	Admit Rate	Size Of First-Year Class Enrolled 2007	Yield Rate (Admits Enrolled)
Amherst	6,680	1,175	18%	474	40%
Barnard	4,574	1,315	29%	559	43%
Bates	4,434	1,312	30%	445	34%
Boston	28,850	7,869	27%	2,291	29%
Bowdoin	5,961	1,130	19%	476	42%
Bryn Mawr	2,106	958	45%	352	37%
Bucknell	8,943	2,673	30%	887	33%
Carleton	4,840	1,444	30%	509	35%
Chicago	10,362	3,597	35%	1,300	36%
Claremont-McKenna	3,778	671	18%	268	40%
Colby	4,679	1,488	32%	467	31%
Colgate	8,759	2,242	26%	746	33%
Colorado	4,826	1,540	32%	524	34%
Davidson	3,992	1,127	28%	467	41%
Duke	17,748	4,077	23%	1,700	42%
Emory	15,366	4,175	27%	1,235	30%
Georgetown	16,163	3,363	21%	1,582	47%
Grinnell	3,077	1,534	50%	427	28%
Hamilton	4,962	1,376	28%	468	34%
Haverford	3,492	877	25%	315	36%
Johns Hopkins	14,848	3,603	24%	1,206	33%
Kenyon	4,626	1,352	29%	458	34%
Lafayette	6,364	2,224	35%	594	27%
Lehigh	12,155	3,882	32%	1,116	29%
Macalester	4,967	2,015	41%	485	24%
Middlebury	7,180	1,479	21%	644	44%
Mount Holyoke	3,194	1,671	52%	522	31%

	Applied	Admitted	Admit Rate	Size Of First-Year Class Enrolled 2007	Yield Rate (Admits Enrolled)
Northwestern	21,930	5,872	27%	1,981	34%
Notre Dame	14,503	3,549	24%	1,991	56%
Oberlin	7,014	2,193	31%	745	34%
Pomona	5,908	964	16%	375	39%
Reed	3,365	1,154	34%	347	30%
Rice	8,968	2,251	25%	742	33%
Richmond	6,649	2,654	40%	810	31%
Rochester	11,676	4,815	41%	1,062	22%
Smith	3,329	1,726	52%	656	38%
Southern California	33,760	8,553	25%	2,963	35%
Stanford	23,958	2,464	10%	1,722	70%
Swarthmore	5,242	930	18%	365	39%
Trinity (Connecticut)	5,950	2,037	34%	576	28%
Tufts	15,365	4,229	28%	1,373	32%
Tulane	16,967	7,526	44%	1,328	18%
Vanderbilt	12,911	4,238	33%	1,673	39%
Vassar	6,393	1,830	29%	681	37%
Wake Forest	7,177	3,041	42%	1,124	37%
Washington and Lee	3,719	1,018	27%	462	45%
Washington	22,428	3,887	17%	1,338	34%
Wellesley	4,017	1,434	36%	590	41%
Wesleyan	7,750	2,123	27%	733	35%
Williams	6,478	1,194	18%	540	45%
Average Admit Rate and Yield Rate			30%		36%
Total Applicants, Admits, and Enrollees	472,383	129,851		44,694	

Data Source: **College Handbook**, The College Board, 2009

the percentage of the total undergraduate student body. We broke out need-based aid from total aid in order to focus on how well colleges are investing in students who have identified financial need.

- Percentage of students graduating within six years: Six-year graduation rates have become the norm, but most students at the Hidden Ivies do graduate in four years. The rates we use here are usually for the first-time, full-time freshmen entering college in 2001 or 2002.

- 2007 endowment: There are some big numbers here and, yes, you can count on the fact that all of them declined in 2008 and 2009, some by as much as a third. Nevertheless, these institutions are in the same boat, so we feel that the relative sizes of the endowments will likely remain stable. New figures for 2008 will be available from NACUBO in 2009.

- Endowment per student: The endowment divided by the total number of students. This figure truly gives you a rule of thumb to compare how much support is available on a per-student basis, and you will note some dramatic differences among the Hidden Ivies.

THREE

Following are individual descriptions of the fifty colleges and universities that are the Hidden Ivies. Based on surveys, questionnaires, interviews, and discussions with students and administrators, as well as our own experience with students going to and coming out of these institutions, the descriptions follow a general conceptual format and are presented in alphabetical order. Some statistical data, following our discussion of key criteria in the previous chapter, and qualitative insights gleaned from individuals and published material, capture the individual personalities of each of the institutions. Unless otherwise noted, most data are taken from the colleges' reported information as part of the Common Data Set initiative (commondataset.org). Endowment data is taken from the National Association of College and University Business Officers (nacubo.org). Tuition, cost, and financial-aid data are also taken from information reported to the College Board (collegeboard.com).

Amherst College

Wilson Admission Center
Box 5000
Amherst, MA 01002-5000

(413) 542-2328

amherst.edu

admission@amherst.edu

[
Number of undergraduates: 1,683
Total number of students: 1,683
Ratio of male to female: 51/49
Tuition and fees for 2008–2009: $37,622
% of students receiving need-based financial aid: 52
% of students graduating within six years: 96
2007 endowment: $1,662,377,000
Endowment per student: $987,746
]

Overall Features

Amherst is one of the most illustrious of the liberal arts colleges in terms of its distinguished faculty and highly selective student body. Amherst was founded in the early years of the nineteenth century. Its reputation for academic excellence and its commitment to undergraduate teaching is second to none. Amherst's leaders consider it one of the preeminent residential liberal arts colleges in the nation. While the arbitrary rankings arrived at by the various national publications are considered flawed, nevertheless Amherst is ranked number one among small liberal arts colleges more frequently than any other institution. Its long history as one of the two or three institutions most competitive for admissions seems only to attract more outstanding candidates every year, as well as world-class faculty who want to teach some of the brightest young men and women.

From its roots as a traditional New England college for men only, Amherst has developed into a highly diverse student body of women and men from all corners of the nation and other countries. Amherst is passionate regarding the advantages that the small residential college provides for students and faculty to learn from and with one another. Every undergraduate lives on campus with the exception of fifty non-freshmen who may apply to live off campus. Amherst has long encouraged bonding for life among its students. Alumni, on the whole, speak with passion about their time at the college, of the friends they made, and of their intellectual development. The distinction of their teachers and their ready accessibility is a constant theme of the advantages of an Amherst education. Every course is led by a professor who oversees all of the work done for the class. There are no teaching assistants on the faculty. Even faculty home telephone numbers are listed in the campus directory so that students may contact them. The student-to-faculty ratio is eight to one, and the average class size is twenty-two students. Amherst is as selective in its admissions as are the Ivy League colleges, which means that the majority of enrolled students are smart, highly motivated, and well prepared to handle an intense workload.

One tradition that continues is the intense athletic competition with Amherst's oldest of rivals, Williams College. It is considered a successful athletic year if the football team beats the Purple Cows. Together with Wesleyan, these colleges comprise the "Little Three," an over-100-year conference rivalry. There are twenty-seven intercollegiate teams for men and women at the NCAA Division III level. Many talented high school athletes are attracted to Amherst for the opportunity to make a varsity team while receiving a first-rate education, especially since the athletic facilities have been upgraded dramatically.

The majority of students join in intramural and club sports. There are some 100 organizations run by students that cover any interest one can imagine. The many music programs, especially the singing groups, are a long and active tradition. A modern student center is well-used by students and adds greatly to activities, since fraternities no longer exist on campus. Theme and special-interest residences are sponsored by the college as well.

Since the marked upheavals in higher education in the late 1960s and early 1970s, Amherst has changed as much as any traditional college in a number of ways. Coeducation and a serious campaign to realize greater diversity in the student body have been dramatic for a formerly all-male, virtually all-white, middle- and upper-middle-class population. Students of color made up 35 percent of the most re-cently enrolled classes. Other changes have had a significant impact on the nature of the college as well. The formerly restrictive and highly traditional nature of the curriculum has been transformed into a very flexible, self-directed program of study for each student. There is but one requirement of all students: a first-year seminar. Beyond that, students choose their courses and even have a good deal of flexibility within their major concentrations.

Operating on the assumption that they are among the best and brightest, the academic leaders trust students to choose wisely and challenge themselves intellectually. Amherst students will readily attest to the demands of the faculty and the challenge to keep up with their talented peers. The change in the makeup of the student body is reflected in the curriculum taught today. Courses that focus on gender, racial, political, and cultural themes are many and are popular with undergraduates. The rise in financial-aid recipients is also a significant indicator of the administration's success in meeting its goal. The once heavily popular fraternities to which almost all men belonged are now a thing of the past. The trustees determined that they defeated the intellectual and social purposes of a small residential community and thus outlawed them from the campus. It is not so easy to block the desires of a smart and independent group of students, however. Even with severe penalties at hand, many have joined underground Greeks off campus. The college has replaced with dormitory and cluster social events and intramural sports the social role the fraternities once played in campus life. Still another significant development since the 1970s is the founding of Hampshire College near Amherst as a highly progressive, student-centered form of education. This then led to the creation of the Five Colleges consortium, which enables a student to choose courses from the other institutions that include Smith and Mount Holyoke Colleges and the University of Massachusetts. The group sponsors specialized majors and faculty that would otherwise be too costly for any one college to support. Social life and cultural opportunities are expanded for many students as well through Five Colleges.

What the College Stands For

Training the minds of young men and women of exceptional talent so that they will think for themselves, understand other people's ideas, write and speak well, and appreciate other cultures are goals the college holds dear. The college shared with us that "Amherst College educates men and women of exceptional potential from all backgrounds so that they may seek, value, and advance knowledge; engage the world around them; and lead principled lives of consequence. Amherst brings together the most promising students, whatever their financial need, in order to promote diversity of experience and ideas within a purposefully small residential community. Working with faculty, staff, and administrators dedicated to intellectual freedom and the highest standards of instruction in the liberal arts, Amherst undergraduates assume substantial responsibility for undertaking inquiry and for shaping their education within and beyond the curriculum." Diversity in all its forms—socioeconomic, ethnic, racial, religious, cultural—is fundamental to Amherst's mission. Here is what the admissions committee states in its literature: "Our commitments to both distinction and inclusion have brought a long line of extraordinarily talented students and scholars to Amherst— people who have enriched our campus, our

country and the world. Today, when we select new students, faculty, or staff, we continue to seek talented people from historically underrepresented groups." Amherst concentrates on the liberal arts and on preparing students for life-long learning and participation in society. As its catalog states, "Whatever the form of academic experience—lecture course, seminar, conference, studio, laboratory, independent study at various levels—intellectual competence and awareness of problems and methods are the goals of the Amherst program, rather than the direct preparation for a profession."

Curriculum, Academic Life, and Unique Programs of Study

Amherst allows its students great freedom in selecting their courses and designing a major that meets with their interests and academic goals. The single requirement is the first-year seminar, but even here there are twenty topics or themes—all of them interdisciplinary in content—to choose from. There are over 800 courses and thirty-four areas of study, with a faculty of 194. Many students create a double major to meet their particular interests. What is called an *open curriculum* refers to the freedom from set distribution requirements. As the college pointed out, "Amherst College is committed to learning through close colloquy and to expanding the realm of knowledge through scholarly research and artistic creation at the highest level. Its graduates link learning with leadership—in service to the college, to their communities, and to the world beyond." The academic content of the courses is of a more traditional nature. The key to carrying off with a rational plan this intellectual cornucopia is the faculty and career advisory system, which assists students in planning their academic program over the four years. An awareness of one's intellectual interests and potential future plans plays an important role in designing a plan of study with one's advisor and departmental leader. The faculty encourages students to undertake independent study, research projects, and a senior thesis. Don't forget that all of the departments of the other colleges in the Five Colleges consortium are also available to Amherst students. About one-third of all students will take advantage of the off-campus and study-abroad programs the college group sponsors. Amherst is also a member of the Twelve College Exchange, which lets a student spend a year at Bowdoin, Connecticut College, Dartmouth, Mount Holyoke, Smith, Trinity, Vassar, Wellesley, Wesleyan, Wheaton, or Williams. The yearly calendar consists of two semesters and a January term. Students are encouraged to utilize this flexible January period to perform community service, undertake an internship with the assistance of the career office and the alumni network, or engage in a project of personal meaning and importance.

Major Admissions Criteria

The selection process is determined by this policy: "Amherst seeks to enroll students who will thrive in our dynamic academic and social environment. A successful Amherst student is someone who will take full advantage of the resources our community offers—such as embracing the academic freedom provided by the open curriculum, learning from our talented and engaged professors, and contributing to a student population which is diverse in both background and perspective." The admissions committee considers the same factors as the other elite colleges in admitting a mere 1,144 men and women from a pool of 7,745 competitive candidates (about 438 of whom enroll). High school grades; the quality and rigor of the academic program undertaken; standardized test scores; recommendations from teachers, advisors, or coaches; extracurricular activities and service to the community; essays on the application; and special talents are all considered. "[B]ut no one of these measures is considered determinative. How they intersect makes the difference." This is how an admissions committee can build a new class each year that will reflect the talent and mix of students that meets the goals of the college. Amherst notes that "we give the greatest weight to your academic transcript. The rigor of courses you've taken, the quality of your grades, and the consistency with which you've worked over four years give us the clearest indication of how well you will do at Amherst. Standardized tests also play an important role in helping us evaluate you in comparison to students taught in very different secondary schools. Recommendations, the quality of your writing, and extra- and cocurricular talents also help the admission committee draw fine distinctions among very talented applicants." A few pertinent statistics will indicate the abilities of those who make it into Amherst: 85 percent rank in the first decile of their high school class and 95 percent in the top quarter. The middle 50 percent had SAT scores of 2,000 to 2,290 combined reading, writing and math. Sixteen percent are valedictorians. Whereas in earlier times a majority of the student body came from a cluster of elite eastern independent schools and suburban high schools, today some 360 different secondary schools are represented in the entering class. About one-third of the places in the entering class are committed to early-decision applicants. Since Amherst competes directly with the other Ivies and Hidden Ivies for top students, a highly qualified candidate can help his or her chances by committing to the college with an early-decision application. Amherst is able to meet the full financial need of all successful applicants who demonstrate need. Thanks to its large endowment funds, the admissions committee has expanded a need-blind policy in choosing which students to admit. Therefore, if you think you can qualify for admission, be sure not to eliminate Amherst from consideration because of its costs. Mirroring what a number of other highly selective and wealthy institutions have been able to accomplish in the last few years, Amherst's financial-aid program is quite generous. When asked what else families should know about the college, Amherst noted the following: "Amherst meets the full demonstrated need of every admitted student. If a family can pay only a small portion of tuition and costs—or maybe none at all—Amherst pays the rest. Last year, Amherst provided more than $28 million in scholarship aid to about half of the student body. The average scholarship award was over $30,000. Beginning in the 2008–09 school year, students will no longer be required to take out student loans as part of their financial-aid awards. Beginning in the 2008–09 admission cycle, Amherst will extend its need-blind admission policy for domestic applicants to international applicants as well. This means Amherst will make your admission decision without considering whether you apply for aid or to what degree you have financial need."

The Ideal Student

The typical Amherst undergraduate of the past is no longer dominant in the community. The diversity of the student body and faculty, the freedom to determine one's own academic ex-perience, the absence of the strong fraternity culture, the multitalented cohort of students, and the intensity of the academic program are the key points in determining if Amherst is an appropriate college for you. Those who are capable of gaining admission to Amherst can have many other outstanding choices. Amherst may be the right college for those who want a very rigorous academic challenge, an opportunity to mix with a wide range of students in a liberal thinking environment, and interaction with faculty of the first level who will push students to perform at their highest capacity. As the college emphasizes, "Amherst is a community that draws its strength from the intelligence and experience of those who come here to learn, to teach, to work. We reaffirm our goal of fashioning the Amherst College community from the broadest and deepest possible range of talents that people of many different backgrounds can bring to us." Even with the Five Colleges program, students who fit best at Amherst enjoy the closeness of a small college community in a small-town environment.

Student Perspectives on Their Experience

Amherst students feel challenged and supported, focused and expanded by their experiences at the college. Students praise the faculty for its intellectual guidance, advice on securing internships and exploring new areas of study, and willingness to stay involved with students beyond the classroom. They admire each other for their diversity of backgrounds, beliefs, and academic and extracurricular pursuits. Amherst is seen as a supportive environment in which to spend one's college years growing intellectually and socially. For some students, the Five Colleges exchange program is a helpful way to expand social, course, major, and resource offerings, while for others the five campuses are far enough apart and the registration process just difficult enough to make the system less workable than would be ideal. Amherst's size and small-town location puts off some students who feel constrained by the fact that news travels fast and everybody knows everybody else. But most appreciate the benefits of close community contact; access to faculty, staff, administrators, and support resources; and small class and advisor group size. Students love the Amherst graduate network and the bonds they develop with each other and with alums past, present, and future.

What Happens after College

A heavy majority of Amherst graduates continue their studies in the top professional and academic graduate schools. Following the trend of students at the other elite colleges and universities, many graduates will explore different opportunities before enrolling in formal studies. Within five years of graduation, 80 percent of Amherst alumni will enroll in graduate school. History and English, which have fabled reputations due to some of the most distinguished teachers in these departments over many years,

ready many students for the premier law and business schools. The sciences also have excellent teachers who provide ample opportunities for research training. A good many students thus gain admission to the major medical colleges or doctoral programs. A combination of student talent and the rigor of the education brings impressive results each year. A well-developed career center at Amherst offers several hundred programs and workshops annually to students, a number of whom go on to win Fulbright, Rhodes, Watson, Goldwater, and other prestigious national and international fellowships with the active support of faculty who know their students well. The college says that "Amherst also offers everything necessary to effectively prepare for study beyond the undergraduate degree: a rigorous curriculum, an emphasis on writing, the opportunity to do graduate-level thesis work, and strong recommendations written by professors who know you well." Amherst fosters a strong alumni network to assist emerging graduates in landing jobs or internships of their choice. If any college in our review attests to the value of an undergraduate education in the liberal arts as a foundation for future opportunities in any field of choice, Amherst serves as a standard-bearer.

Barnard College

3009 Broadway
New York, NY 10027-6598

(212) 854-2014

barnard.edu

admissions@barnard.edu

[
Number of undergraduates: 2,295
Total number of students: 2,295
Ratio of male to female: 100% female
Tuition and fees for 2008–2009: $37,538
% of students receiving need-based financial aid: 44
% of students graduating within six years: 89
2007 endowment: $207,301,000
Endowment per student: $90,327
]

Overall Features

Barnard enjoys a unique situation within the corps of elite women's colleges. While it is a small liberal arts college dedicated to educating talented women for future roles of leadership, it is a distinct component of one of the great research universities, Columbia. As a result of this relationship, Barnard students have a "best of both worlds" opportunity. The college is dedicated to first-class undergraduate teaching, personal engagement with students, and a special concern for the needs of young women. A dean says, "As a college for women, Barnard embraces its responsibility to address issues of gender in all of their complexity and urgency, and to help students achieve the personal strength that will enable them to meet the challenges they will encounter throughout their lives." At the same time, the extraordinary resources for learning at a leading research institution are readily available. It would be rare for a Barnard student to complain about limited course offerings, laboratory or library facilities, or extracurricular options. A student can determine for herself to what extent she wants a small-college experience or a broader, more diverse learning experience by taking courses of her choosing within the Columbia curriculum. Some students even choose to live in coed Columbia College dormitories. Barnard takes pride in the quality of its faculty, selective student body, and available learning resources. It makes certain that such support services as personal, physical, and academic counseling are tuned in to the needs of young women. An assistant provost told us that "close advising relationships and close relationships with faculty help students a lot. The small, personal setting helps students feel comfortable and therefore make them more willing to take intellectual risks. Small classes allow them to make connections more easily with faculty and with each other. Students have a genuine sense that faculty and staff are there to help them and advocate for them, and this allows them the sense of security to excel and succeed." Barnard students have the most dynamic city in the world outside the walls of their campus. Virtually every kind of interest can be met. Extracurricular activities are coordinated with Columbia College so that all clubs, committees, athletics, publications, affinity groups, and artistic endeavors are available to all undergraduates on both campuses. Yet Barnard is not every student's cup of tea. Because of its location in an urban setting and its relationship to a very large university, there is less of the intimate residential culture one would find at many of the other selective liberal arts colleges. Barnard's diverse student body and political and academic climate is a natural extension of these factors.

What the College Stands For

A dean of students succinctly captures Barnard's educational goals: "To ensure that they read critically, write powerfully, speak eloquently, and think clearly." Barnard is a standard-bearer for the education of women and their preparation for future success in all walks of life, plus an exposure and appreciation of international, cultural, religious, gender, social, racial, ethnic, and intellectual diversity. Its student body is composed of one of the largest racial minority populations of any private college in the country: some 35 percent are women of color. The college accomplishes this commitment by virtue of its very generous financial-aid program.

Barnard's president has written to prospective students about the college's philosophy and important attributes: "As a residential liberal arts college, Barnard offers students a faculty of distinguished scholars who remain accessible to undergraduates, along with a dedicated and responsive student services staff. Barnard's New York City setting offers students a world of museums, theater and music, as well as possibilities for year-round internships in institutions that stand at the center of the fields of commerce, publishing, science, medicine, education, the arts, and finance. As members of one of the four colleges of Columbia University—and the only one to remain independent—students are part of a vibrant 'academic acropolis' on Morningside Heights, which also includes the university's graduate and professional schools and a number of neighboring institutions, including Teachers College, the Jewish Theological Seminary, and the Manhattan School of Music. And, as a women's college, Barnard is a place where women never take second place." Barnard is pushing its students to be aware of and able to succeed in the global environment. As a dean shared with us, "Barnard is a nurturing community that offers an excellent education which is unique in that it is individually tailored to meet the needs of a curious, creative, smart student body. It has a great history of having educated great women, known and unknown historically." Barnard has become more selective and challenging academically, and at the same time is seeking to become more internationally focused in its educational approach and outreach.

Curriculum, Academic Life, and Unique Programs of Study

Barnard's major educational goal, according to a dean, is "to inculcate analytical and critical skills by exposure to a liberal arts curriculum. This exposure puts a premium on breadth in balance with the depth that grows out of the study of a major subject." There are general education distribution requirements in place at Barnard, with a good deal of flexibility in course selection to fulfill the four areas of study. The workload is quite heavy, and since the majority of classes are small, the faculty requires of their students a good deal of writing and independent projects. The humanities and social sciences are the stronger areas of study and the favored disciplines of Barnard women. A good number of premed students add to the mix and they frequently take their science courses at Columbia College. In keeping with its high expectations for its students, Barnard requires that all seniors complete a senior thesis on a focused topic of interest within their major or take a comprehensive examination to demonstrate mastery of the discipline. Another special advantage of the college's location in New York City is reflected in a number of special combined majors. For example, students can achieve a major in music with courses from both Barnard and Juilliard or the Manhattan School of Music. A student can create any combination of dance, music, theater, visual arts, or writing concentration within and without the Barnard curriculum. A talented young woman can look at Barnard as a school with many options for study, and therefore not have to seek out a more specialized arts environment. Through opportunities like the Civic Engagement program, students are able to take advantage of Barnard's surroundings. As an administrator points out, "Our location in New York City is one of the primary reasons that students choose Barnard; it allows them a context in which to test out theory from the classroom in real-world situations through community service, internships, on-campus jobs, and more. Barnard stu-

dents have a love of learning. They want it all." Barnard has recently focused attention on building up campus traditions, spirit, and a more residentially focused college community through initiatives like Spirit Day, New Student Convocation, and a community-service event called First-Year Reach Out.

Major Admissions Criteria

Barnard's dean of admissions tells us, "We are looking for academically motivated young women who want to take on the rigors of a challenging academic environment. They need to be independent enough to handle an urban environment and want to delve into the cultural life and internships that will be present. Students who do well at Barnard are all of the above plus individuals who are secure with themselves, thrive on being active and stimulated by those around them." In its materials, Barnard notes that "the college seeks women who will benefit most from the Barnard experience: a diverse group of motivated and curious young women who will draw from its deep well of opportunity and contribute to its stimulating community." A dean looks for "smart, self-motivated students who have a good preparatory education." And, more formally, the college states, "The Committee on Admissions selects young women of proven academic strength who exhibit the potential for further intellectual growth. In addition to their high school records, recommendations, and standardized test scores, the candidates' special abilities and interests are also given careful consideration. While admission is highly selective, no one criterion determines acceptance. Each applicant is considered in terms of her individual qualities of mind and spirit and her potential for successfully completing the course of study at Barnard." According to the dean of admissions, the key admissions criteria at Barnard are "proven academic achievement and signs of potential to succeed at Barnard as assessed through transcripts, standardized testing, recommendations, and writing; evidence of motivation and desire to learn, experience difference, and leave one's 'comfort zone' as assessed through recommendations, an optional interview, and writing; and a track record of outside involvement and sustained interest in nonacademic pursuits."

The Ideal Student

A young woman who is truly in search of a number of new challenges will do best at Barnard. The many types of personalities, backgrounds, and points of view that make up the undergraduate student body will stimulate the right person. "Brains, hard work, organizational skills, and the willingness to ask questions" will help students succeed at Barnard, according to one dean. The college cannot but help take on some of the tone of its environs: There is a competitive, bustling, urban tone exuding a scent of brashness and independence. Barnard women do not hesitate to offer their opinions on issues of importance to them. The workload and quality required of all students can be daunting for someone looking for a less than intense intellectual program of study. With the infinite array of activities and distractions to be found in the Big Apple, Barnard has less of an intimate, cohesive community feel to it. Even so, one dean says that students "should know that Barnard really cares about its students and that faculty, administration, and staff consider them almost like family." The college offers this portrait of its students: "Independent. Intellectual. Interested in ideas. Eclectic. Feisty. Serious. Talkative. Inquisitive. Cultured. Self-assured.

Hardworking. Enthusiastic. Politically aware. Adventurous. These are a few of the ways Barnard students collectively describe themselves. Some of these qualities—seriousness, eclecticism, independence—are evident in the experiences and achievements they bring to Barnard. They have graduated from many different kinds of secondary schools and they come from all over the United States and more than forty countries. Their accomplishments range widely. They have been editors of high school newspapers, scientific researchers, published poets, gifted athletes and performers, tutors of disadvantaged children, activists, and committed students. Barnard students are superb resources for one another. They make Barnard an exceptional, exciting institution. But for all of their diversity—ethnic, cultural, social, economic, academic—there is common ground among them. They have great talent and they are open to challenges. They take on new ideas, a rigorous academic program, an international city, and one another with an ease and self-confidence that is distinctly Barnard." Many of Barnard's students spend the greater part of their free time in the canyons of Manhattan. Since most students self-select the type of environment they wish to experience, Barnard women are independent, assertive, creative, mature, and willing to experiment, on the whole. There are many other colleges of excellence that would better suit the young woman not yet ready for this stimulating and challenging campus.

Student Perspectives on Their Experience

Why do students choose Barnard? Because of its location in Manhattan, its academic reputation and intellectual caliber, the diversity of its student body, its connection to Columbia University, and, given all of this, its continuing small-college feel, which is supported by strong faculty contacts. Says one student, "I chose Barnard because of its easy access to Columbia, because of New York, and because what they say in the brochures is pretty true: Students at Barnard get the best of both worlds. A small college and an academic advisor (unheard of at Columbia) in a huge university and city."

Students appreciate the level of education at Barnard, and are sometimes surprised at the rigor of the academic demands placed upon them. As one graduate notes, "Barnard offers a great education. When I first arrived here I was hung up on the fact that I was probably better prepared than other first-year students. This may have been true, but my grades my first semester were subpar and this was because I had slacked off. Barnard is not a slacker's college. I think it has a real hang-up in that it can be seen as the 'college across the street.' In general, I have had to work much harder in my Barnard classes than in my Columbia ones. It has made me excited about learning and I know I am getting a great education."

Complaints about Barnard have tended to focus on a lack of strong sense of school spirit or identity, on lengthy administrative processes, and on facilities. These are all areas the college has been working on, for example with its First-Year Focus residential program. As a student puts it, "In general, I don't view myself as a Barnard student. If you want a pretty much all-girl's environment, it's possible, but I don't, so I do identify more with Columbia."

Barnard students say that those who do best at the college are into the city environment, academics, and their independence. This is not a school for those strongly interested in sports or Greek parties. Those who do well are "motivated," "independent," and "smart," and are comfortable making new friends in what is a new environment for most. Barnard is a fairly competitive place, according to students, and those who enter ought to be prepared to work, to strike out on their own, and to be proactive in meeting new people.

Praise for Barnard centers around its faculty, its academic support and attention to individual students, its more flexible degree requirements, its relationship with Columbia, and its connections to the world of opportunities that is New York. Says a student, "I think it's great that BC [Barnard] has advisors for all their undergrads. All of my friends at Columbia complain about being anonymous and BC kids have someone to address any educational issue to." Barnard is "incredibly diverse . . . the kids aren't dressed similarly. . . . there are clubs for basically all groups and ethnicities. . . ." A sophomore says, "It's hard to talk about Barnard College without also talking about Columbia College. They are, if you want them to be, incredibly intertwined. That's what makes it sometimes difficult to identify and be proud of BC. Also, going to school in New York is an amazing opportunity, especially now with it being so much safer and such a trendy spot, but it does have its downside. It's not a typical college experience. But once you go to school here for a year you're hooked. I can't imagine going to school in the country. And it is a lot of work."

What Happens after College

One dean says of Barnard graduates, "Barnard students develop the intellectual resources to take advantage of opportunities as new fields, new ideas, and new technologies emerge. They graduate prepared to lead lives that are professionally satisfying and successful, personally fulfilling, and enriched by love of learning."

Barnard has a particularly fine reputation and history of sending its students on to top graduate programs. In fact, the college reports that it ranks third among private undergraduate colleges in the total number of students earning Ph.D.'s and in the top five in preparing future female physicians. Some two-thirds of Barnard students graduate having completed internships, many of them with prestigious firms and facilities in their field of interest in New York. These help down the road with career and graduate placement possibilities. Almost one-third of Barnard graduates will go directly on to an advanced degree, most in medicine, law, or business. The college lists the popular employment fields of the remainder as "the arts, communications, teaching, social services, and many other fields."

Barnard offers a counseling-oriented approach in its career development office: "Our department's philosophy as a developmental service, rather than a placement service, plays a critical role in assisting students as they transition from the structured experience of Barnard to the decidedly unstructured life as a college graduate. Not only does this mean that we provide resources and support to students in sorting and valuing their interests and passions in order to locate career options, but it also means that rather than placing students in internships or creating community service opportunities, we assist students in clarifying their values and giving them the tools necessary to locate venues to gain the knowledge, skills, and experiences that they need and desire."

Bates College

23 Campus Avenue
Lewiston, ME 04240

(207) 786-6000

bates.edu

admissions@bates.edu

[
Number of undergraduates: 1,660
Total number of students: 1,660
Ratio of male to female: 48/52
Comprehensive fee for 2008–2009: $49,350
(includes room and board, tuition, and fees)
% of students receiving need-based financial aid: 40
% of students graduating within six years: 89
2007 endowment: $275,557,000
Endowment per student: $165,998
]

Overall Features

Its founding in 1855 by Free Will (Northern) Baptists has left a significant imprint on Bates's social and academic environment. This early religious denomination was in the forefront of the abolitionist movement in the nineteenth century. From its beginning, women and minorities were actively encouraged to matriculate, and the college never had fraternities or sororities. Bates continues its commitment to embrace all voices, points of view, and types of students. Thus for a college of small size located in an old industrial town in rural Maine, Bates attracts a fairly mixed student population. It has become more international and national in recent years, enrolling half the class from outside six New England states and maintaining a proportion of the incoming class of 16 percent American students of color or international students. On the whole, students and faculty care about social issues and the environment. This is reflected in the many campus social organizations and various fields of study available to students. Bates has never had to deal with the issue of fraternities and sororities because of its founding philosophy. Consequently, students feel a close sense of community and a concern for their fellow learners, which influences the overall tone of the college. Bates has never had the image of a traditional, preppy, or cliquish college. Faculty are recognized for the quality of their teaching and commitment to inspiring their students. Classes are kept quite small and interactive.

What the College Stands For

Bates is a total undergraduate educational community that focuses on educating motivated, intelligent young men and women in the liberal arts tradition. Its mission is to provide access to individuals of all backgrounds and points of view who wish to learn. It has a quality of openness that is reflective of its mission statement. As with the majority of the colleges in *The Hidden Ivies*, its goal is to prepare responsible young men and women for positions of leadership in their future personal, professional, and civic lives. It emphasizes its concern for social action and protection of the environment through its teaching and programs. Bates has never sought the status of a socially or economically elite institution. A high percentage of students receive financial aid at a college with a smaller endowment than that of the majority of its peer institutions. This is a full measure of its commitment to its purposes. One administrator noted, "The aid budget at Bates has often gone up by between 2 percent and 5 percent more than the fees each year, so we are able to meet the full need of all students for four years at Bates, and with sharply reduced loan expectations." Another added, "There is a strong sense of community at Bates, both among our students and between students and faculty and students and administrative staff. There is a collegial spirit of excellence here. Our academic atmosphere drives students to succeed without being consumed by a competitive, cutthroat mentality."

Bates celebrated its sesquicentennial in 2005, focusing its vision on four principal educational goals as part of an extensive planning process:

- Academic rigor and achievement through teaching, learning, and scholarship.

- A distinctive ethos and culture of engagement, civility, and service.

- Connections and an integrating cohesion expressing the promise and value of a liberal education.

- A flexible, principled residential community that is attentive to individual learners.

These goals clearly reflect Bates's two core founding principles of academic rigor and egalitarianism. Bates plans to stay true to its mission and its educational direction in the future. The college requires that all organizations be open to everyone and sees its students as open, friendly, and interested in others. As one dean noted, "Bates is an academic community that is unpretentious and genuine in its interest in the well-being of constituencies within and without. Bates is an involving place—inspiring more students to become engaged in activities outside the classroom that are meaningful to them, add texture to campus life, and make a difference in the lives of others."

Bates has significantly invested in itself over recent decades, tripling its faculty and doubling its student body since the 1960s, adding seven new buildings and many rebuilt buildings in the last twenty years (including new student housing and a new dining commons), and expanding its women's athletic program. Future plans include building a student center, expanding endowment, continuing to try to recruit top students, and bringing more diversity to the college. Bates's Multicultural Center is a "palpable addition" to the campus, improving an established situation of cooperation and communication among diverse groups on campus. Of the faculty, almost 15 percent are people of color and 48 percent are women. A dean comments that "technology will challenge us to make the residential nature of the college more meaningful. We need to continue to focus on faculty, students, and facilities (including technological facilities)." Overall, Bates will continue to focus on emphasizing rising academic achievement levels and the development of the "transformative nature of a liberal education."

Curriculum, Academic Life, and Unique Programs of Study

Bates provides a flexible curriculum with broad distribution requirements. It has a special academic calendar that is made up of four-four-one terms (four months, four months, and one month). A student can choose a specially designed course of interest, often a nonmainstream topic, for the one-month spring term. Among the over 100 independent study projects completed each year are: alternative education, Horace, ceramics, and bacterial plasmid profiles. Given its mission, Bates's course selection includes a number of multicultural, cross-discipline courses. These include interdisciplinary majors in neuroscience, environmental studies, biological chemistry, East Asian studies, women's studies, African American studies, American cultural studies, and classics and medieval studies. Classes are small and interactive, and faculty are regarded as available and interested in their teaching. As one administrator noted, "The faculty and staff at Bates come to the college for the same reasons the students do—they get to fire at close range and don't lose people in large lecture halls or between the cracks." It is fairly easy to arrange for a tutorial or special project with a professor in a field of interest, and it is not uncommon for student-faculty research to be published. Bates is well regarded for the

quality of its science programs. A large number of students major in the sciences, especially premedical courses and environmental sciences. The track record of acceptance into professional graduate schools is excellent. Competitive debating on a national level is an active mark of distinction at Bates. Many students are attracted to this demanding endeavor for its rigorous training in logical thinking and communication skills.

Bates prides itself on its service learning program and its international study offerings. With a population of about 65,000, the twin cities of Lewiston-Auburn present communities that have assistance needs that Bates students can address. The Harward Center for Community Partnerships "uses a collaborative model in their work with social service agencies, involving them as partners in needs assessment and the development of service opportunities for students. As a result, Bates students, through their service, are engaged in reciprocal relationships with those they are working in partnership with." Bates consistently ranks in the top ten among all colleges in terms of student participation in international study. Almost all seniors at Bates write a thesis, a program on which the college puts a great deal of emphasis.

Major Admissions Criteria

Like its neighbor, Bowdoin, Bates makes standardized testing an optional requirement for admission. Because of its emphasis on thinking, writing, independent study, and active class participation, the college has found standardized test scores to be a less important predictive factor than a good high school academic record, "extracurricular engagement of substance," and "values," as demonstrated in essays and letters of recommendation.

The admissions office looks for "demonstrated academic achievement in high school, a broad pattern of commitment or interest in others, and particular skills or talents that have been brought to a high level of ability." Bates seeks students who are inquisitive, open, and perhaps with "rough edges." Applications are at an historically high level at Bates, with a 29.6 percent acceptance rate and a 33.7 percent yield rate. Over half of Bates admitted students were in the top 10 percent of their high school class and a third were in the top 5 percent.

The Ideal Student

It is obvious that Bates is not the logical college choice for the urban-oriented student. While Portland, a medium-size city, is within comfortable driving range, Boston is 140 miles to the south and not too easy to reach during the long winter months. For the motivated, smart student in search of a warm, close-knit, socially open environment with an emphasis on outdoor life and Division III and intramural sports, Bates will welcome you with open arms. In the parlance of the contemporary student world, Bates is a very suitable place for an "L. L. Bean" rather than a "J. Crew" type. "Students are apt to be involved in four or five different activities, 'sea anchors' that keep them involved and committed. There are lots of trip wires and support mechanisms: faculty advising; dormitory junior advisors and resident coordinators; mentors, coaches, and the like."

A dean notes that "students who have genuine intellectual curiosity will probably flourish at Bates, and students who have honest and curious interests in others who don't share all

of their ideas and background. It's not the right school for someone who wants to sit comfortably in a fraternity or sorority with thirty like-minded people for four years. Bates is not perfect at getting people to move outside their assumptions or backgrounds, but is active in encouraging students to do so." Another adds, "Bates students, for the most part, are modest in their sense of self but clearly recognize that they are a part of a rigorous academic community and have the ability to excel academically, meet the expectations of our first-rate faculty, and positively contribute to campus life."

Student Perspectives on Their Experience

Says one student, "The major reasons that I chose Bates are the friendliness of the student body, the fabulous academics, and Maine!! Bates appreciates and respects the notion of being an individual." Students appreciate Bates's sense of self and its advocacy of inclusion. They value the academic level of challenge at the college and their access to faculty. "Academics are great, with small classes and cool, really interesting profs. Furthermore, they are challenging and definitely make you re-evaluate the way you see the world." Student diversity in its many forms is both appreciated and lamented. "The neat thing about Bates is that you have people with as different post-graduation plans as Harvard Law and working on an organic farm all living and interacting on campus together." Yet "Bates is in Maine, which is not a bastion of worldly culture or ethnic diversity. . . . The college is full of interesting and bright kids, and not just a bunch of rich kids either. The kid who lived next to me last year was from Ethiopia and had never seen snow before." A senior notes, however, that "the college strives to bring in as many diverse students as possible and make them feel welcome and comfortable. I have to say Bates is still very white, upper-middle-class, though."

While the absence of fraternities and sororities is appreciated, there is still alcohol on campus. A junior student advisor points out that "Bates has numerous activities planned each weekend that do not involve the consumption of alcohol, as well as during the week. There are many clubs and activities worthy of joining." Another negative that students mentioned was the small size of the college, which they felt facilitated "gossip" on campus: "It almost feels like what is your business is going to become, at least, your neighbors'." The flip side of this is that students help each other to succeed: "We all help each other out; without that we would not be able to advance."

Students are divided in terms of feeling they have an impact on campus decision making and would like more communication with the administration, but they expressed the sense that the college, and the faculty in particular, are attentive to their needs. "The college is small enough that we have personal contact with professors and deans, and even the president. If a student wants to get something done, has an idea or whatever, it is possible to get it done. Everyone has a voice, all the time!" A junior resident coordinator notes that "the school has a talented and enthusiastic faculty, with an intense and demanding academic program. It emphasizes the importance of learning to think critically and also of balancing strong academics with a healthy extracurricular activity. The school offers a vast amount of things to be involved with."

Students report that the "well-rounded" student does well at Bates, "a motivated, well-balanced student who knows what it takes to be successful, but also knows how to have fun."

Clearly, students should want and appreciate the small college community that Bates offers. "I am impressed with the connection that occurs between students and faculty. The faculty is available to meet whenever needed and is always highly approachable." Another student adds, "Bates makes a family type of community where you can feel comfortable being yourself so that you can meet new people. There is no competition, which makes it even better. The competition is with yourself."

Lewiston does not receive rave reviews, but as one student notes, "I also chose Bates for its location. Yes, some may think this is strange, as the Lewiston-Auburn area is not the most desirable area to dwell in. But I saw incredible opportunities for community service work. I am heavily involved in volunteering with the public school system in looking for ways to boost the aspirations of children in this area." Additionally, students take advantage of the Maine outdoors environment surrounding Lewiston. Overall, "Bates is a great place to spend four years learning and growing. It is a special place." It is special because of "the people and the love and passion they feel for Bates College and the community here, and the support the community offers its students."

What Happens after College

One administrator commented on some alumni survey findings: "For a college with a longstanding reputation in the sciences, we were surprised to discover we had more graduates whose title was *entrepreneur*, having founded their own businesses, than graduates who were M.D.s." Some two-thirds of Bates graduates earn graduate degrees, and there has been an increase in the proportion of graduates who are pursuing business, law, and technology professions, as compared to education, for example. Bates provides "broad preparation for service as well as leadership." A student reports that "the career center is invaluable." It helps students go on to a broad area of jobs and graduate programs. The highest percentage of alumni go on to careers in business and industry, education, health and medical services, and full-time graduate study. As Bates aptly notes, "Alumni frequently cite the capacities they developed at Bates for critical assessment, analysis, expression, aesthetic sensibility, and independent thought."

Boston College

140 Commonwealth Avenue
Devlin Hall 208
Chestnut Hill, MA 02467-3809

(617) 552-3100

(800) 360-2522

bc.edu

[
Number of undergraduates: 9,060
Total number of students: 13,087
Ratio of male to female: 48/52
Tuition and fees for 2008–2009: $37,950
% of students receiving need-based financial aid: 40
% of students graduating within six years: 91
2007 endowment: $1,670,092,000
Endowment per student: $127,614
]

Overall Features

Extraordinary changes have occurred in recent years at this traditional, Jesuit-founded university. Boston College has transformed itself from a local commuter school of limited resources for mostly working-class Catholic men to a national university with global aspirations. In the words of the dean for enrollment, "As someone who has worked as an instructor and an administrator at the college for over thirty years, I can give witness to the massive increase in available resources to expand opportunities in research and student learning." The leaders of the College made several strategic decisions in the 1970s that have borne fruit in the ensuing years. As they concentrated on making BC into a fully coeducational, residential institution, they also made the decision to maintain the on-campus enrollment at 9,000 students. Rather than expand the size in the face of a constantly increasing application pool, the commitment was made to increasing the quality of the academic environment. The dean for enrollment told us that "by maintaining a constant undergraduate enrollment and by graduating a high percentage of students together as a class after four years, we also encouraged the same sense of community and esprit de corps that has characterized a Boston College educational experience for many decades." The goal has been to raise the bar for each entering class, a fact that is clearly reflected in the greater selectivity of recent entering classes.

The contemporary environment on the Heights, as BC students and faculty refer to their campus, reflects more than ever this sense of a close community of students and faculty, unusual for an institution of its size. The strategic decision to create a national-level intercollegiate athletic program has resulted in a significantly broader geographic and a somewhat more diverse student body. Given the spirit on campus that derives from a nationally prominent athletic program, the broader outreach by the admissions staff, and its improved endowment, BC has achieved its goal of joining the ranks of the very selective, nationally recognized universities. While maintaining its historic identification as an undergraduate college, BC has developed into a full-blown university due to the growth of graduate programs that include law, business, education, nursing, and a number of academic disciplines.

Boston College is unique in a number of ways, starting with its historic and still active role as a Roman Catholic institution, its midrange size, its location nearby a major metropolitan city, and yet in other ways it has taken on the features of some of the outstanding state universities. BC offers minors in Jewish and Islamic studies, yet recently added Christian symbols (mostly crucifixes) to all its classrooms, as well as statuary and other Catholic elements to its campus and programs. It offers a broad range of major concentrations that include arts and sciences and career-related fields across nine different colleges; the presence of a number of significant professional and academic graduate schools; the commitment to Division I-A intercollegiate athletics, and the intense level of support for its teams are more akin to the public universities than to the smaller, strictly liberal arts colleges included in the Hidden Ivies.

BC's success in NCAA competition in the highly visible sports of football, basketball, and ice hockey have led to increased applications from the more traditional student who wants to combine a strong academic experience with the opportunity to follow top athletic teams. This factor, in combination with the growing reputation as offering challenging academics, has made the school very competitive

for admissions in recent years. Another result of these powerful factors is increased fundraising from alumni and friends of the college. BC has thus launched a ten-year strategic plan to build new residential, academic, and athletic facilities to replace outdated facilities.

The college's location is one of its great attractions. While situated on a distinctive campus in an attractive suburb, it is only twenty minutes away by trolley from Boston, the mecca for thousands of collegians from all parts of the country and abroad. Parents as well as students are attracted to its sense of place with its green campus, many new buildings, and its safe environment. There are two campuses on which students reside as a result of the closing of a women's college and the purchase of its property by BC some years ago. Students differ in their attitude toward living on the smaller, more intimate campus that is separated from the main campus. Over the past five years, the college has doubled the physical size of the campus while intentionally not increasing its enrollment. It has a defined plan to remodel older residence halls and build new ones in order to guarantee 100 percent housing for all its undergraduates.

An observer of the college world could be forgiven for referring to the miracle on a hilltop in Newton, Massachusetts. From a financially fragile college on the edge of bankruptcy that historically served first-generation students of limited means who commuted to a campus with limited resources, Boston College's leaders in the 1970s determined to create a new university by dint of a focused vision of what could become a highly selective, multidimensional, community of students and teacher/scholars. Its mission as a Jesuit institution that educated young men and women in the religious and social responsibilities of this tradition has been maintained through this dramatic transformation. This is reflected in the curriculum and the diverse social service activities on and off the campus.

What the College Stands For

BC believes strongly that it has a distinctive mission as a Jesuit and Catholic university. It pursues its mission to serve society in three ways: by fostering the intellectual development and the religious, ethical, and personal formation of its students in order to prepare them for leadership in a complex, global society. The Jesuit concept of *cura personalis* (care for the individual) means that the faculty are expected to take an active interest their students. Several prominent teachers spoke to this spirit; a professor in the Carroll School of Management put it this way: "After teaching at Duke and Harvard, and studying at Cal Berkeley, it is an honor to be a professor at Boston College. There is nowhere else I would rather teach. BC is unique because of its focus on truly caring for its students and seeking the best from them and for them in everything we do." Said a senior administrator, "The faculty works at accomplishing the college's goals by producing significant research that advances insight and understanding, addressing important social issues, and creating an active dialogue between religious belief and other formative elements of culture through the intellectual inquiry, teaching and learning in a residential community of teacher/scholars and students." A professor of chemistry underscored the priorities of the college and thus the tone of the campus this way: "Among my goals at the undergraduate level are to produce students who successfully compete on the national stage, which in many cases means getting into first-rate graduate and pro-

fessional schools. I push my students to excel in class, which is what many are capable of doing, and what BC is all about." This same teacher echoed the view of a number of her colleagues that the caliber of the student BC enrolls today is brighter and better academically prepared and thus prepared to handle a higher level of expectations from their teachers.

BC began more than a century ago as a college to serve upwardly mobile immigrant populations. It continues its mission to provide opportunities for emerging populations that seek an advanced education. In the past five years the college has guaranteed to meet the full demonstrated financial need of all its undergraduates. A policy of need-blind admissions is in place and thanks to the generosity of alumni and friends of the college, scholarship funds are available to all students who are accepted. A commitment to expanding the racial and ethnic diversity of the student body is in place, fitting with the mission of the college.

Building on its tradition as a Jesuit institution, BC has a vision of creating an international dimension to its teaching and learning experiences. It plans to reach out to the long-established Jesuit universities around the globe to broaden the scope of its programs to undergraduate and graduate students. Between its ambitious capital improvements to the campus, and its additional faculty and programs, BC conveys a sense of excitement, confidence, and energy for the next phase of its remarkable development into a world-class institution.

Curriculum, Academic Life, and Unique Programs of Study

Boston College may not appeal to the student who seeks a highly flexible, self-directed undergraduate experience. True to its vision of what it wants students to learn and consider in their adult lives, the college has a required core curriculum. BC believes that the educated citizen and professional in all walks of life will profit from exposure to the ideas and principles of both Western culture and the other major world cultures. Students indicate that there is a wide range of subjects in the various disciplines that can appeal to the individual, and that fulfilling the core requirements is not too onerous.

The undergraduate college is comprised of four distinct schools: the College of Arts and Sciences, the Carroll School of Management, the Lynch School of Education, and the Connell School of Nursing. While each school has its own specific course requirements, students are expected to explore subjects in the social sciences, humanities, physical sciences, math, arts, and theology. In addition, all freshmen are required to take a writing course and a course that focuses on cultural diversity. In the senior year a series of seminars on specific topics are offered in the Capstone program. High school students interested in BC should review carefully the general core requirements, as well as those in each of the four schools, to be aware of what will be required of them if they enroll in any one of them.

The college offers an honors program that provides advanced-level seminars that integrate related areas of learning. There are many cross-disciplinary majors and minors that are popular with undergraduates. Many courses combine work and service off campus that fulfill various core requirements together with the goal of expanding the intellectual and social horizons of the students. There are opportunities to work with faculty on advanced research projects as undergraduates, something that is usually reserved for graduate students at most research-driven universities. Some 20 percent of undergraduates have worked on a project with a faculty member and some have claimed joint authorship with a professor on a research project. BC makes available to its undergraduates seventy study-abroad programs, in which over 40 percent of the student body participates. The college believes that its deep Jesuit network around the world provides a standard of quality for its students that is frequently lacking on many campuses, often affecting recognition of credits by the home school in fulfilling its degree requirements.

In keeping with its religious tradition, volunteerism is a major activity among a large number of undergraduates. The college's Appalachia Volunteers program is particularly popular. A wide range of religious and nonreligious retreats are available and, according to one dean, virtually every student has participated in multiple retreats. The college has purchased a dedicated retreat house located nearby on the Charles River to meet the demand for retreat space. The PULSE program combines elements of the core curriculum, specifically the theological and philosophical courses, with a service component. The program is meant to introduce students to social injustice in action and give them the opportunity to actively combat such injustices. The program focuses on analysis and reflection of the service component.

Reflective of the greater faculty attitude toward the students they teach, a senior professor described the qualities he looks for in his students this way: "I look for intellectual curiosity, enthusiasm, creativity, and breadth. Students who do well at BC (and after graduating from BC) are those who are very bright, have broad interests inside and outside the classroom, and seek (and take advantage of) opportunities to learn and grow in all aspects of their lives." This same professor perceives the faculty, staff, and administration as all working together to create a truly caring and enriching atmosphere that encourages, seeks, and expects (and gets) the best from each student at BC. This attitude plays a significant role in the unusually high rate of retention and successful graduation by BC students. In a recent class, 91 percent who entered as first-year students graduated within six years, a statistic that matches the most selective colleges in the nation.

Major Admissions Criteria

Here is how the dean for enrollment describes the kind of students it looks for in a large and competitive pool of candidates: "We look for students with a passion to accomplish something worthwhile, who are well-rounded, inquisitive, articulate, creative, and show promise of becoming future leaders of the world." While this statement will resonate with the majority of the selective colleges and universities included in *The Hidden Ivies*, BC reviews its applicants within the context of its unique environment. Students must embrace the overlying philosophy of the institution in terms of its social consciousness and faith; be prepared to work hard in their studies; and participate wholeheartedly in the community through various social, athletic, political, artistic, and service activities.

Accordingly, the admission committee reviews the level of academic challenge students have undertaken within the context of their high schools' curricula. The strength of the students' performance in tackling the academic resources available to them is the most important factor in evaluating candidates. The dean goes on to say, "Admissions officers seek to become experts on the high schools that they visit, and they pore over transcripts and recommendations to see who has that passion to learn and to lead. We of course also consider test scores, writing samples, and other credentials that a student may offer, but primarily these are used to fill out the picture of what drives and motivates each student." The college's commitment to expanding its representation of students of color, different ethnicities, and cultures is reflected in its outreach efforts. While the numbers are not large, recent admitted classes indicate increases in the number of black, Latino, and Asian American students.

The Ideal Student

Students who are happy and successful at BC possess traditional religious values, a strong work ethic, a solid academic foundation and executive learning skills, social and emotional maturity, professional career aspirations, and an enjoyment of community life. With 9,000 undergraduates on campus and a host of available organizations and activities, an individual should find it easy to form close friendships, engage in meaningful activities, and identify an academic field of significant interest. Because BC is a larger environment than the majority of selective liberal arts colleges and with so many options on all fronts, a student has to have a good sense of herself in terms of interests, values, and priorities. Good time-management skills and consistent focus are key to a successful experience. BC has worked hard to provide small-size classes and faculty advising, but an independent and assertive attitude is necessary to make one's way through the myriad academic and social options. The payoff is the opportunity to make new friends, engage in new activities, and explore new fields of study throughout the undergraduate years.

BC is not the most amenable environment for a reclusive or socially insecure student. The campus is alive with activities and social relationships. A student who needs a good deal of direction and encouragement is likely to be more comfortable on a smaller campus. The atmosphere on campus is influenced by the major athletic programs and many students follow the teams by attending games and the social events that revolve around these events. BC students are physically active as well. Nearly one out of ten students participates in varsity sports and a greater number in intramural programs. An interest in participating in some area, while not required, does make for a successful experience. There is a large cohort of students who are not necessarily interested in the social/athletic side of campus life. They commit their energies to volunteering through the many social service options made available to them; so for them, BC can be a good place as well.

Student Perspectives on Their Experience

BC undergraduates are among the most enthusiastic collegians one will encounter. They speak of the demanding, but not oppressive or overwhelming, academics; the knowledgeable faculty who make themselves available for advice and direction; the upbeat campus spirit and pride in their college; the multitude of activities on the social, athletic, and community service fronts; and the easy access to Boston and Cambridge with their offerings of good music and arts events, eateries and watering holes, and professional athletic teams. With its emphasis on building a comprehensive college community, the college has stepped up dra-

matically the number of activities on campus, from artistic offerings to socials and, of course, the intercollegiate sporting events. One junior who is engaged in a number of campus organizations summed up the appeal of BC to a majority of its students this way: "The atmosphere in Chestnut Hill is so young, vibrant, and exciting. It feels like a college environment should. To me, BC holds the quintessential collegiate experience: the fanatic sports fans (both for the college and professional teams around Boston), an outstanding academic (and Jesuit) reputation, the excitement of the nearby city, and the whitewash brick buildings." The opportunities for volunteering and community service through the various organizations with a spirit that incorporates the Jesuit ideals and outreach locally, nationally, and internationally are considered another of the outstanding features of the college. Many students are drawn to the off-campus retreats and service trips at some point in their undergraduate years.

Students find the advisory system very helpful for those who take advantage of it. The faculty are expected to connect with their advisor to review their course programs and potential fields of study. The faculty is there for you if you reach out to them, according to students who found this to be a valuable factor in their success.

Several caveats expressed by many students center on the expectations of their faculty and their peers. The workload is relatively heavy, good grades are not easy to come by, and there is an assumption that virtually everyone is preparing to enroll in graduate school, especially in the professional schools of law, medicine, business, and education. The other concern is for the lack of diversity within the student body, notwithstanding the efforts of the administration to broaden minority representation. Viewed from a historic perspective, it is ironic that students perceive their community as heavily dominated socially and economically by a white, middle/upper-middle-class, socially sophisticated, preppy cohort. In a kind of social ying-yang, many students would like to see more diverse students on campus while at the same time being very comfortable in their world of peers with whom they share similar backgrounds and exposures.

A number of students comment on the lack of access to some of the outstanding courses taught by top professors due to their popularity and the ceiling on the size of the classes. Some students have expressed their difficulty in adjusting to the size of the college, and others complain about the availability and quality of the housing (an area the administration is actively working to remedy). But this represents a minority of students who express tremendous enthusiasm for their college, from the faculty to their peers to the opportunities put before them.

What Happens after College

In a recent survey, 95 percent of undergraduates indicated they planned to pursue advanced degrees. Over a third of students continue directly on to further studies, particularly law-, medical-, and other career-related graduate programs. A large number of students enter the job world and then enroll in advanced degree programs in business, social work, education, nursing, and medicine. The many opportunities for internships available in the Boston area give undergraduates a leg up in the job market at graduation time. A significant portion of

graduates commit to volunteer service for several years in popular programs such as the Peace Corps, Teach for America, the Jesuit Volunteer Corps, and AmeriCorps.

BC has put a great deal of resources into its sciences, and the results show. An indicator of the attitude of the college as a whole toward achieving a reputation for preparing students for future educational opportunities second to none is expressed by a professor in the physical sciences who offered this view of her work with undergraduates: "Among my goals are to produce students who successfully compete on the national stage, which in many cases means getting into first-rate graduate and professional schools. I push my students to excel in class, which is what many are capable of doing, and what BC is all about. I expect the students in my classes to more than hold their own with students at other top schools on GREs and MCATs. And they do." BC has staked its growth as a leading college and university, in part, on the preparation of its students for admission into the top graduate schools and successful completion of the degree programs. A student who has aspirations for a professional career and is willing to work hard in his or her studies, with a proper balance of social and extracurricular time, will be recognized and supported by the faculty in meeting these aspirations. A meaningful indicator of success on the part of faculty and student teams is the remarkable number of BC graduates who have won the prestigious Fulbright scholarship for advanced studies: in the most recent graduating class, fourteen students were awarded Fulbright scholarships.

Bowdoin College

5700 College Station
Brunswick, ME 04011-8441

(207) 725-3100

bowdoin.edu

admissions@bowdoin.edu

Number of undergraduates: 1,716
Total number of students: 1,716
Ratio of male to female: 47/53
Tuition and fees for 2008–2009: $38,190
% of students receiving need-based financial aid: 42
% of students graduating within six years: 89
2007 endowment: $827,714,000
Endowment per student: $482,350

Overall Features

Founded in 1794 to educate leaders for service to the larger community, Bowdoin has continued its mission in modern terms. It continues to focus on a strong liberal arts education that will prepare students for virtually all fields of endeavor. Its large endowment for a small institution enables it to present a broad range of courses and physical resources to deliver a first-rate education. Its historical emphasis on teaching and interaction between faculty and students in an intimate setting are the overriding features of the college. Its location on the Atlantic coast in a small town in rural Maine impacts the overall environment and, to some extent, the curriculum. "The subtleties and the harshness of Maine, combined with the pride and driven practicality of northern New England set Bowdoin apart from most competitive institutions around the country," says one member of the college community. Students

enjoy the out-of-doors lifestyle, with the Bowdoin Outing Club being one of the most popular activities. Students experience an intimate sense of community in a totally residential campus. Faculty are known for caring about students and interacting with them in and out of the classroom. At the heart of Bowdoin is its tradition of academic excellence and participation in a total living-learning environment. Its healthy financial resources allow Bowdoin to attract talented students of all backgrounds irrespective of their ability to pay the costly tuition. Bowdoin is one of the handful of colleges still able to follow a "need-blind" admissions policy.

Diversity and technology development are major themes of the administrative members as they want to prepare the next generation of community and professional leaders. Since 2000, Bowdoin's student body has grown by just over 100 students, but the student-faculty ratio has remained quite low. Bowdoin has the resources to accomplish this, in part due to one of the largest single gifts ever given to a small college: $36 million in 1997. The college hopes to continue its path to becoming more national, rather than regional, in outlook and reputation. These days, only about 40 percent of Bowdoin's students come from the New England states, and about 5 percent are internationals from twenty-six different countries. Almost every state of the United States is regularly represented in the Bowdoin student body. Bowdoin's campus is physically beautiful and immaculately kept up. Several new buildings have added substantially to the campus, including Kanbar Hall, which houses Bowdoin's programs in psychology, education, and teaching and learning; and new LEED-certified residence halls. Bowdoin's art museum received a dramatic expansion and renovation in 2007.

What the College Stands For

Bowdoin's mission statement reaffirms old principles and continues its commitment to them in the future. It concludes, "The purpose of a Bowdoin education—the mission of the College—is therefore to assist a student to deepen and broaden intellectual capacities that are also attributes of maturity and wisdom: self-knowledge, intellectual honesty, clarity of thought, depth of knowledge, an independent capacity to learn, mental courage, self-discipline, tolerance of and interest in differences of cultural beliefs, and a willingness to serve the common good and subordinate self to higher goals." Enunciated by the first president of Bowdoin, Joseph McKeen, in his inaugural speech, serving the common good is still present in the minds of all on campus. As one administrator put it, "We expect our students

and graduates to serve the common good. We prepare them to do this by teaching them to think for themselves, to learn to speak well publicly, to learn to read critically and write well, to tackle complex research problems and projects, to understand cultures and places other than their home, to develop a sense of history, to develop an appreciation for the arts and music, to understand science and the scientific method, to use information technology confidently and to its full potential."

Words that are used frequently by students and staff alike are *leadership*, *relations*, and *involved*. Students are expected to take the initiative in their learning, to lead their peers and communities, and to give back. Students establish significant relationships with faculty and each other and involve themselves in the myriad activities available on and off campus. The administration and trustees removed the traditional Greek system from the campus as a means for creating a more unified student body. The class of 2000 contained the last fraternity and sorority members. Similar to other small residential colleges, Bowdoin does not perceive a justification for social units that define and separate the student body. In place of the fraternities, a system of affiliation with a particular dorm or cluster, called the College House System, was established. "This is a creative and brave step for any institution, and the CH system takes a former strength of the institution and modifies it for today's more inclusive objectives. It continues to give students a sense of place, greater residential options, and opportunities for leadership." First-year students at Bowdoin live in the "Bricks," the traditional first-year dorms, forming the relationships and learning experiences that have been referred to as "the building blocks" of their Bowdoin experience during college and afterward.

A commitment to the faculty and to teaching comes first at Bowdoin. "Faculty are hired and promoted with teaching, as well as research, paramount and the culture of the college fosters strong faculty-student relationships." Students and faculty are said to openly share an enthusiasm for learning, while the atmosphere on campus is described as "open, down to earth, and rarely cynical."

Curriculum, Academic Life, and Unique Programs of Study

"The academic honor code and social code is at the heart of a Bowdoin student's experience. It sets forth a fundamental value of the institution: honesty in all things. We take our students seriously and expect them to approach their education seriously," says a college official. The college does not report GPA or class rank, and "students report that group learning is encouraged and valuable." A large percentage of each class majors in the sciences, particularly the life sciences, as a preparation for medical school or environmental studies. Another significant portion concentrates in the social sciences, such as history and economics, or in languages. The requirements for graduation are relatively flexible in that students must complete courses in each of the four general fields of learning, but they have an extensive range of choices. Bowdoin offers many interdisciplinary majors so there is ample opportunity to create a major concentration around one's personal interests. Such programs include Asian studies, African studies, women's studies, and environmental studies. The college continues its tradition of faculty mentoring students by means of collaborative research in the sciences and liberal arts. Many enthusiastic undergraduates speak of the significance of

working on a science project with a knowledgeable teacher or writing a creative novel or a paper on an extensive historical or literary topic. First-year students can choose from more than sixty seminars covering a wide range of topics. Over 60 percent of students engage in one-on-one independent study at some point in their college careers. Over half of Bowdoin seniors complete senior honors projects, involving extensive research, careful writing and revision, and an oral defense. As one administrator puts it, "These projects test many of the things we teach." Another says, "Such courses provide students with wonderful opportunities to test their own wings and to take control of their own education. The independent study promotes intellectual and/or artistic growth, independence of perspective, and confidence." Bowdoin is a member of the Twelve College Exchange program, which lets students spend a year at any of the other colleges. For some this presents a wonderful opportunity to experience a different setting and community and, in many instances, additional academic studies. Bowdoin students also have ample opportunity to study abroad for a semester or a full year. The college raves about its programs in Asia, South Africa, and Ecuador. "Study away has an enormous impact on Bowdoin students, providing over half of our graduates with opportunities to expand their horizons beyond the campus. Study away permits students to test themselves and their intellects in different cultures and institutions and to return more mature, with a wider vision, and almost always a deeper appreciation of Bowdoin and the education it makes available." For international students on campus, the Host Family Program provides a home away from home and an introduction to the United States. One of the college's most unique, special programs is the

Coastal Studies program, a marine biology and ecology laboratory located twelve miles from the campus. The Outing Club supports Bowdoin's preorientation trips, which help incoming students to appreciate the beauty of the college's surroundings. Bowdoin's performing arts series "brings the world of theater, music, visual, and performing arts to the campus and local community. One of the challenges of rural institutions is to expose students to the world outside the campus (a.k.a., the "Bowdoin Bubble"). By bringing the arts to the campus, you help expand the bubble and give students a lifelong appreciation of the arts—which is often felt to be a part of the liberal arts vision."

The Student Activity Fee Committee (SAFC) has changed the funding of student activities on campus. An event-by-event funding system has allowed more student organizations to flourish, so that the now more than 100 clubs and organizations have more flexibility in programming and funding.

Major Admissions Criteria

Bowdoin turned the tightly knit world of selective liberal arts colleges on its head almost forty years ago when it made optional the submission of the SAT. The perception of the admissions committee then, and now, is that other factors weigh more significantly in judging qualified candidates. This option has encouraged large numbers of talented and motivated high school students to apply to Bowdoin, allowing the committee to evaluate and accept candidates on more relevant information. "We were the first institution in the country to institute this policy, and it reflects our value of a concern for the individual and for individuality. We recognize that not all students fit a rough mold of scores, grades, and extracurricular achievement. Over the years, many, many extremely gifted individuals who would not have come had SATs been required, have come to Bowdoin and taken full advantage of this place and gone on to do impressive things in the world (including some of our young trustees)." One staff member recently described the resulting admissions process as a "transcript-heavy selection process." Bowdoin believes that "character should count," looking to recommendations and essays, for example, for evidence of leadership and concern for others.

Teacher recommendations, "particularly as they relate to attitude about learning," student essays, and a student's record of involvement are all particularly important to Bowdoin's selection process. In fact, Bowdoin's supplemental essay requires students to identify the high school teacher who has had the most significant positive impact on their development and to describe that relationship.

Important admissions criteria at Bowdoin are academic ability and potential, "leadership and the demonstrated potential to make a difference in the world," and a "commitment to serving the common good." Creativity and "independence of mind" are other factors mentioned. Students are expected to show "excitement about learning, interest in the world of ideas, openness to difference and to new areas of knowledge, courage and a willingness to grow and change, the desire to be part of a residential community."

Bowdoin is seeking to increase the diversity of the college, even as it reports about 25 percent students of color from all over the world, including America. Utilizing its needblind admissions and resources, the college continues to try to expand socioeconomic diversity, particularly in terms of maintaining its historic mission to provide a top education to citizens of

Maine, the poorest New England state, whose students represent the third highest proportion of students on campus by state origin, behind Massachusetts and almost even with New York. Former secretary of defense William Cohen and former senator George Mitchell both attended Bowdoin as scholarship students from Maine in the 1950s and early 1960s.

The Ideal Student

The qualities that Bowdoin looks for, and that are said to help students succeed, are self-motivation, high academic ability, leadership in multiple arenas, social conscience, and "the ability to think for oneself, not simply to do well on a multiple-choice exam, but to be able to write, compose, and create an original work." Students are curious, active, and independent. They are not competitive with each other, but rather supportive of one another's success and learning. They are problem solvers, critical thinkers, adventurous, and confident. One admissions officer noted, "They are incredibly adept at juggling numerous activities, and they are rarely spectators."

As one dean put it, "Maine is known for its physical beauty and the friendliness and rugged individualism of its people. Bowdoin reflects that. When you enter Maine on Interstate 95, the sign says *Maine: the way life should be.* That applies to Bowdoin, too: the way college should be." Bowdoin takes pride in its location and sees it as a major strength. Students should be looking for a personal, friendly campus, nestled in a rugged outdoor environment. "Some families may feel that Bowdoin is too isolated, but I contend that around this coastal location, which is full of natural, accessible beauty, one can develop and profit from exposure to a sense of wonder. It is hard to put into words, but Bowdoin and Maine are magical places because of their proximity to the sea, and this very unusual coastline, which is rocky and inhabited by mast pines and all sorts of unusual creatures."

A long-serving academic dean observed the ways in which Bowdoin students grow during their education at the college: "I see students who have found their own voices. Engagement and confidence, shaped by some sense of humility about the unknown, are key indicators of the personal and intellectual growth of Bowdoin students. All of these qualities were exemplified, for example, by the five environmental-studies coordinate majors I met at a lunch yesterday where they made presentations about their summer internships. The clarity and enthusiasm of their talks, the thoughtful and careful ways that they answered questions, their confidence in taking on challenging tasks with great responsibility, and the breadth of their perspectives both on their particular tasks and the larger enterprises in which they are engaged helped to exemplify for me the success of our teaching/learning community at Bowdoin."

From the admissions office comes this point: "Bowdoin is not for everyone. If a student desires an urban setting, if he or she is driven by grades or is content to be a spectator, Bowdoin would not be a good college choice. On the other hand, if he or she is unabashedly excited about learning, is not only unafraid of new challenges but embraces them, Bowdoin may be a good fit."

Student Perspectives on Their Experience

Students show evident pride in the college, its personal qualities, the relationships they have developed, and the access to faculty that they enjoy. They are somewhat more concerned over a perceived lack of racial, ethnic, and income diversity on campus, although this is not a universal feeling. What comes through most clearly is their sense of Bowdoin's excellent teaching, familial environment, and beautiful location. "What Bowdoin does best is to truly let the students value and appreciate our teachers. They play a very active role in the students' lives not only in the classroom, but outside the classroom as well," says one graduate. Another student offers, "The school is tremendously strong academically. Its size also allows students to develop mentor-type relationships with faculty, staff, administrators, trustees, and alums. These relationships complement the academic education by offering a personal education about life. These relationships, which I couldn't have imagined before attending, are perhaps even more valuable than the formal education I've received (which is very valuable in its own right)." About the classroom, we hear this: "On average, class size is about sixteen, which is very comfortable. This maximizes class discussion, and classes are less about taking notes, more about intellectual discourse." An international student says, "The relationship with professors has been extremely rewarding for me. Several of my professors have become more than professors to me. I consider some of them my friends. Bowdoin has given me more than education: I believe it has changed my personality in a positive way."

Students succeed at Bowdoin, says one junior, because of "the laid-back academic atmosphere and the student support. Yes, the work is hard and challenging, but we don't let it get to us, we don't stress out. We don't let the work control us. In addition, there is no competition among students. You don't have to be afraid to tell people you got a C on a test, it's OK to share notes. Students here help other students with work as much as possible. It's not a battle for grades here, it's a joint effort to learn!" Students caution that "the only people who have a hard time at Bowdoin are shy and unwilling to get involved right away in class discussions and extracurricular activities. These things are critical at Bowdoin for both social and academic reasons." A recent graduate adds, "Bowdoin promotes individual thought which reaches beyond a set of lectures and a textbook to make learning a reformation and enrichment of thought processes. It allows the opportunity for extracurricular activity, which is pivotal to a well-rounded, academically enhanced, and generally happy education."

Academically, students reflect the value of the broad liberal arts education at Bowdoin, especially the flexibility but also the extent of the distribution requirements. "One of the best traits about my education at Bowdoin has been the diversity of fields that I've taken courses from. While I'm an economics major, roughly half my courses are in the humanities. I don't think I would have had that opportunity at a larger university." Students value their academic opportunities and their faculty, and, if they are critical at all, it is about the need to promote more diversity on campus, to open the administration to student concerns, and to try to weed out from the campus those students who do not take their education seriously or appreciate the opportunity they have been given to attend Bowdoin.

Students we heard from clearly chose Bowdoin in part because of its physical environment. "Bowdoin could possibly be in the best place for any college to be. First of all, we are nestled amidst the beautiful Bowdoin Pines. The town of Brunswick is a terrific small town with restaurants, bars, theaters, stores, banks, mini-malls—everything you could want in a college town, most of the things within walking distance. Freeport, one of the best shopping towns on the East Coast, is only fifteen minutes away, while a major city, Portland, is only twenty-five minutes away. Bowdoin is also minutes away from the coast and within an hour or two you could take advantage of the great outdoors, be it skiing, hiking, climbing, rafting, etc. Even though there is an abundance of things to do on campus, the number of things to do off campus is endless."

Students leave Bowdoin with a sense of the place and their connection to it and each other. One student going on to teaching says, "Part of teaching is also learning, and I'm sure that with a liberal arts education, I will be ready to teach in almost any subject within reason. I like the challenge of being able to shape minds and challenge them. I had a rough time as a teenager, and my teachers were among the few people who I adored, so I hope to have that kind of impact on my own students." Another says, "Bowdoin is special in many ways. The community feel at Bowdoin is unparalleled. You know the checker at the dining hall, the lady at the mailroom, and the security officers on a first-name basis. Professors are there to help educate, not to lecture for an hour then return to research before passing out poor grades. You are a person at Bowdoin, not a statistic. You are a face and a person who contributes and learns from the entire Bowdoin community."

Another student tried to capture the essence of the college's history and its impact on her: "I have yet to find the words to describe the feeling that I get when I am here. I knew from the very first time I saw Bowdoin that this was the place for me. Maybe it has something to do with the history of the college. Each incoming freshman is invited into the president's office to sign 'the book' that the likes of Nathaniel Hawthorne, Henry Wadsworth Longfellow, General Joshua Chamberlain, former senator George Mitchell, and Olympic gold-medalist Joan Benoit have signed during their first few days at Bowdoin. It is an inspiring feeling to sit in a classroom where Harriet Beecher Stowe sat, or to look up at the same huge oak trees that Nathaniel Hawthorne saw when he stepped out of his classroom. The students at Bowdoin will always aspire to greatness, and maybe that's what makes it so special. It is an indescribable feeling of honor to be a Bowdoin student; it is the past, present, and future students of this college that make it so special."

What Happens after College

An inordinate number of Bowdoin graduates pursue a career in public service, following their sense to serve the common good. About 80 percent of students major in the humanities, fine arts, or social and behavioral sciences, vastly outnumbering those in the traditional math and science fields, though a quarter of graduates have double-majored. Government and economics are consistently the two most popular majors.

About 75 percent of Bowdoin graduates enter graduate or professional school within five years of graduation, from Ph.D. programs in the sciences to medical colleges to economics. Many Bowdoin students go into teaching at the secondary school level. Other areas include the film industry, the business world, advertising, law, and volunteer programs like the Peace Corps and AmeriCorps.

With the personalized attention that Bowdoin students receive and their opportunities for extracurricular and cocurricular study, projects, involvements, and internships, they develop their potential at the college. "A Bowdoin education," says one administrator, "enables students to develop many of the critical abilities needed to succeed in today's workplace: problem-solving, analytical reasoning, written communication, public speaking, and leadership skills." Bowdoin alumni are "fiercely loyal" and willing "to bend over backward in helping students to make connections through networking; locate internships, summer jobs, and full-time positions; as well as relocate to different parts of the country. When you come to Bowdoin as a student, you become part of a very large extended family for the rest of your life. And wherever you may travel in the world, you will run into other Bowdoin people, and when you do, you will be bound by a common thread that is somewhat inexplicable to others."

Bryn Mawr College

101 North Merion Avenue
Bryn Mawr, PA 19010-2899

(610) 526-5152

brynmawr.edu

admissions@brynmawr.edu

Number of undergraduates: 1,287
Total number of students: 1,745
Ratio of male to female: 100% female
Tuition and fees for 2008-2009: $36,540
% of students receiving need-based financial aid: 50
% of students graduating within six years: 84
2007 endowment: $663,626,000
Endowment per student: $380,301

Overall Features

While Bryn Mawr is no longer officially affiliated with the Society of Friends, which founded the college in 1885, the philosophy of the Quakers is instilled in the community, as witnessed by the diversity of the student body, the high percentage of scholarship recipients, the emphasis on intellectual exploration and the freedom to voice one's opinion at all times, and the honor code. Bryn Mawr was one of the early colleges founded to educate women in an academically and intellectually rigorous curriculum equal to that which men received at the Ivy League universities. Bryn Mawr puts a great deal of emphasis on its small-classroom, teacher-led approach to educating young women. As one senior administrator points out, "The college remains passionately committed to the proposition that women can do and be anything they choose, a commitment that originated in the vision of founding dean and second president M. Carey Thomas that Bryn Mawr would offer women the best education available." Bryn Mawr prides itself on its sense of an intimate community with strong faculty-student interaction and opportunities to explore virtually all fields of interest despite its small size. Enjoying a sizable endowment for a small college and with an open relationship with its neighbor, Haverford College, Bryn Mawr enables its students to experiment with myriad academic and extracurricular programs. If a course of a more exotic nature is sought, a young scholar can take advantage of the college's relationships with nearby Swarthmore, Villanova, and the University of Pennsylvania. Bryn Mawr is a picturesque suburban campus located only fifteen minutes from Philadelphia. The honor code that permeates both the academic and social life of the college is a key element in defining the environment. As one student describes it, "The honor code that we have in place has been very important to me. The professors trust us and in my opinion, respect us more. Professors leave the classroom when we are taking tests and best of all, we have self-scheduled exams, which makes it possible for us to take our finals when we feel prepared to take them. The honor code also extends to social life at Bryn Mawr. The social honor code stands for general respect for your peers who you are living with. Students leave their shoes in the hallway, their food in the pantry fridge, laundry detergent in the laundry rooms, and toiletries in the bathroom with the trust that things won't be stolen. The honor code is also made possible by our Student Governance Association (SGA). Each Sunday there are meetings which everyone is welcome to attend. Also, twice a year SGA hosts plenary, our form of a town meeting where everyone comes together to vote on resolutions written by students to change aspects about the college."

What the College Stands For

A senior administrator provides an excellent summary of Bryn Mawr's educational goals: "Bryn Mawr endeavors to encourage and develop a student's ability to think creatively, critically, independently; to express herself articulately; to experience the excitement and joy of learning; and to acquire the flexibility of intellect to continue to learn and adapt throughout her life." A dean notes of the college's expectations for students' learning that "in the end, one hopes that they have learned how to seek answers for themselves, as well as the pleasures and satisfaction which come in finding them. One also hopes that they learn, both in the residence halls and classrooms, on the stage and in the gym, a great deal about negotiating differences of value and opinion in pro-

ductive ways, and about working cooperatively." The college tries to promote a diverse and international student body, with about 6 percent of students coming from 43 countries abroad and about 25 percent being American women of color. Bryn Mawr is small, only about half the size of the other women's colleges in the Hidden Ivies (Barnard, Mount Holyoke, Smith, and Wellesley), and sees itself as having "a strong sense of student community and a very accessible and engaged faculty, deans, coaches and other student life administrators." Students are helped to succeed by, as one dean notes, "a supportive environment in which the students have close contact with the faculty who are enthusiastic and supportive and in which students are encouraged to work collaboratively with other students."

The college's future priorities include increasing its emphasis on international studies and interdisciplinary programs, promoting increased faculty diversity (16 percent currently are persons of color), and developing the use of technology and computer science in instruction, research, and linkages to neighboring colleges. The college will seek to make more internships available to students and build on its current strengths in undergraduate research. Bryn Mawr expects to increase its academic and athletic competitiveness. And the college "has an excellent arts faculty in areas ranging from dance to creative writing to theater, with a strong music program available to Bryn Mawr students through Haverford College. We expect to support students' increasing interest in both curricular and cocurricular opportunities in the arts."

Bryn Mawr's library holds over 1 million volumes, quite a large number for any undergraduate college. The college sees its links to Swarthmore and Haverford as being integral to its mission and availability of resources. "The college has long been renowned for the rigor of the education it offers, and the culture of learning on campus means that it is possible to learn as much if not more from conversations and discussions with one's peers as one learns from the faculty." Students from all four classes—freshmen through seniors—are mixed in Bryn Mawr's dormitories, encouraging diversity and communication in the college community. A final word from one dean captures the college well: "Bryn Mawr is a very open, congenial, supportive, collaborative, and diverse institution, which at the same time is rigorous with many longstanding traditions. The mix of old and new, small (small college) and large (large urban environment), American and international together with a student body and faculty from a wide variety of backgrounds makes Bryn Mawr an extremely interesting and exciting place."

Curriculum, Academic Life, and Unique Programs of Study

Students feel the honor code helps to cut down on the academic competition and create a real sense of intimacy among the community members. The tone is one of working hard in a challenging program and deriving satisfaction from a job well done. This is an important factor as Bryn Mawr insists on completion of a fairly strong set of distribution requirements in three divisions: the natural sciences, social sciences, and humanities. First-year students take a seminar of choice focusing on analytical and writing skills. All students must complete the equivalent of two years of foreign language studies. Bryn Mawr has been long known for outstanding science programs and faculty, and one-third of the students major in one of the

sciences. As one administrator notes, "Bryn Mawr is being recognized increasingly as the most successful college in the United States in encouraging women to pursue further study and careers in the basic sciences." There are ample opportunities to work side by side with a teacher in a research project that provides an advantage for graduate school admission and success. "By learning in a purposely small institution, students are able to test themselves both in the classroom and in a culturally diverse community, to learn both broadly and in depth, and to ultimately take a leadership role in their field," says one administrator. A significantly high proportion of graduates go on to medical school or doctoral programs. Here, too, students are encouraged to work with their faculty mentors in research projects and to do independent study. Bryn Mawr also has excellent departments in the humanities, such as English, art history, languages, classics, and archaeology. It has ranked first in the nation in the percentage of graduates completing doctoral degrees in the humanities. In all fields, "the curriculum relies on hands-on research, primary sources, and few or no textbooks." Although graduate students generally do not teach courses, "the tradition of small graduate programs in many of the disciplines makes the academic setting more complex and interesting," states one academic dean.

In addition to the student-administered honor code, the Bryn Mawr Self Government Association—the first such association for students—"reinforces the independence and integrity" of the students. "There is no dean's list, no Phi Beta Kappa, no published grades, even though Bryn Mawr is one of the most selective colleges in the United States. Every student establishes her own standard of excellence and the measures of success are one's own," says an admissions official. The college encourages study abroad, community service, and research or career-oriented internships and externships. Students are supported by a network of health services, a system of "peer tutorials," and class deans. Each student is assigned to an assistant dean for academic advising, and "the faculty and dean's office work closely in providing a seamless program that promotes academic persistence and campus adjustment," according to one administrator. The collaboration with Haverford, only about a mile away, is the closest of that with any other neighboring college and is accessible and well-utilized in terms of students taking classes and even majoring at the Haverford campus. Additionally, the college participates with Haverford and Swarthmore in the Tri-College Summer Institute for entering students of color to help make the college more accessible. The college continues this effort with an affirmative-action compliance plan for faculty searches, a full-time admissions assistant director to coordinate minority student recruitment, a Diversity Leadership Group and Diversity Council, and multiple clubs and cultural organizations on campus.

Major Admissions Criteria

The Bryn Mawr admissions offices ask questions in three main areas in their evaluation of applicants:

- Has the student challenged herself in the most academically demanding program available to her? Has her performance distinguished her from her peers?

- Has the student made the best use of the resources available to her? Has she been

recognized for exceptional academic promise and do her teachers and school recognize her intellectual creativity?

- Has she made contributions outside the classroom?

Thus the college focuses primarily on a student's high school curriculum and performance, personal qualities (such as "independence, responsibility, integrity, maturity") as shown in interviews, essays, and recommendations, and evidence of key talents and interests. As one student describes her admissions experience, "The people who come to Bryn Mawr are a self-selecting pool of exceptional women. We rely on one another, grow with one another, and learn from one another. I felt very much at home in this environment and welcomed. I had never visited another campus where all the students met my eye level and actually smiled back. It was comforting."

The Ideal Student

There is a set of characteristics that defines the successful and satisfied Bryn Mawr woman: focused on her education, intellectually curious, intelligent, motivated to achieve, and open-minded. Students use words like *intellectual, passionate, self-directed, supportive,* and *sassy* to describe the college and its students. One dean refers to the college as a "community of serious student scholars." An admissions officer notes that the Bryn Mawr Book Award is given to "an outstanding female student in the junior class who demonstrates a love of learning and a curiosity about the world around her." He comments, "Bryn Mawr students are generally those who have distinguished themselves in their secondary school and stand apart from the crowd in talents, interest, and academic seriousness. The most successful students are those who are unafraid to tap into the array of resources available to them." Another dean describes the following qualities that help students to succeed at Bryn Mawr: "intelligence, intellectual curiosity, independence, willingness to work hard, ability to manage one's time and to balance serious academic commitment with engagement in the community, and openness to and curiosity about other people from very different cultures and backgrounds." With its commitment to embrace all ideas and points of view, together with its socially diverse community, the college encourages open dialogue on political, intellectual, cultural, and gender issues. This is not an environment for the faint-of-heart student who is more comfortable with a conservatively inclined, homogeneous population. The social life revolves around campus activities, lots of sports at the intercollegiate and intramural levels, and time with one's friends and studies. The bright young woman who is looking for a large, socially driven, traditional college environment will not take well to Bryn Mawr.

Student Perspectives on Their Experience

Students work hard at Bryn Mawr, and they appreciate the college's intellectual focus. The college "makes it OK to be smart, to work hard, to stay up night after night reworking your thesis. You're not dorky, not nerdy—you're a Mawrtyr. There is tremendous potential in a women's college. Bryn Mawr exploits that potential in a wonderful way—women (of course) do everything here!" Another student says, "Bryn Mawr is best at giving its students confidence in their abilities and helping them learn to respect themselves and the other students around them." Students feel supported as female schol-

ars, and also "know how to have fun and enjoy ourselves while we're working hard." Successful students are "open-minded, hardworking, and dedicated to succeeding," and are involved academically and socially. "Bryn Mawr is full of multitalented, fun, and intelligent women."

Traditions at Bryn Mawr are felt to be very important. "BMC's traditions really give the students a sense of being part of a group and belonging at BMC. The honor code has also been really important to me because it gives me the freedom to be myself academically and socially, because I know that my peers respect me and will treat me honestly and honorably." A junior says, "There are awesome traditions, beginning with Parade Night, where the classes get to know each other at the first Step Sing of the year, moving on to Lantern Night, where the frosh choose sophomore 'hellers' for one week as initiation into Bryn Mawr, and finally, May Day, where the year wraps up with celebration, music, and fun. The traditions really bring students together. Each class has their own set of songs that they sing at Step Sings." Another student says, "Every tradition is an event. They are fun and they are a great way for the different classes to interact. Every class year lives together so there isn't a solely designated freshmen or upper classmen dorm. You get to know people throughout all the years and it really broadens your experience and makes it more worthwhile."

Academics are important to the students: "If you work hard and love learning, you will pass. You have to be good to get good grades, really good. You have to be on top of things, hardcore in to what you study. But that's how we are." Deans help students to succeed, and older students help the younger classes. "That's why the mixed-classes dorms are wonderful. The older students have taken the classes, dealt with the profs. They can help oh so much." Continuing the theme of student cooperation, another student says successful Bryn Mawr students are "people who are willing to work hard, but not just to compete with other students. People who are interested in learning for the sake of their own knowledge and aren't interested in grade-grubbing. The honor code really helps students to focus on challenging their own personal limits and not worrying about how they compare to other students." "I would say that Bryn Mawr's strong suit is its academics. It was established at a time when most women's colleges were nothing more than finishing schools, Bryn Mawr was formed to counter this and to offer a strong and complete education to women. This tradition has not died off. The classes tend to be small and have engaging discussions with excellent professors. Professors are always willing to talk with students and are not only helpful, but genuinely interested in students' input and own ideas. Because of this professors love having students help them with their research and there are lots of research opportunities for undergraduates." Some students, however, note that there is a "facade" of a "noncompetitive and low-stress atmosphere," while "this place is a pressure-cooker. . . . Students are fiercely competitive and driven. Bryn Mawr is intense." Students are advised to "learn how to study and what works best for you . . . The secret of success is to study what you love. Find something you are passionate about. Liberal arts offers so many options."

Another student notes, "I wish that the Bryn Mawr students talked more about what they do both within the classroom and in the outside community. Because of the noncompetitive atmosphere here, I feel that students don't like to talk about what they do because it might appear like they are bragging. I would like people to be able to share their stories, because that is often the most inspiring thing about coming here."

There are some calls to make the linkage with Swarthmore more usable, to improve athletics, and to push students to get involved in more off-campus programs. Perhaps one senior sums up the campus life best: "BMC is a school/community. It is difficult to be flat or one-sided or to hide. It is demanding academically but in a good way that makes learning more concentrated and bonds very intense. It is getting through something together while opening the world to individual goals at the same time. The word that comes to mind is *balance*. BMC teaches balance while also supplying it. There are pressures and there is support."

According to one student, "Lesbians have a very strong voice on campus. And it makes sense, doesn't it?" Students do recognize the diversity on campus, but see the college as moving in the right direction, meaning that there is work to be done. Within the diversity, one student sees most students with similar personalities "conservatively minded, politically liberal, studious, and introverted. Students not of this type tend to hate Bryn Mawr and transfer out." Along similar lines, another student says, "I would change how everyone feels the need to be so politically correct all the time that the slightest thing can sometimes offend a person. I don't think that means that people can say whatever they are thinking especially if it could hurt someone, but there are some things that offend that shouldn't for any reason." The Customs program, however, is seen as extremely important by students in helping them to make the transition from high school to college. Regarding Bryn Mawr's single-sex environment, a student says, "The main concern that many prospective students have is in a word: men. They are worried they will step on this campus and never see a boy again for four years. That is definitely not the case. Haverford boys are here all the time, and you can take classes at Haverford as well if that is a prime concern." Another says, "While Bryn Mawr does not have the same kinds of diversity you may find at a larger state school, considering the fact that Bryn Mawr is a small private school, the diversity here is outstanding. The most important aspect of diversity at Bryn Mawr, in my opinion, is the diversity of thought and our ability to share and learn from our peers. There is no pressure to conform to a norm because there is no norm here."

The faculty are strongly valued by students. Says one, "Since I have been at Bryn Mawr, there is something that has helped me a lot, and that is the following: The profs are just people. They are not here only for me, but they are here to help. They mess up sometimes, too. They burn toast on occasion. Talk to them. See what kind of people they are. Profs can help you only as much as you let them (really) and don't be too defensive—they are really smart, you know!" Another states, "Professors are incredibly supportive and helpful for each student in the class." And, "You really get to know your professors. It's not uncommon to eat over at your professor's house or babysit their children."

"I would have to say that what Bryn Mawr really does best," says a junior, "is make your college years memorable—with beautiful traditions, amazing academics, and lasting friendships." Says another, "Bryn Mawr is a unique and special place . . . it offers the best of everything— academics, traditions, small social functions like coffee houses and then bigger parties at Haverford, the adventure and excitement of a big city, the ability to meet lots of new people from so many places." Women are empowered at Bryn Mawr, that is for sure. Says one graduating senior, "As a senior I have been thinking about this a lot and in May I won't feel like a flower that's opened. That's high school. This May I will feel much stronger, like a firecracker— everything at Bryn Mawr is intensified."

What Happens after College

Some 37 percent of Bryn Mawr students major in math or natural sciences, leading many graduates to pursue continuing studies in those areas. Bryn Mawr has ranked in the top ten in the *Wall Street Journal*'s list of liberal arts colleges sending students to top law, business, and medical schools, and the college is also in the top ten of all colleges and universities nationally in the percentage of students going on to complete a Ph.D. Regading the alumnae community, one student says, "I'm constantly meeting Bryn Mawr women who are the grandmothers and aunts of family friends. Because we have this history as a women's school, there is a greater bond between women you meet and great stories about the school from years ago." An administrator says, "In addition to their commitment to their academic work and the vibrancy and vitality of the campus community, Bryn Mawr students are passionate about improving the world. As the school that educated the first woman to win the Nobel Peace Prize, Emily Balch, the college has attracted and gained from the presence of students whose ambitions, beyond their own successful careers, have included living lives that contribute to the social and common good." Education, law, and medicine attract many alumnae, but so do business, management consulting, and international careers. The majority of Bryn Mawr graduates will earn graduate degrees. As one student and future veterinarian/ Ph.D. says, "I want to show the world that Bryn Mawr women are smart, strong, and can do anything they put their minds to."

Bucknell University

Freas Hall
Bucknell University
Lewisburg, PA 17837-3538

(570) 577-1101

bucknell.edu

admissions@bucknell.edu

[
Number of undergraduates: 3,560
Total number of students: 3,696
Ratio of male to female: 45/55
Tuition and fees for 2008–2009: $39,652
% of students receiving need-based financial aid: 47
% of students graduating within six years: 89
2007 endowment: $599,399,000
Endowment per student: $162,175
]

Overall Features

At first viewing, Bucknell may appear to fit snugly into the category of a small, traditional, liberal arts college situated on an attractive 450-acre campus in a rural setting with an intimate social connection among its students and between students and faculty. It is this and more. Bucknell is a unique combination of a traditional residential liberal arts college and a mini-version of a university. Defined primarily by its curricular offerings across the arts and sciences like its peer schools, it also incorporates a strong school of engineering and sciences and a school of management. Bucknell is the largest of its peer schools with 3,500 undergraduates. This allows it to offer a broader range and depth in a number of academic fields than the more typical liberal arts colleges with an average of 2,000 undergraduates. The university takes great pride in its campus aesthetics and facilities, which influence students' experience in their four years

there; it is one of the most handsome campuses in the country and is laid out in a manner that encourages interaction among the student body and faculty.

Virtually all undergraduates live on campus (first-year students are required to) which adds to the strong sense of community. A plan is underway to study the facilities in order to enhance the school's social and academic programs in the future. New buildings for residential living and teaching, and renovation of older facilities, are geared to maintaining the aesthetic look of the campus and enhancing the mission of the university. The goal is to expand the science, engineering, and management facilities for the professional programs. An arts complex is also part of the plan.

Academic programs and a traditional student body combine to give Bucknell a reputation as heavily weighted toward a preprofessional environment. However, there is another deep tradition that defines the school, and that is its emphasis on social and public service. A considerable portion of the student body (85 percent of the seniors) participate in community activities on and off campus. There are active religious organizations (these include a Christian Fellowship, Catholic Campus Ministry, and a Hillel center) on campus that foster community engagement and social outreach.

Given its rural location, life revolves around campus activities. The Greek system drives the social life for a majority of students with over 50 percent of upperclassmen belonging to a fraternity or sorority. While the administration emphasizes their role in community service and educational role, they are also popular as party and drinking centers. Students who are not part of the Greek system can find social outlets limited and travel to major cities inconvenient. There are 175 student organizations that can meet a student's interests. Students are encouraged, and readily agree, that a part of their educational development should be exploring life in other environments. Over 40 percent of students study abroad in keeping with the mission of the college to develop globally aware graduates. Bucknell competes with great success in a wide range of sports in the Patriot League. There are more than twenty-five men's and women's varsity athletic teams and impressive facilities to support their winning teams. Because of the size of its enrollment, Bucknell competes at the Division I level in most of its sports. School spirit is built, in large measure, on the strength of the athletic competition on campus.

What the College Stands For

Bucknell is strongly committed to educating its students for future productive lives for themselves and the world at large. It believes this happens best by bringing excellent teachers together with intelligent students. The emphasis for the faculty is on teaching and mentoring their students accordingly. Research is an important factor in bringing expertise into the classroom, while conveying their knowledge and accessibility to their students are the college's priorities. The college's leaders state their determination to maintain its commitment to small classes and academic excellence across its liberal arts and professional programs. Accordingly, there is a Teaching and Learning Center to build and sustain the quality of what transpires in the classrooms.

The Bucknell Forum series brings to campus well-known speakers to address important national and international issues and to interact with students. An unusually broad range of formal colloquia on current social, political, economic, and cultural concerns are held each year as well. Among its peer colleges, Bucknell ranks near the top in the number of graduates who join the Peace Corps. Its president has articulated the college's mission in response to its recognition as one of the major sources of Peace Corps volunteers: "We believe in enrolling outstanding students who have a passion for both learning and life. We are proud that as alumni they engage at the highest levels in the world around them to help make a difference. Their high involvement in the Peace Corps is proof of that."

Bucknell is also committed to educating students of all socioeconomic, cultural, and racial backgrounds, and has committed a significant portion of its funds to accomplish this goal. Sixty percent of the students in the most recent graduating class received some form of financial aid. A large number of Bucknell Community College Scholars and Posse Scholars are enrolled annually. This is necesssary in a college that is largely composed of a white, upper-middle-class student body from suburban high schools and independent schools.

Curriculum, Academic Life, and Unique Programs of Study

Not many of the small liberal arts colleges offer professional schools of engineering and science and management at the undergraduate level. For many students the opportunity to combine studies in these disciplines with a wide range of offerings in the arts, humanities, and social sciences is a key attraction to enrolling in Bucknell. There are core requirements that include courses in English, math, and a thematic seminar intended to build analytic thinking and writing skills, lab science, the humanities, social sciences, and subjects that emphasize writing skills. Several unique concentrations include animal behavior, environmental relations, international relations, international engineering studies, combined management and technology studies, and neuroscience. To encourage students to pursue their special academic interests, independent study and honors programs are available. There are a number of specialized centers intended to foster advanced exposure for the committed student that include the Writing Center, which encourages writers to share their interests and support one another's creative writing efforts; the Stadler Center for Poetry, which supports the work of emerging and established poets and writers; the Institute for Leadership in Technology and Management; the Environmental Center; and the Center for the Study of Race, Ethnicity, and Gender.

Major Admissions Criteria

Bucknell has joined the ranks of the highly selective liberal arts colleges in a steady climb over the years. Over 9,000 students apply each year and only a third are accepted. SAT scores are in the mid- to high-600 range and ACT scores are in the 28 to 32 range for the midrange of accepted students. The admissions committee, in a fashion similar to the other selective liberal arts colleges, focuses its review of applicants on the quality of the high school curriculum (the level of demanding courses across the areas of English, math, science, social studies, and languages); participation in leadership, community service, or athletic activities; admissions test scores; and a compelling personal application that indicates maturity, curiosity, energy, and strong communication skills. Bucknell has two rounds of early-decision application programs.

The Ideal Student

This is an ideal environment for the individual who wants a complete campus collegiate experience. The academic demands are heavy and the expectations of the faculty in an intimate teaching structure call for motivation to learn, interest in being exposed to a broad range of subjects, and the ability to manage time efficiently. An outgoing personality and interest in either or both athletic and community service are important to finding a place within the community. An interest in Greek life is a plus for many students. The student who wants the advantages of an urban college exposure will be disappointed by the location of Bucknell and the emphasis on campus-centered life. Although the college is committed to diversifying its student body through active use of its financial aid funding and active recruiting, the student body remains predominantly white and middle and upper middle class. Bucknell students have an exceptionally high rate of graduation over a four- and six-year period, an indicator of their satisfaction with their collegiate experience.

Student Perspectives on Their Experience

One undergraduate put it this way: "If you know what Bucknell has to offer as a social community and a learning environment up front and you find this appeals to you, you will not be disappointed. The college makes no apologies for what it is and what it has to offer, and it comes through on what it promises." Other students who declare their love for the college nevertheless encourage studying abroad at some stage in their studies in order to break up the experience of life in the "Bucknell Bubble," as well as to gain broader exposure. Several varsity-level athletes are happy at Bucknell because they can play at a high level of intercollegiate competition and still experience a high level of education in a relatively small college. One young woman recently transferred to a large urban university in order to expand her social experience after two years on campus. A thread common to all the students interviewed is the quality and care of the faculty in their teaching and advising.

A number of students we have counseled about Bucknell over the years have found themselves surprised by the rigor of the academics they encounter there. In some cases, prospective athletes have needed to drop their sport in order to maintain their studies in engineering. When students are well matched with Bucknell they are "happy," "comfortable," "challenged," and "involved." Such students are often outgoing, social, and smart.

What Happens after College

Bucknell claims that 96 percent of all recent graduates either are employed or are pursuing advanced degrees, approximately one-quarter in the latter category. In keeping with the preprofessional goals and the fields of study available to them, Bucknellians do well in landing jobs in traditional fields of business, engineering, education, and social service. Many return to graduate school after a period of working. The strong community experience while on campus binds students for life, it seems, and leads to a good deal of social and job networking as alumni.

Carleton College

100 South College Street
Northfield, MN 55057-4016

(800) 995-2275/(507) 646-4190

carleton.edu

admissions@carleton.edu

[
Number of undergraduates: 1,986
Total number of students: 1,986
Ratio of male to female: 50/50
Tuition and fees for 2008–2009: $38,046
% of students receiving need-based financial aid: 55
% of students graduating within six years: 93
2007 endowment: $663,500,000
Endowment per student: $334,088
]

Overall Features

Carleton was founded in 1866 to provide for outstanding students in its region of the country a liberal arts education comparable to the very best of the New England colleges. To this day the college maintains its focus on training young men and women in the intellectual skills and moral and ethical values that a first-rate liberal arts education provides. It has continued and expanded its commitment to a wide mix of talented students of all backgrounds. It enjoys a significant endowment for a college of its size, which it puts to good use in providing financial aid for its students. Carleton also offers a large number of national merit scholarships out of its own funds that can cover all or most of the tuition. It can truly state that it seeks talented students who can gain admission irrespective of ability to pay. Carleton is a socially, politically, and intellectually liberal environment. It stimulates students to think

critically and open their minds to many points of view. Students are attracted to the college for its reputation for academic excellence, a major commitment by the faculty to teaching and mentoring students, its spirited and close community, and its success in graduate school admission.

Faculty members are generally known for excellence in teaching and knowledge of their disciplines. Many have achieved renown through research and publishing. The student body is serious about both studies and active participation in extracurricular organizations. While its location is known for the frigid weather that prevails, the tenor of the community is one of warmth and friendly engagement among students and faculty. Reflecting the tradition of training students for future intellectual and leadership roles, Carleton graduates have won an unusually high number of Rhodes scholarships over the years. Over one-half of any graduating class goes on to leading graduate schools. In fact, Carleton has ranked number one in the nation among liberal arts colleges in the number of graduates who attain doctoral degrees in the sciences. It ranks among the top five in doctoral recipients in all disciplines. There is no Greek system in place, so the social emphasis is on dorm and college organizations for social and leadership life. An unusually large percentage of students get involved in community service organizations in the larger community. Carleton has a strong tradition of providing the best campus facilities possible, including a $10 million recreational center. While Carleton is a beautiful campus with a total residential population in a small town, the twin cities of Minneapolis and St. Paul are only forty miles away. This allows many outlets for social fun and community service. Athletics are active but not intensely competitive.

What the College Stands For

Carleton's former president, Stephen Lewis, succinctly described the college's mission in a letter sent to parents of incoming students: "A Carleton education is first and foremost an experience that helps young people learn to think clearly, analyze critically, and express themselves effectively. And, as a residential, liberal arts college, Carleton has always been concerned with the development of the entire person." Another administrator describes Carleton's goals as "to assist students in their intellectual and emotional growth and to encourage their physical and spiritual health by providing a wide array of activities and support; to equip students with the academic and personal skills to become independent, successful adults; and to encourage students to lead balanced lives."

Carleton sees itself as a place where issues are taken seriously, but people at the college, including students, do not take themselves too seriously. There seems to be a sense of humor to the place, as illustrated by the student tradition of displaying a bust of Friedrich Schiller, which was "liberated" from the president's office in the 1950s, at major events. The bust is kept in secret by a group of students and occasionally wrestled away by other students during a showing. It drops in by helicopter, is paddled down the river in canoes, and is ushered into auditoriums by ninja-clad students! A Carleton administrator describes the college's uniqueness by the degree to which faculty are involved with students outside the classroom; "there is support and shared activity among the students themselves; competition of students against one another is absent from the culture; there is physical activity and life outside the classroom relative to the hard work within it; friendliness and decency are present; myths about the weather have an impact upon the ethos of the campus; and there is a sense of humor in an otherwise serious-minded campus population."

Future initiatives include a continuing effort to promote racial diversity, curricular development within and among academic disciplines, and several key building projects. One dean notes progress on developing the student body: "Always geographically diverse (over 60 percent come from homes more than 500 miles from campus), students are increasingly diverse (different from one another) ethnically, politically, and religiously. It is the ideal setting for students to learn from (in the words of Carleton's president, Rob Oden) 'meaningful encounters with difference.' Carleton has a mission statement that features diversity as an institutional principle; a separate statement on diversity . . . Carleton is an institution that values inclusion and diversity, and makes those values central in the operation of all of its divisions and departments."

Carleton has already built many excellent campus facilities and programs, with one dean pointing out "an extensive recreation center with indoor track, climbing wall, fitness center, racquetball courts, and convertible courts for tennis, volleyball, basketball, etc. We also added more acreage to the Carleton arboretum, bringing total area to nearly 900 acres. Most recently, we are constructing a new student residence hall and have plans underway for a new arts center which would combine several departments within humanities and arts under a single roof with extensive facilities for dance, theater, studio arts, and cinema study. The core of this enterprise is a new exploration of creativity across disciplines, especially in humanities and the arts. We have expanded programs like environmental studies and cinema and media studies and continue to review and change curricular offerings within our tradition of offering strong departments across the physical sciences, humanities, and social sciences."

Curriculum, Academic Life, and Unique Programs of Study

The distribution requirements necessary to graduate are both flexible and rigorous in nature. There is a wide choice of courses to fulfill the goal of the liberal arts education, but every student must also complete the equivalent of four years of a foreign language as well as a six-credit course to complete the Recognition and Affirmation of Difference Requirement. Students are required to complete their education in twelve terms, and there is a four-year residency requirement. A senior project that demonstrates a comprehensive understanding of one's academic concentration is required. Many interdisciplinary and major/minors are available to students, and the college's music programs are seen as particularly strong. Many take advantage of the outstanding science departments and collaborative research with faculty. Newer buildings include the Center for Math & Computing, the biology center, and the Olin Hall for science. Two-thirds of Carleton students seize the opportunity to study abroad for a semester or a year. It is the rare young scholar who has not been able to build a particular field of study around his or her interests with the support and encouragement of the faculty. Although people choose to enroll in Carleton for its academic excellence, there is not an intense competitive quality to the community. Collaboration in learning, support for one another, and individualism in the intellectual enterprise are encouraged. A senior faculty member tells us, "There is a strong and growing culture of peer teaching and learning at Carleton. Students comment on the lack of competition (coupled, though, with a strong work ethic and an intense, sometimes 'hothouse'-like academic environment). In this culture, students know that the success of a group will improve their individual success. Several faculty and departments give students in their classes formal instruction in working in groups; some have adapted assignments for courses with a range of student experience, so that the more advanced students help the younger students."

Carleton puts a great deal of emphasis on faculty development, "hewing teachers out of Ph.D.'s," and sees a "symbiotic relationship between students and faculty." There is faculty review via student letters; there is a senior faculty mentoring program, a Learning and Teaching Center for teacher training, and new faculty workshops make for a fine continuum of training and support for faculty. The benefits of this emphasis for Carleton students are clear. Students are helped to succeed by "small classes and an outstanding commitment on the part of faculty; strong academic support programs, especially in modern languages and math; the students themselves—a deeply caring, very friendly, and supportive group of people; a tradition of intellectual engagement; a student life staff committed to helping students learn how to live in a complex, diverse world; and literally hundreds of activities to choose from."

Students have a voice in the college through the College Council, which includes five elected students, five elected faculty, two staff, a trustee and an alumni observer, and the president. This is a voting body that impacts college policy and represents the participatory nature of Carleton. A student senate controls a large budget for student organizations. Eighty percent of students work at least ten hours per week on campus, and some 30 to 40 percent are active as volunteers at any given time.

Carleton currently teaches Chinese, Japanese, Arabic, and Russian, and sees their Asian studies program as a key strength in this area. Carleton has a 20 percent minority and 7 percent international student population, with an undergraduate grant and research program for minorities. There are over a dozen multicultural student organizations, an Office of Multicultural Affairs, a summer math and science program for minorities, and other support programs on campus.

Major Admissions Criteria

Applications, and yield, are up at Carleton. A student's academic record and intellectual interest are clearly important in admissions, as are involvement in school and community and character traits. Carleton enrollees are "lively, friendly, and active." A senior administrator says, "First, Carleton seeks curious students who naturally are eager to learn. While ability to succeed in a challenging environment is gauged by test scores and high school records, the Carleton admissions office examines evidence beyond this data to assess students' engagement with learning. This may come from course selection and performance, teacher recommendations, and reflections in student essays, among other things. Second, Carleton seeks individuals that will come together to create a vibrant, engaged class. This is assessed through meaningful participation in activities, evidence of seeking opportunities outside the classroom to learn and grow, and demonstrating a willingness to share and examine one's own views and perspectives while listening to the perspectives of others. Third, Carleton seeks a diverse class, admitting students across a broad spectrum of experiences and backgrounds and with a breadth of viewpoints and interests." The dean of admissions says that "Carleton seeks students who have demonstrated a passion to learn through their high school careers. . . . Intellectual curiosity is the chief characteristic possessed by most Carleton students. It is an interest that makes students desire to take courses across the curriculum and to pursue unusual questions within their major fields. We also seek students who have lots of energy to participate in a residential college. We look for students who will populate our orchestra, join an athletic team, or toss a Frisbee. There is no particular talent or interest that we seek but the drive to contribute to a small community in constructive ways. A spirit of adventure is helpful and a willingness to take risks are strong Carleton characteristics. . . . What we seek is evidence that study for a student is more than a duty or a ticket to the next certification or job; a love of learning and questioning attitude are the elements that we seek in essays and recommendations."

The Ideal Student

Individuals who love learning in an intimate and interactive academic environment will thrive at Carleton. This is not the right environment for the student who prefers to remain anonymous in the classroom or who does not like to engage in active learning. Studies come first, but a high premium is put on community involvement and service. Those who care about social problems and taking action will find a host of opportunities and an encouraging environment at Carleton, and so will those who like to participate in leadership activities or low-key athletics at the intercollegiate or intramural level. Given its diverse student body and generally liberal-minded tone, a candidate for Carleton should be open to new people, new ideas, active debate, and discourse. One should have concern for social issues and should be willing to work hard at one's studies.

At the same time, Carleton students are not seen as being competitive with each other, or elitist. As the president of the college said at a recent talk for new students, "As part of our Minnesota roots, Carleton people tend to be pretty self-effacing—lots of Garrison Keillor 'shy people.' You'll find very few who are self-promoters, and that distinguishes us from a lot of other places. This is a diverse community of strongly individualistic people. . . . A typical Carleton student is one who would be outraged by the notion that there was a typical Carleton student!" Students who are "freer spirits" and "fierce individualists," "curious, active, seekers and risk-takers" will feel at home at Carleton.

An administrator notes that "the campus culture has moved significantly over the past decade. There is less insistence on independence and autonomy for their own sakes, and more acceptance of the responsibility that accompanies freedom and independence. Students are more open to guidance—but not interference—in their personal lives and their groups. Students have continued keen interests in engaging in multiple activities, physical recreation, and intensive study, and seem very open to working in groups." "If you come to Carleton," says one administrator, "bring your walking shoes. Exploring the arboretum in the changing seasons and all times of day is a never ending discovery of sights, sounds, smells, vistas, wildlife encounters, and immersion in the grandeur of the natural world."

Student Perspectives on Their Experience

One student tells a revealing story about how he chose Carleton. "In selecting a college to attend, I worked from larger concerns to more specific school decisions. After realizing that a liberal arts experience could only be had during undergraduate education, I decided to attend a small school where I, rather than the literary journal, would be the focus of a professor's attention. Once this was determined, I asked a high school guidance counselor for the best liberal arts school in the country. She said the best learning environment was at a place called Carleton College, so I applied, was accepted, and arrived for new-student week six months later." He went on to become student body president.

This student and others find the open, involved learning environment highly rewarding. "At its best, Carleton fosters an environment where students are truly free to examine differing ideologies and confront new ideas under the guidance of an involved, student-driven course. But this is done outside the classroom as well, with significant intellectual challenges arising anywhere from the library to the dining hall to the weight room. Also, I recall going over to my professor's home with my classmates, who numbered seven total, and discussing the implications of the cold war on current U.S.–China policy over dinner and after-dinner treats." Students find help at the Write Place for editing papers and at the Math Skills Center.

One graduate says, "Carleton is an exceptional school for students who do not know what career path they want to take immediately after college, but who are highly self-motivated, like to explore many different opportunities, and are smart and focused on academic excellence. The school provides what I would say is an ideal environment for this type of student to develop to his or her full potential (insofar as a college education can help)." Nevertheless, "while students at Carleton are encouraged to push academic boundaries, the school is not a good environment for those who are not certain of their political, social, or religious beliefs. The 'liberal' sociopolitical environment is not helpful to those who are still exploring their social, political, or religious beliefs, regardless of whether or not they end up agreeing with the prevailing college culture. Especially at risk are students from nonmainstream backgrounds (racial minorities and the like) who are not firmly committed to getting a good academic education."

"Students who are smart, study hard, keep goals in mind while maintaining a flexible and explorative mind, are self-motivated, willing to take the initiative to make use of college resources, explore new things with the guidance

of professors and fellow students, and who are able to keep in mind that the world is a lot larger than a small college campus in rural Minnesota do best at Carleton." Students are said to benefit greatly from informal faculty friends and mentors, but one student says, "The formal academic advising system is a shambles, and students are really on their own. The corollary to being free to explore your options is that the college really can't step in to help out until it's too late." A junior notes, however, that "the school excels in academics and maintaining high standards of excellence not only with students, but also with professors and faculty. There really isn't anyone at Carleton who doesn't belong there, and that goes for both students and faculty. The school does best in making its students succeed in every aspect and seeing that nobody falls through the cracks. You have to try very hard to be a failure at Carleton." The professors "are truly amazing. Not only are they brilliant, they are so anxious to talk to students about their subject that students can't help but become excited about their studies."

"There isn't a way to construct a quintessential Carleton student and the same goes for who does the best. Everyone who goes through Carleton finds something unique about himself or herself," says one student. Another adds, "Carleton is a place where students can experiment both academically and socially, in a beautiful setting with exciting, intriguing people."

A note on the environment: "A really warm coat makes walking across campus to go to class in minus-fifty-degree weather a whole lot easier." And one will need to go out, since the students who do best are "people who take advantage of everything! There are so many clubs, concerts, talks, and events going on every single day, that those who don't go are really missing out." And "who else has late-night trivia, ice sculptures, concerts, plays, events, and so many truly fun opportunities. There are a lot of outgoing, creative people here."

Complaints focus on alcohol use and abuse, with requests for the college to take a more active role in prevention, and there is the wish to improve multicultural recruitment, retention, and involvement. "Even having one multicultural student drop out is devastating to our already small community, let alone half a dozen. It seems to me that there aren't enough multicultural faculty and that the college isn't trying terribly hard to overcome this problem." On the other hand, a junior says, "Some of the requirements Carleton imposes that are supposed to ensure diversity overdo it a bit. . . . Carleton *loves* diversity. While we don't have many racial minorities, we do have a wide range of students from different socioeconomic positions and just a lot of really neat, different, interesting people." And a senior, while arguing for improving the diversity on campus, adds, "I hope as we try to increase our diversity we don't do it for numbers sake, but for enriching the lives and educational experiences of young people."

One student sums up his experience by saying, "Carleton, in my opinion, provides the best, most challenging undergraduate education in the world. It is not a practical education, but students graduate with the ability to do pretty much anything. This is because of one thing: While Carleton does not have anywhere near the resources of the more traditional Ivies, the compensation is that Carleton professors love what they do, love to teach, and are darn good at it. Carleton's faculty are consistently ranked some of the best in the nation." Says another, "Even if I didn't want to succeed, Carleton would make me succeed and the faculty would, and does, personally see to it that I do

just that." And another maintains, "I couldn't have picked a better college. I love the fact that I can walk across campus at 11:00 p.m. after a Model UN meeting to find friends studying in Sayles, people in the science and computer labs, and students even playing Frisbee or hockey long after dark. The energy around Carleton is invigorating and I adore it."

What Happens after College

One administrator notes, "Alumni continually comment that Carleton gave them the confidence to try many different things and, in fact, that is what happens. There is no typical career path, so alumni scatter around the world in business, law, medicine, education, research, government, community service, religious vocations, journalism, and politics." Carleton has an extensive career center to serve students: "Special programs and workshops teach students how to write résumés, interview successfully, select a major, search for jobs and internships, and many other valuable skills. Students may tap into a vast and supportive alumni network to conduct informational interviews and arrange career-shadowing experiences." Some 20 percent of Carleton graduates are enrolled in advanced degree programs six months after graduation, including academic study, law, medicine, and other areas. Six months after graduation, 66 percent of Carleton graduates are in the workforce. The largest numbes are in business-related fields and education. Service programs, such as the Peace Corps and AmeriCorps, are also popular. "A Carleton education prepares students to excel in many different fields. Students learn excellent communication and analytical skills, and have ample opportunities to develop their leadership skills. Students are also encouraged to develop a commitment to their communities, diversity, and global awareness."

Carleton's institutional research director reports, "Carleton alumni continue on for graduate study at very high rates. The overall number of doctoral degrees earned by Carleton graduates is up 22.4 percent from the previous decade, and by 51.1 percent from the decade before that. Among institutions whose highest degree is the bachelor's degree, Carleton ranks first in the physical sciences and geosciences, third in life sciences, fourth in the social sciences, and fifth in the humanities. Carleton also ranks well compared to this group in the smaller disciplines (less than 100 doctorates earned by Carleton graduates): third in business and management fields; fifth in communications and librarianship; seventh in arts and music; seventh in mathematics and computer sciences; eighth in engineering; ninth in religion and theology; and fifteenth in education. Carleton ranks sixth (per hundred graduates) among all U.S. colleges and universities in the proportion of bachelor's degree recipients who earned doctoral degrees in all science and engineering (STEM) fields. A group of private small liberal arts colleges (known as the "Oberlin 50") has consistently and significantly outperformed all other Carnegie institution types in the proportion of STEM field doctorates produced per hundred graduates." Supporting one of our premises for attending a Hidden Ivy, a number of which are part of a study group called the Oberlin 50, he notes, "From 1996–2006, the Oberlin 50 group averaged between five to six doctoral degrees per 100 graduates, compared to about three degrees per 100 graduates at the next highest category—the very-high-research-activity group of research universities. Carleton, an Oberlin 50 member, averaged 11.7 STEM discipline degrees per hundred graduates, a ratio that ranked only behind California Institute of Technology, Harvey Mudd, MIT, Reed, and Swarthmore. Members of the Oberlin 50 group also outperformed all other institution types in the social and behavioral sciences disciplines, with just over two degrees per 100 graduates over the past decade."

University of Chicago

1101 East 58th Street
Rosenwald Hall, Suite 105
Chicago, IL 60637

(773) 702-8650

uchicago.edu

collegeadmissions@uchicago.edu

Number of undergraduates: 5,027
Total number of students: 11,650
Ratio of male to female: 50/50
Tuition and fees for 2008–2009: $37,632
% of students receiving need-based financial aid: 49
(Class of 2012)
% of students graduating within six years: 90
2007 endowment: $6,204,189,000
Endowment per student: $532,548

Overall Features

The University of Chicago states straight out that it is one of the great intellectual communities in higher education. No one in the academic profession would argue this point, nor would any student who experiences a Chicago education. There are many outstanding features that define this famous institution. It starts with its reputation as an intellectual stronghold for both undergraduate and graduate education. Over the years eighty-one Nobel Prize winners have taught here, with seven presently on the faculty. In a powerful looping effect, intellectually gifted students who put the life of the mind above all other factors in choosing a college are attracted to the faculty and atmosphere that define Chicago, and outstanding teachers and scholars are drawn to this kind of student body.

Chicago is no place for a bright but lazily inclined student. From its inception in 1890, which makes it a relative newcomer among the nation's most prestigious universities, Chicago has been an innovator in a number of key areas, starting with its commitment to educating young women on an equal plane with young men, creating a core curriculum to implement its belief in the vital importance of interdisciplinary learning and research, and maintaining a small undergraduate enrollment with an emphasis on productive interaction between students and their teachers, while at the same time building a world-renowned graduate faculty and eliminating big-time athletic competition in order to put the greater part of its resources into its faculty, facilities, and students. This decision was made at a time when football was becoming the nation's passion and the university had nationally ranked teams and the first Heisman Trophy winner.

Founded with a generous gift from John D. Rockefeller, the university was able to build a self-contained, collegiate, Gothic campus in the Hyde Park section of Chicago. Its resources and, most of all, its stated approach to educating students attracted a series of outstanding academic leaders through its history. Today there are on campus a number of innovatively designed buildings by some of the world's leading architects. While Chicago holds true to its intellectual traditions, it continues to push out new frontiers of knowledge and new ways of teaching and learning. In 1998 its new curriculum adjusted some of the core requirements for greater flexibility but continued the tradition of requiring an interdisciplinary or core education for all its undergraduates. Chicago is committed to maintaining its small and strictly liberal arts curriculum at the undergraduate college while building its graduate professional and academic schools.

In recent years, the university has made a concerted effort to lighten up its campus environment and image in order to make the experience more appealing to the smart, intellectually oriented students who apply to Chicago in fewer numbers than to their peer universities, particularly some of the Ivies and Stanford, where the vibrant academic life is joined by a thriving social and athletic scene. A new athletic facility, new and revitalized dormitories, more on-campus social events, and even the resuscitation of intercollegiate football have been instituted to this end. Chicago students now compete in a wide array of intercollegiate athletics at the Division III level.

In order to create an intimate social/academic experience for its students, Chicago has put in place a house system of small units within the dormitories. Students report that the houses take on a distinct personality or culture, and that they provide a comfortable social transition into the community in the first year. Housing is guaranteed to all who want to live on campus, though many upperclassmen opt for the independence of apartment life on the edge of the campus. The city of Chicago is another great asset. Students come to enjoy its many offerings, from excellent multiethnic foods to theaters, movie houses, museums, art galleries, orchestras, professional sports teams, and entertainment bars, all of which are close to Hyde Park. In recent years there has been a concerted effort to revitalize the neighborhoods that surround the beautiful campus. This appears to be making considerable progress.

What the College Stands For

Its underlying mission and philosophy of education is summed up in the words of the sitting president: "The University of Chicago, from its very inception, has been driven by a singular focus on inquiry . . . with a firm belief in the value of open, rigorous, and intense inquiry and a common understanding that this must be the defining feature of this university. Everything about the University of Chicago that we recognize as distinctive flows from this commitment." One has the strong impression that the University of Chicago, of all the major research universities, would be the last to surrender its commitment to a grounded, intensive liberal arts education. While holding to the tradition of the core curriculum across the major arts and sciences, Chicago presents an intellectually liberal, open-minded, free-thinking culture, in which it takes considerable pride. It wants its students to develop their analytic skills, their passion for exploring ideas with an open mind, and to assume roles in society where they can exercise their gifts and their learning in meaningful ways. The physical symbol of what Chicago considers its priorities is the massive library located in the center of the campus and which is in the midst of expansion. Books and intellectual resources matter at this serious learning place.

Curriculum, Academic Life, and Unique Programs of Study

The university operates on a quarterly academic schedule, each quarter eleven weeks in length. There are three distinct components to the curriculum: the general education requirements, a concentration in a particular discipline, and elective courses. There is no way of working around the general education requirements if one is to graduate from the university; exceptions are rarely granted. The core consists of an integrated, interdisciplinary sequence of courses and they must be completed by the end of the second year. With forty-two courses required to graduate, the general education requirements take up a sizable portion of a student's program. They are divided into three distinct disciplines: humanities, civilization studies, and the arts; natural and mathematical sciences; and social sciences. Much of the work in the core courses centers on exploring original texts in order to build critical reading, analysis, and writing skills. The depth and the breadth of the curriculum is impressive, making it possible for students to select courses to fulfill the core requirements as well as a major that suits their interests. About one-third of an undergraduate program will consist of elective courses. Students must also demonstrate reading, writing, and speaking competency in a foreign language equivalent to one year of college-level study. Students can fulfill this requirement by placement tests such as the AP exam or a college-administered competency exam. Intensive summer language studies can also reduce the number of quarters of required course credits. Chicago continues to value the concept of sound mind in a sound body; thus there is a physical education requirement that can be met by involvement in a variety of activities that range widely from archery to yoga.

Major Admissions Criteria

For a university of such national and international reputation, Chicago receives a much smaller number of applications and admits a much higher percentage than the leading universities with which it is compared. This is primarily the result of the self-selection process on the part of high school students. Knowing Chicago's rigorous academic standards, its demanding core curriculum, and its image as an all-business-and-no-fun environment, and facing what has been a very challenging, uniquely styled application, many qualified students choose not to apply. The university accepts both the SAT and ACT testing formats. Test scores on the SAT for enrolled students cluster from 670 to 750, and in the 28 to 33 range on the ACT. SAT subject tests are optional.

The admissions committee pays the most attention to signs of intellectual ability and interests as demonstrated through the quality of subjects studied in high school and participation in enriched learning experiences such as independent study, research projects, and academic experiences outside of the high school setting. Recommendations from teachers and mentors play a major role in the review of applicants. The admissions office has recently adopted the Common Application, which will simplify the process of applying. However, students must still complete particularly unique (and fairly offbeat) supplemental essays, continuing the tradition of Chicago's "uncommon application" of recent years. Chicago is committed to building and maintaining its diverse population of students, and it uses its generous financial-aid funding to make this possible. It is need-blind in considering whom to admit each year. Merit-based awards are also used to attract unusually talented individuals.

The Ideal Student

The type of student who prospers at the University of Chicago is very clear: those young men and women who see themselves as serious about their academic career, intellectually curious, hardworking, desiring to be stimulated by demanding faculty, and who embrace the underlying principles of the traditional core curriculum. Chicago is not a setting in which to hang out or do enough to get by and feel a part of the mainstream. If national-level sports, either as a participant or spectator, are an important piece of your college experience, this is not the right school to attend. Engaging in intellectual and artistic issues in the classroom, the dorm, the social scene, and even on the athletic teams is what makes people glow. The general social and political environment on campus is liberal with a sprinkling of conservatives who seem not to hesitate to speak up. Students who take advantage of the opportunities of a major city will thrive on the campus and in the entire city of Chicago. For some, there is the appeal of receiving a first-class academic education while enjoying participation in Division III athletics. A respect and appreciation for engagement with students who represent a diverse number of cultures and races goes a long way toward helping one feel at home at Chicago. Satisfaction with their education is extremely high among students, as

indicated by the return for their sophomore year of 98 percent of first-year students and the graduation rate of over 90 percent of the student body annually.

Student Perspectives on Their Experience

One young man, an outstanding competitive swimmer in high school, was surprised, happily in his case, to find that a majority of his Chicago teammates traveled to their meets with a book bag so that they could read or study between their heats. Every student we encounter immediately refers to the caliber of the faculty and the fact that they teach undergraduates with a passion for their subject. One young man about to graduate and head to law school stated that "the image of the college as filled only with techno-nerds and hyper students who do not know how to relax and have fun is simply not true. Sure, everyone here is serious about his work, but we share so many interests that this may be the easiest campus to make friends and have a good time with them." A junior exclaimed that "unlike what I hear from many of my friends who are at other top colleges, the faculty is really here for us and most of my classes are small in size." A young woman who recently arrived on campus from a small, semirural town exclaimed that "the city of Chicago is a gold mine of wonders to be explored and savored for four years." Students report feeling challenged but mentored by their faculty, especially in the higher-level classes, and exceptionally well prepared for rigorous graduate level studies.

What Happens after College

Reviewing a roster of Chicago's distinguished alumni in its literature underscores the talented people who study at the university, the advantages of the core curriculum, and the wide range of areas in which students are prepared to excel in their future lives. While exact numbers are not available, a majority of undergraduates continue their education at the graduate level in the sciences and humanities and in the professional schools of business, law, education, and medicine. The reputation of the demanding, broad-based curriculum makes Chicago students attractive candidates for graduate studies. The university maintains an active alumni network of volunteer mentors and potential employers. The career and placement services runs a number of helpful workshops on choosing a career and preparing for life after Chicago.

Claremont McKenna College

890 Columbia Avenue
Claremont, CA 91711-6425

(909) 621-8088

claremontmckenna.edu

admission@claremontmckenna.edu

[
Number of undergraduates: 1,135
Total number of students: 1,135
Ratio of male to female: 56/44
Tuition and fees for 2008–2009: $37,060
% of students receiving need-based financial aid: 45
% of students graduating within six years: 90
2007 endowment: $474,022,000
Endowment per student: $417,640
]

Overall Features

From a chronological perspective, Claremont McKenna is a youngster among the select cohort of liberal arts colleges founded one to two centuries earlier. Nevertheless, it has come of age at a rapid pace due to its quality of faculty, facilities, focused curriculum, and setting. Its sizable endowment is remarkable for so young an institution. The commitment to assisting talented students to attend the college is made possible by this endowment. A small college as measured by its enrollment, it presents the course offerings of a much larger institution. Claremont McKenna was one of the five founding colleges that comprise the Claremont group. The others are Pomona, Pitzer, Harvey Mudd, and Scripps Colleges. Each has its own unique style, campus tone, and emphasis in educating its students, but in combination they provide a marvelous range of learning options for the serious student. All classes at the five colleges

are available to all students. There does not appear to be much red tape involved in registering for courses outside one's home institution.

Claremont McKenna's academic content, from the start, has been centered on government, economics, management, international relations, and public policy. As one dean describes it, it is a liberal arts college on the crease between the academic and real worlds. Students are attracted to Claremont McKenna for the opportunity to prepare for future roles in the business, consulting, legal, and governmental arenas in a strong teaching-oriented program. The college attracts top faculty in these particular disciplines who are interested in inspiring bright students to master their subjects. The student-to-faculty ratio is quite impressive and thus classes are small, allowing for a good deal of interaction. Students are expected to speak up and speak out in an intelligent and articulate fashion. A great many students will undertake independent projects and assist faculty in their research throughout their four years. With an eye on their future in national and global leadership positions, a significant number will study abroad. The college environment has a traditional and pragmatic tone, given its particular academic emphases and its students' professional ambitions. There seem to be very few liberal arts colleges where a substantial portion of the student body are registered Republicans. Claremont McKenna encourages exploration on and off campus, and leadership experience is available through the many clubs and organizations. There are no fraternities or sororities at Claremont McKenna, so social life is centered in the dormitories and in campus organizations. For so small a college there is a very active athletic program, with about one-third of students involved. Students join with Harvey Mudd and Scripps Colleges for Division III intercollegiate competition. The college is located only forty miles from Los Angeles, so there is ample opportunity to experience a large urban center.

What the College Stands For

Claremont's mission statement emphasizes the college's major goals: "Its mission within the mutually supportive framework of the Claremont Colleges is to educate its students for thoughtful and productive lives and responsible leadership in business, government and the professions and to support faculty and student scholarship that contributes to intellectual vitality and the understanding of public policy issues." With an average class size of seventeen and about 93 percent of students living on campus, Claremont carries out its mission in the midst of an intimate residential college setting. "We receive incredible support from trustees, alumni, parents, and faculty. Our tight-knit community provides an environment in which students may learn and grow as future leaders," says one administrator. Another institutional leader notes, "Claremont McKenna College provides an education that is both highly pragmatic and deeply philosophical: We want our students to learn how to think independently, creatively, and analytically; to learn how to utilize their thinking skills and apply them to real-world matters; and to possess both the ability to recognize opportunities and the courage to take risks."

The college sees itself as providing "an environment in which it is easy to maintain a high level of morale—regardless of the fact that the academic demands are intense. CMC students genuinely enjoy learning—in classes and from one another. Student life is campus-focused, with virtually all students living on campus and freshmen, sophomores, juniors, and seniors all living in the same dorms." The college emphasizes that "teamwork at CMC is a given," and that "support for CMC students is everywhere." One dean notes, "This is a first-rate faculty dedicated to undergraduate education," which helps the students succeed. A well-known historian, Jonathan Petropoulos, who was hired by Claremont, spoke clearly of his and the college's commitment to both teaching and research when he was quoted as saying, "This is the paragon of a liberal arts college. It's an atmosphere ideal for learning, yet it has the resources of a research university."

Curriculum, Academic Life, and Unique Programs of Study

Given its academic emphases, Claremont Mc-Kenna can best be described as a traditional model of education, what one senior administrator called a "focused liberal arts college . . . I believe we are better defined and focused than most liberal arts colleges." Both faculty and students tend to put their energy and thought into the subject matter at hand, which means learning the nature and dynamics of social, economic, political, global, and managerial material. The distribution requirements are traditional as a consequence of the curriculum. Students must gain exposure to the full range of arts and sciences with more specific courses than those required at a majority of Claremont's peer institutions. The curriculum consists of one-third core and distribution requirements, one-third major requirements, and one-third full electives. All students must complete an in-depth senior thesis in their field in order to graduate. With its collegiate neighbors, Claremont McKenna provides students with fifty different study-abroad programs to choose from. A majority of the student body will take advantage of this opportunity at some point in their four years. For the student who determines that he/she wants to attend medical school or other science programs, the excellent level of science courses and facilities at several of the other Claremont colleges are at hand. Another unique feature is the presence on the same campus of ten research institutes, all of which are focused on various areas of professional leadership, public policy, and the law. Ambitious undergraduates can avail themselves of their presence to gain research and internship experience. In 2007, the Robert

Day School of Economics and Finance was established with a $200 million gift. According to the college, "This was the largest recorded gift to a liberal arts institution, the largest gift in the field of finance and economics, and among the top twenty largest gifts ever given to a college or university." Students may now pursue an M.A. in finance, and as undergraduates may participate in the Robert Day Scholars Program in business, finance, government, and nonprofit leadership. In sum, Claremont McKenna holds a special position within the world of small liberal arts colleges. It presents the same range of liberal arts curricula as other top colleges but with a concentration in the applied social sciences and a much wider course selection, thanks to its four affiliated colleges and its graduate management program and the research institutes.

Other programs of note include the Athenaeum speakers program, which brings well-known "historians, authors, poets, politicians, musicians, professors, economists, and more" to campus four evenings a week; a 3/2 management/engineering program, in which students complete three years at CMC and two years at a school of engineering; the Washington, D.C. semester; and the Practicum program. In terms of student support, an administrator points out that "faculty advisors offer guidance in course selection, career paths, majors, and general life issues. Student sponsors help students during freshman year with everything from class selection to adjustment to college life. And the career services department helps immensely in providing internship and job opportunity information and guidance."

We can expect CMC to stay true to its founding mission and identity in the future. As the college welcomed only its fourth president in

1999, it emphasized the recruitment of "superlative" students and faculty and technology. Says one administrator, "We will seek to build our name recognition on the East Coast and seize an unassailable position on the top-ten list of private U.S. liberal arts colleges." A dean says, "'Steady as she goes.' CMC's mission is still fresh although some adjustments will occur," notably in increasing commitment to and enrollment in the sciences and modern languages. Still leading Claremont today, President Pamela Brooks Gann is one of the longer-serving college presidents nationwide.

Major Admissions Criteria

Claremont looks for students who are "intellectually able, ambitious, and self-disciplined." One administrator lists the most important admissions criteria as "high school grades, SAT score, and record of activities." Another says, "CMC attracts students who take the initiative, whether in a student leadership position, as captain of an athletic team, or president of a club. We want students to be involved and to stand out as leaders." The major admissions criteria are:

- High school transcript—challenging courses along with high grade point average.

- Extracurricular activities—demonstrated leadership.

- Scores—SAT or ACT.

A dean notes a particular interest in students' participation in student government. The admissions office says, "The admission commit-

tee considers the whole person and assesses each applicant's potential for success, both during college and after graduation. Before making our selection, we weigh the following: academic performance and promise; personal characteristics, achievements, and goals; participation in activities and organizations; interests, talents, and leadership potential."

The Ideal Student

It should be obvious that Claremont McKenna provides a particular brand of education in a unique environment in comparison to the majority of liberal arts colleges. The student who is passionate about the social sciences, management studies, and leadership training in the public, private, or international fields will find this college most attractive. Conservative in nature, Claremont McKenna best serves those who are either conservatives themselves or who welcome the opportunity to challenge themselves in such a setting. The workload is demanding but there is an emphasis on collaboration and support among its student body and with the faculty. An intelligent young man or woman who might be considering an Ivy League institution with a vision of preparing for a career in politics, government, law, or business should seriously consider the educational foundation to be acquired at Claremont McKenna within a more personalized environment. Here is one college where a young man can bring his blazer, white shirt, and tan bucks.

In this environment, however, the student body is relatively diverse, with about 8 percent international students, 4 percent African American students, and, reflecting its California environment, about 11 percent Latino students and 12 percent Asian Americans. (One-third of students are Californians.) There are a number of organizations on campus to work with diverse populations, and CMC seeks to promote diversity by offering substantial financial aid packages. Overall, however, "CMC students are themselves focused, ambitious, and driven. They're also friendly and open to working with other students in groups, and eager to support each other's achievement and success," says an administrator. Students who do well are "ambitious, self-confident, intellectually curious, and socially outgoing." They are "positive and upbeat with a can-do attitude. CMC challenges and pushes students to the limit, but at the same time, students are given the tools and resources to succeed. CMC combines thought-provoking classes with a warm, family-type environment."

Student Perspectives on Their Experience

"I think that our best attribute as a college," says one sophomore, "is the competitiveness to be the best and to achieve excellence. It is dynamic and always trying to progress. . . . It is also a great networking opportunity. CMC puts its students first and takes care of us."

The students who do best are "those who are the most driven to do the best. There is a heavy workload, so those who can organize their time and stay focused are the most successful." Small class sizes and "a great faculty" help. "The whole environment is conducive to a family-like atmosphere where students and faculty help one another. With small classes, you can't get lost in the shuffle. Also, having all of the finest and latest resources (technological and others) is greatly appreciated." Within this environment "the freedom that the school gives you to

conduct yourself as an adult without lots of personal restrictions helps students to grow as people. And the willingness to help out in every way possible with any activities one wishes to plan makes living here enjoyable." A senior notes, "It's very helpful to be able to reach our professors at home during weekends and around finals time." And "anyone who puts in enough time and effort is able to perform well at CMC. No class is extremely difficult if the student is eager enough to visit the profs and/or tutors."

Students succeed "who really like to get involved (academically and socially). Those who take their classroom learning beyond the academics (with internships, research, on-campus organizations) find themselves intellectually stimulated and *busy*!" Echoing this, a sophomore says, "Really smart students with no personality have no place at this school. The school trains leaders, not midlevel management. If you have no voice, no opinion, and don't get involved, you probably won't get a whole lot from CMC." Students help each other to do well. "There's a strong sense of solidarity among our student body (particularly when it comes to surviving classes). We help each other out a lot." Another says, "Takers do best at CMC—students who know what they want and take it. Vocal, forward, strong-willed, self-confident students succeed here; it is not a school for the passive." This sentiment is echoed by other students, both positively and negatively.

A senior campus leader planning to go into information technology consulting says, "The management-engineering program offered me the opportunity to meet a lot of the science department faculty, who are important to my physics major interest." Another notes the attraction of the Claremont system: "The five college program is really neat. I get to live on a government-, economics-, and foreign policy–based campus (one that, because it has concentrated its funds in the above list, has no music department) and still can be a music major by taking those classes at Pomona College. The system also tremendously changes social life; because all the campuses are adjoining, I essentially go to school with over six thousand people. But I share my deans and administration at CMC with only one thousand. I have the community bits of a small school with the social options of a larger university."

Some students complain about wanting to make the "study material a little more culturally diverse," although "the other Claremont Colleges make up for that." One quips that the college "trains corporate attack dogs quite well" and that "it lacks the typical leftist slant, and thus has little truck with ethnic studies or feminism. Whether it studies these fields in a 'worse' manner than others is not necessarily bad." To illustrate the contrast, another student says, "Strong support of the ROTC [Reserve Officers' Training Corps] program has been important to me. Support like that goes against the grain of increasing 'politically correct' and overly liberalized schools." General campus diversity, however, is seen as positive, particularly when including the other Claremont campuses. Students would like to see CMC more well known and publicized on the East Coast, and a visit is clearly crucial in evaluating the college; potential students have given their campus visits glowing reviews. Some students do not see every area of the college administration (i.e., the registrar) as following through on the small-school model where all students are treated as individuals. But "student government has an incredible budget and is well respected by the administration."

"CMC is a true home away from home. Personally, I think that all the staff here have been nothing but helpful, especially for incoming freshmen, who often have a rough transition time." Students appreciate access to faculty, the small campus size, and the campus living situation, which "promotes good friendships" among students of all classes. "If I had the choice of going to Harvard, Princeton, or Yale, I would still have chosen CMC," says one student. Another, headed eventually to law school, makes the point that "the school helps its students succeed."

What Happens after College

After Claremont, about two-thirds of students go on to graduate or professional education over time, with two-thirds of these to law school and business school, and one-third to other academic programs. Immediately after graduating, however, some two-thirds of students enter the professional employment market. Consulting and investment banking are the most popular early career areas, but the list of choices is eclectic and includes nonprofit options like the Peace Corps and Outward Bound and academic fields from education to social psychology. Nevertheless, some 70 percent of CMC graduates have "economics, government (political science), or both in their diplomas." An administrator notes, "We focus strongly on the development of leadership skills—a fact that is reflected in our students' career outcomes: One in eight CMC alumni now holds an executive management or key leadership position in business, industry, the professions, the sciences, government, or public service. CMC alumni achieve the success that they do because they take from campus the ability, skills, and confidence to succeed, as well as the ability to apply information—whether mathematical, technical, scientific, philosophical, or moral—to problems they encounter in their work as well as in their daily lives." Going further, we hear that "CMC graduates have pride in themselves and confidence in their ability to 'take on the world' in many arenas: business, medicine, government, law, and much more. Our students are leaders, not only in the organizations in which they are building their careers, but also in city, state, and national government, and in their local communities."

Colby College

4800 Mayflower Hill
Waterville, ME 04901-8848

(800) 723-3032/(207) 859-4828

colby.edu

admissions@colby.edu

[
Number of undergraduates: 1,867
Total number of students: 1,867
Ratio of male to female: 43/57
Comprehensive fee for 2008–2009: $48,520
(includes room and board, tuition, and fees)
% of students receiving need-based financial aid: 38
% of students graduating within six years: 87
2007 endowment: $598,729,000
Endowment per student: $320,690
]

Overall Features

Here is another example of a small, independent college founded in the early 1800s to educate its students in the classic liberal arts tradition and encourage them to assume positions of leadership in the larger world. The northern Baptists who founded the college made sure to include a clause in its charter that promised religious freedom to all who attended. This attitude was expressed later in its admission of African Americans and women more than a century ago. Colby had the first campus-based antislavery society in the nation (1833), was the first previously all-male college in New England to admit women (1871), pioneered the January Plan (1962) (see "Curriculum, Academic Life, and Unique Programs of Study"), and encouraged outdoor orientation trips for all new students. The college's former president made it an active policy to attract students of differing racial, religious, and ethnic backgrounds. Still a small, entirely residential college, Colby places a premium on its faculty teaching and advising students in a personal and friendly environment. There is also a heavy emphasis on community interaction and sharing. This philosophy led to the elimination of the fraternity system on campus in 1984 and, in its place, the development of a residential commons plan. The dormitories were configured into four commons, each with its own dining hall, social activity center, and student-led activities and regulation. Faculty are affiliated with the commons and some even live in a dormitory. Colby students take advantage of the isolated, wintry setting through skiing, club outing activities, and cheering for their Division III intercollegiate teams. A majority of students play on a varsity or intramural team. A good number are attracted to Colby for its strong environmental science studies, which also reflect its country setting.

What the College Stands For

Colby sticks closely to "The Colby Plan," which consists of "ten educational precepts that reflect the principal elements of a liberal education and serve as a guide for making reflective course choices, for measuring educational growth, and for planning for education beyond college. Students are urged to pursue these objectives not only in their course work but also through educational and cultural events, campus organizations and activities, and service to others." These precepts, which the college believes are at the heart of a liberal arts education, are:

- To develop one's capability for critical thinking, to learn to articulate ideas both orally and in writing, to develop a capacity for independent work, and to exercise the imagination through direct, disciplined involvement in the creative process.

- To become knowledgeable about American culture and the current and historical interrelationships among peoples and nations.

- To become acquainted with other cultures by learning a foreign language and by living and studying in another country or by closely examining a culture other than one's own.

- To learn how people different from oneself have contributed to the richness and diversity of society, how prejudice limits such personal and cultural enrichment, and how each individual can confront intolerance.

- To understand and reflect searchingly upon one's own values and the values of others.

- To become familiar with the art and literature of a wide range of cultures and historical periods.

- To explore in some detail one or more scientific disciplines, including experimental methods, and to examine the interconnections between developments in science and technology and the quality of human life.

- To study the ways in which natural and social phenomena can be portrayed in quantitative terms and to understand the effects and limits of the use of quantitative data in forming policies and making decisions.

- To study one discipline in depth, to gain an understanding of that discipline's methodologies and modes of thought, areas of application, and relationship to other areas of knowledge.

- To explore the relationships between academic work and one's responsibility to contribute to the world beyond the campus.

These precepts are another excellent summary of the goals and tenets of a liberal education, forming the core of Colby's mission. Like other Hidden Ivies, Colby has seen its student body increase in qualifications and recognition. "Colby was a fine regional college; it is, today, a fine college with a national and, indeed, international reputation. It has changed by an ever strengthening faculty which has brought an increasingly stronger and more diverse student body," says one long-serving dean.

Another administrator notes Colby's perception of itself and its bucolic environment: "Colby's classic, small-town, liberal arts campus creates a community of learning and a rich intellectual climate. The resources dedicated to cultural programming are extraordinary and there is always something stimulating going on. . . . The location also makes Colby one of the safest environments in which to study, as campus crime statistics prove." But what makes Colby most unique, raves one administrator, are the people: "It's the personal touch, whether it comes from the president and his wife visiting a sick student in the hospital or the physical plant department helping students build a climbing wall in the field house or secretaries serving cookies at 'Midnight Munchies' exam breaks. When a late-April snowstorm hit while the admissions office had a big group of recruits visiting, students loaned enough Bean boots so the prospective students *and* their parents and siblings all had proper footwear for the campus tour; they all enrolled. . . . When the downtown arts center ran into financial difficulty, Colby stepped up with some funding and volunteer support of key administrators to bail it out and make it a nonprofit institution. When, at a local restaurant, our Rhodes scholar told our president he was taking an unsuccessful finalist for the scholarship out to dinner, the president surreptitiously intervened and picked up the tab."

Curriculum, Academic Life, and Unique Programs of Study

Colby insists on student exposure to a full range of the arts and science offerings. While it has a relatively traditional set of requirements that

must, as a result, be fulfilled, it also gives students some highly appealing means for meeting them. "We have resisted specializing and emphasize the broad liberal arts," says one administrator. "Unlike many colleges which eschewed traditional programs (i.e., foreign language requirements) in years past, we have clung to the broad basics while adding many of the programs that are demanded by the times." For example, to complete the language requirement a student can spend a semester in an intensive language course in Mexico or France. A large number of study-abroad programs are available throughout the four years, and on campus, the college has recently built the Oak Institute for the Study of International Human Rights, which brings a human rights practitioner to campus for a semester and includes a lecture and scholarship program. The Green Colby programs highlight Colby's interest in sustainability and conservation. The Goldfarb Center for Public Affairs and Civic Engagement, opened in 2003, connects students and faculty with national and local leaders. On campus, new buildings have been built and others are planned to complete the development of a "Colby Green expansion." Colby maintains the January Plan, which many other colleges dropped a few years ago for reasons not fully understood. Over the course of the Jan Plan a student can pursue a specific topic in considerable depth without the pressures of the regular academic semester. This can take the form of intense study of a subject with a professor, an internship, an artistic project, or study abroad. Colby students tend to favor concentrations in history, government, international relationships, English, and environmental studies. The Olin Science Center attracts faculty of note and students to science studies. Colby's Museum of Art houses major collections, and received a significant gift from the Lunder Collection in 2007. Another interesting program is Integrated Studies, which teaches three courses in different disciplines as an interdisciplinary cluster. The administration and trustees have been hard at work on behalf of Colby students in raising money to enhance the learning and living opportunities on campus.

One dean expects to see "increased emphasis on student services. There is no college with a better residential life program." The college sees itself as having strong student services, faculty advising, specialist deans for multicultural and international students, a Writers' Center, and a tutoring program. Colby is a "hands-on" kind of place, with "a great deal of individual faculty attention in and out of class and with a student service program that emphasizes close attention and follow-up, which our students appreciate." Another administrator notes that "Colby's faculty has always been good. Today it is excellent. Ninety-nine percent of teaching faculty have terminal degrees in their fields and scholarship is high. A common thread is that teaching comes first, and Colby distinguishes itself with the close interaction between faculty mentors and students. Those relationships are part of a long-standing culture at the college, but they also are supported by programs including a books-in-the-residence-halls program and a take-a-professor-to-lunch program. Senior exit interviews continue to rate student-faculty interaction and collaboration as one of Colby's strengths."

Colby has consistently been an early adopter of new technology. "The college's location in central Maine has changed from a mixed blessing to a strength, in part thanks to technology. Fax, e-mail, the World Wide Web, the proliferation of media (including cable channels), satellite television links (up and down), and videoconferencing all keep students and profes-

sors intimately linked with the larger world. We have legitimate national experts on various aspects of the federal government, for example, who are wired into Washington despite living six hundred miles away. We still have the benefits of Maine's marvelous environment but we now have many more opportunities to interact with the larger world." An alum and current administrator notes that "Colby's campus is often cited as one of the nation's most beautiful and, though the enrollment has grown only slightly, twenty new buildings (more than one-third of the buildings on campus) have been built in the twenty-five years since I graduated. Science students who go on to graduate programs marvel at the access to sophisticated lab equipment (electron microscopes and MRI, for example) that undergraduates here enjoy when, as grad students, they have to wait in line for equipment at major universities. The library has improved from a somewhat neglected facility and collection twenty-five years ago to a larger, better stocked facility that is a technology leader among its peers. The increase in Colby's resources (i.e., endowment) has ensured top-notch facilities that are well maintained."

Major Admissions Criteria

Colby, like its peer colleges, looks at each individual application carefully to find out as much as it can about what makes its students tick. Unlike the other two Maine colleges in the Hidden Ivies, Colby does require some standardized testing, which it views in the context of an applicant's overall profile. Colby will, for the first time, allow applicants for the fall of 2010 to submit SAT subject test scores in place of SAT or ACT. An admissions official describes the criteria that Colby prioritizes in admitting its students: "Colby College looks closely, in a personalized way, at each application. Factors that weigh most heavily in our decisions include a student's high school record, though not only the record of achievement one has accumulated, but, just as important, the degree to which an applicant has challenged himself or herself in the courses he or she has elected to take. The majority of students admitted to Colby place near the top of their graduating classes, after having taken advantage of their high school's most demanding courses. While we have found that a student's academic performance in a rigorous college preparatory curriculum is the single best indication of his or her promise for success at Colby, standardized test scores also may indicate academic potential and will enable us to measure along a common yardstick students applying from all over the world. Colby requires the SAT i. But SAT ii subject test scores are optional. Writing that is required on the application helps us to understand each applicant's ability to think critically and communicate effectively through a medium that will be of the utmost importance throughout each student's college career. Required recommendations from college counselors and teachers (2) help us to gain a more comprehensive perspective on each applicant and often influence our judgment. In making admissions decisions, Colby seeks excellence—in academics, art, music, theater, work experience, publications, leadership, public service, or athletics. We value racial, cultural, and socioeconomic diversity throughout the college and seek candidates from all parts of the country and the world."

The Ideal Student

Prospective students who visit the campus seem to know almost immediately if this is the right place for them to spend four years. With its handsome set of buildings on a pristine campus, Colby appeals to a mainstream, socially outgoing individual who seeks an intimate community and culture. A dean describes Colby students as "nice and smart." Those with an interest in an active learning format and the out-of-doors will thrive in this setting. Those looking for an urban experience with lots of off-campus options and a very diverse student population are not likely to choose Colby. "There's still a sense that Colby has a way to go before its student body reflects American demographics, but in the last twenty years minority enrollment has increased from two percent to about thirteen percent and next year's freshman class will be the most diverse ever. In addition there is another ten-percent cohort of international students from 62 countries. It remains a priority to recruit more students and faculty of color. A multicultural center was opened a decade ago and Colby is aggressively pursuing initiatives to make the campus welcoming and comfortable for all students." Most students (two-thirds) will spend a semester or year in one of the many off-campus programs and return to Colby refreshed and eager to reunite with friends and certain teachers. This is not a community made up of tree-huggers, to use the vernacular of college students today. It is more a moderately traditional, socially and physically active group of collegians who thrive in a close-knit community and appreciate good teaching. It is not the best place for loners, urbanites, and social reformers.

Student Perspectives on Their Experience

Students choose Colby because of its academic reputation, location, and relaxed, active environment. "I chose Colby because of the great reputation it had as a work-hard, play-hard kind of school. The close connection with professors and the friendly atmosphere also played a major role in my decision. Colby's campus was one that I fell in love with when I came up to visit as a pre-frosh. It is absolutely beautiful." A sophomore notes, "Unsure of what major I wanted to pursue in college, I was attracted to a liberal arts school. I also knew that I wanted a small school. Visiting the campus is what really persuaded me—I had a great time; all the students were so friendly and the campus is beautiful. Each time I visited I just had a good feeling about the school." Another graduate says, "Not only did Colby fit all of my requirements, but I fell in love with its white columns, brick buildings, ivy, and spacious campus." What students find is a quality academic program, great faculty, and a high-achieving peer group. "I think camaraderie among students pushes others to succeed. The academic environment is competitive, and the competition helps students to succeed."

Says one graduate, "Colby does a lot to attract some fantastic professors who really have an impact on one's life. All professors and members of the administration have an open-door policy which really helps a student feel like they belong here. The professors do their own outside research and there is ample opportunity to apply for a research-assistant position. I never would have thought that I would be working for a professor after my first year of college, but it happened." And, "The close attention from the professors really helps a student succeed. If you are struggling, the professor's door is open to you, along with possible tutoring from a TA. Furthermore, the open-mindedness of most of the students is a big plus." Another student says, "Colby offers academic help to students in many ways. I can offer personal examples. Struggling in our biol-

ogy class, my roommate and I found it easy to meet with a tutor who was very helpful and accessible. My calculus class met with two math majors once a week to clear up any questions dealing with our homework. I frequently met with a TA for guidance in economics projects and my English comp. professor scheduled regular one-on-one meetings to discuss individual writing issues. Help is readily available to students and they are encouraged to take advantage of this help."

Students love the Jan Plan options and the COOT^2 program: "This is Colby Outdoor Orientation Trip. Each incoming student is put into a group of about ten other first-years and two or three upperclassmen. These groups hike, canoe, bike, and/or camp out in the beautiful Maine woods for four days before on-campus orientation and classes begin. COOT^2 provides new Colby students with a core group they come to depend on in the first few weeks of school, which is usually a very critical time. Most students remain in contact with their COOT^2 friends throughout all their years at Colby."

Students recognize the need for more diversity on campus, and, in line with their administrators, the importance of continued endowment building to support the college's programs. "In addition to fund-raising, Colby has tried to bring in more minority students. Although the numbers are increasing, it is a slow process that will take many more years to get to the level they want to be at." And, "Colby remains very homogenous with its predominately straight, white, middle-to-upper-class population—despite efforts to diversify. Offshoots of this problem include poor town-gown relationships, self-image problems among students (eating disorders and alcohol abuse are two symptoms of these problems), and a generally elitist and apathetic student body." Nevertheless, students report an accepting environment for racial/ethnic minorities, international students, and gay and lesbian students. And "there are plenty of cultural activities on campus—theatrical performances, symphony and choir concerts, festivals celebrating various cultures, and lectures."

Who does well at Colby? "The students who work hard and play hard do the best," says one student. Another reflects, "It is difficult to say who does best because there are many ways to succeed at Colby. My grades are pretty good, but I am not a straight-A student, and I'm a pretty good runner but not a huge point-scorer for the track team. If grades or athletics are what is important then you could say that I am doing just pretty well. But I did declare a major, accomplished many goals along the way, and made friends and memories that will last a lifetime. Everyone does best at what they feel is important." A graduate adds, "Colby provides incredible opportunities. I was able to play three sports, work for the school newspaper, write Web pages, and obtain interesting internships. I also was able to study abroad. So Colby offers opportunities that couldn't be found at larger institutions."

"There are some exceptional and very special people who work on Colby's staff [custodians and secretaries]; the food and dining hall are far superior to any I've experienced elsewhere; some members of the faculty and administration are exceptional people and mentors; and the library is phenomenal," says a recent grad.

What Happens after College

Insists one dean, "Eighty-five percent will go to grad school. They are prepared to do and be anything they want to do or be. I can prove it." Another administrator notes, "Colby's commitment to the liberal arts embraces the firm belief that the breadth and quality of a Colby education should be extended to include an equally broad choice of meaningful and rewarding career opportunities. The Office of Career Services strives to acquaint students with career options, offers insight into various professions, and assists in preparation for the actual career search." About 17 percent of Colby grads begin graduate school within three months, and within three years some 40 percent have entered or completed graduate school. Choices include study in a broad variety of arts and sciences programs, law school, medical and veterinary school, and business school. Employment directions are also broad. Students go on to work in education (15 percent); economics, sales, and consulting positions (11 percent); science and health (13 percent); financial services (6 percent); social services and government (16 percent); communications (9 percent); and law (4 percent). "The Peace Corps and AmeriCorps are also popular destinations," notes an administrator, "perhaps in part stemming from an active student-run Volunteer Center." She goes on to say that "the assistance of alumni and friends of the college is particularly important. With the generous support of Colby graduates and parents of current students, a broad network of persons in various professions and widespread geographical locations has been established to assist students and alumni in career exploration. Parents and alumni have agreed to conduct informational interviews, be hosts for on-site visits, sponsor internships for January and the summer, and provide housing for interns and job seekers in their areas."

Colgate University

13 Oak Drive
Hamilton, NY 13346-1383

(315) 228-7401

colgate.edu

admission@mail.colgate.edu

[
Number of undergraduates: 2,806
Total number of students: 2,814
Ratio of male to female: 48/52
Tuition and fees for 2008–2009: $39,545
% of students receiving need-based financial aid: 37
(class of 2012)
% of students graduating within six years: 90
2007 endowment: $709,047,000
Endowment per student: $251,971
]

Overall Features

Colgate may call itself a university, but it is in reality an undergraduate, residential liberal arts college. Here is another of the outstanding colleges established by the northern Baptist church, this one in 1819 to train young men for the ministry and civic leadership. Colgate reflects the more traditional model of the private, residential college that prevailed prior to the social revolution of the late sixties. It admitted women in 1970, as did many other all-male colleges at that time. While it has expanded its curricular offerings extensively over the years, Colgate still requires students to undertake a core curriculum and fulfill a number of subject distributions in order to graduate. The strong commitment to outstanding teaching is a hallmark of the school. Here is a refreshing example of a faculty whose members are evaluated regularly on the quality of their teaching as well as on original research in their respective disciplines. Teachers are expected to engage with students out of the classroom as well in this intimate, rural campus setting. The workload is pretty intense and students are expected to participate actively in the classroom. The Greek tradition has been a signature feature of campus life for many years. Close to one-half of the men and a third of the women belong to fraternities and sororities respectively. Many of them live in their chapter houses instead of dormitories. Not unlike many other institutions, Colgate finds itself in the throes of heated debate on the role of Greeks on campus life. The faculty has voiced its opinion that they should be removed since they seem to be a socially divisive element and the main source of anti-intellectualism on campus. Both undergraduates and alumni are highly emotional on the value of the system and will not let the trustees and administration eliminate them without a fight.

Athletic competition at the Division I level is a significant component of the Colgate experience. Talented high school athletes who want to combine an excellent education of a traditional nature with strong athletic competition find Colgate particularly attractive. Its location in rural, upstate New York has created a self-contained community of smart, physically active, socially engaged students who can thrive on its beautiful campus. It has also resulted in a mostly middle- and upper-middle-class cohort of students who feel comfortable in this traditional environment. It is a challenge for Colgate to attract a critical mass of minority students to the campus. The great majority of students prepare for professional careers in business, law, and medicine, befitting their traditional backgrounds.

What the College Stands For

A dean describes Colgate's educational goals: "We are dedicated to the total education of students, both academically and personally. We are proud that students learn in the classrooms, labs, studios, study groups, and other traditional academic settings. But at Colgate we recognize that significant learning also takes place in the residence halls, on the athletic fields, in student activities, and other nonacademic settings. We work hard to create vibrant learning environments in all of the college's programs, and we think we do a pretty good job." Another dean notes that "Colgate is committed to providing a rigorous and far-reaching liberal arts experience for undergraduate students. A Colgate education demands that students stretch their limits and engage with ideas in a variety of different settings."

An admissions dean summarizes Colgate's goals as follows: "Colgate's mission is to provide a demanding, expansive, educational experience to a select group of diverse, talented, intellectually sophisticated students who are capable of challenging themselves, their peers, and their teachers in a setting that brings together living and learning. The purpose of the university is to develop wise, thoughtful, critical thinkers and perceptive leaders by challenging young men and women to fulfill their potential through residence in a community that values intellectual rigor and respects the complexity of human understanding."

Residential life at Colgate has been improved, even to the point of the college advertising for volunteers to move off campus at times to relieve a temporary housing crunch. The college has added and enhanced facilities, moved fraternity and sorority rush to fall of sophomore year (students may not live in Greek houses until junior year), and increased to almost 25 percent the proportion of students of color and 5 percent internationals on campus through recruitment and retention efforts. "Improving our village's economy and services to residents and students is a high priority for Colgate," says one dean, noting the college's goal to help alleviate the results of a sagging economy in upstate New York. What makes Colgate unique is "our commitment to the overall growth and education of students, study groups run by faculty, a world-class career services center, Division I-AA sports, our location in upstate New York, our willingness to experiment with change, and the amount of technology available to students." With 2,800 students, Colgate is larger than most other small liberal arts colleges, yet still retains the "feel and personal touch" of its smaller cousins. The college sees its students as having become more academically serious and qualified over the past decade and less inclined to drink. "We still have a ways to go in terms of reducing the role of alcohol use in student life, but it is far less central than it was years ago. Accordingly, there is a more vibrant intellectual life outside the classroom than there was at that time, which makes Colgate more attractive and comfortable for students with social interests that vary from what has long been the 'norm' here."

Several administrators stressed their belief in both the college's excellence and its refusal to ever "be satisfied with the status quo." A dean illustrates Colgate's view of itself, as well as one of the central arguments of the Hidden Ivies: "I have worked at two schools ranked higher than Colgate University by a news magazine, and my children attended two others. Having worked at two other places and being a consumer at two more, I seriously question numerical rankings. Overall, I believe Colgate University is superior to three of the other colleges and on par with the fourth. We care deeply about our students and our programs reflect that attitude. I feel honored and humbled to be the dean of such a fine college."

Curriculum, Academic Life, and Unique Programs of Study

Colgate does not apologize for retaining its traditional, Western civilization–oriented curriculum and requirements. The administration instituted a set of core requirements for first- and second-year students that include these four major courses: Challenge of Modernity; Western Traditions with Literature; Cultures of Asia and the Americas; Scientific Perspectives of the World. In addition, two courses each in the humanities, sciences, and social sciences are required. This is an effort to make certain all graduates will have been exposed to the major disciplines of a liberal arts program. Says a dean, Colgate's core curriculum enables students "to learn how to communicate effectively in many ways and different contexts. . . . Students are asked to confront the question of 'Who am I?' and cultivate habits of mind that form the lifelong legacy of a liberal arts education. Students learn to think in an integrative manner that crosses boundaries, to see disparate ideas in context, and to communicate effectively as they examine basic questions on the nature of their individual and cultural identities." The most popular majors once these foundation requirements have been fulfilled are history, political science, economics, and government. Says one dean, "We expect students to be well versed in the history and traditions of Western thought and require them to engage in the study of at least one non-Western culture. Our core curriculum also requires students to explore the role of science in contemporary society, and the distribution requirement ensures that students may not be too narrow in the way in which they think about and engage in the academic endeavor."

The college has an honor code that echoes those of many other Hidden Ivies. Other key programs are study groups and student research. Numerous support programs and organizations are available through the ALANA Cultural Center. "Theme houses" like the Harlem Renaissance Center, Bunche House (peace studies), and Ecology House appeal to students with particular interests and affinities. "Clearly," says one dean, "supporting students who are different from the typical Colgate-bound student is a high priority for us." The college has an office of Undergraduate Studies and Skin Deep retreats to help students become acclimated to campus. All faculty at Colgate, a point of pride for the college, are involved in student advising, as are a network of deans and administrative advisors. Again emphasizing Colgate's interest in having its students succeed outside as well as inside the classroom, some 75 percent of students are involved in athletics of some sort, and the college supports over 100 student activities and more than 20 off-campus study programs.

Major Admissions Criteria

Colgate's admissions officers are clear about their most important criteria in admissions:

- A strong high school transcript. It records a student's academic achievement, ambition (seen in course selection), and pattern of success, advancement and diversity of interests over 3 to 4 years of high school. From our experience, a student's high school records (GPA) and the level of difficulty in their course selection combine as the most powerful predictor of academic success in the first year at Colgate.

- Recommendations and testing. These items "tie" for second place because they really do counterbalance each other as elements that broaden our understanding of a student's achievement—in the eyes of those who know them at school and according to a standardized measure of achievement. . . . We're looking for the stories about a student in the context of their high school, their classes and the school community. We want to know what

contributions the student makes, in and out of class, and whether their ideas, involvement, and example have an impact on their peers. . . .

- Standardized testing is also important. That does not mean that the SAT is *the* measure of who gets into college, just as our saying "the transcript is more important" does not mean that SATs or ACTs can be discounted.

- A student's activities and overall "personal profile" of honors, involvements, and summer pursuits. We attract (of necessity) students who are "doers," who are accustomed to getting involved, making things happen and helping to create the life and activity around them. Ours is not a passive campus, though I would also note that students need not be "super achievers" in order to be admitted.

One dean summarizes the admissions committee's goals: "Priorities in the admission process reflect the priorities and values of the Colgate community. Students admitted to Colgate have an ability to respond to challenges, develop their own ideas, use their imagination and challenge themselves in and out of the classroom. In this dynamic, interactive learning environment, students and faculty learn from one another, and a willingness to contribute is paramount. The word *contribution* is often used to refer to involvement outside the classroom, but the greatest contribution a student can make at Colgate is participating in the intellectual life of the college. Many applicants excel academically. Many excel in their activities and as leaders in the community. The admission staff looks for excellence in all areas, starting with the academic record."

The Ideal Student

Those high school students who are seeking an academically rigorous, more traditional orientation to the study of the arts and sciences where good teaching is recognized and rewarded will find Colgate very appealing. They must also feel at home on a campus that favors athletic engagement and lots of social interaction through Greek and dormitory life, as there are few alternatives to be found off-campus in Hamilton and its environs. Colgate students describe themselves as intelligent, confident, energetic, moderate to conservative, career-oriented, and in search of a good time to balance the hours spent on school work.

It is hard to do better in matching the college to its ideal students than the following statement from a senior dean: "We seek students who are demonstrated academic achievers and involved community citizens . . . who are excited by an atmosphere that emphasizes participation—in and out of class—and expect students to engage with one another and their professors. We seek bright, intelligent, engaged learners. I sometimes use the phrase 'practical intellectuals' to suggest the joining together of intellectual life with 'real-life' settings for learning. . . . Colgate appeals to the best and brightest students who also seek a down-to-earth and rather lively environment . . . I don't mean purely social involvements, but rather a lively sense of involvement and direct contribution to the college community. Every student at Colgate can make a difference. In a small-town setting, every voice matters. We appeal to students who see the merit in that atmosphere versus a more

anonymous, typically larger or more urban setting. Colgate students often speak of the balance in their lives—not a 'perfect' fifty-fifty proportion of academic and personal pursuits, but rather a sense that both sides of that equation really do matter and are worthy of attention. Colgate students and Colgate as an institution are nothing if not 'ambitious.' We sometimes use the word 'scrappy' to describe the institution—suggesting a sense of 'not settling for second-best,' or the 'we-try-harder ethic,' or in our athletics program, the image of a 'giant-killer'—the small school that competes against—and beats—institutions more than twice its size. There is a strong spirit and personality to Colgate, a vibrant, positive, upbeat style. That 'can-do' attitude (corny as it sounds) permeates much of what we do, and characterizes the kind of student who will do well here."

Another dean adds, "We are looking for academically motivated students who have a proven ability to succeed in a rigorous environment, students who are intrigued by ideas and display genuine intellectual curiosity. Additionally, given our small, rural setting, we want students who are motivated to create opportunities for themselves and their peers outside the classroom. We provide resources and guidance, but we want students who will seize the opportunity to edit the student newspaper, or start a programming board, or engage in community service, or direct a play. The students who do best here are those who extend their zeal for learning beyond the classroom and make meaningful commitments to cocurricular involvement."

Student Perspectives on Their Experience

Students tend to feel "right" on Colgate's campus fairly quickly. "When I stepped onto Colgate's campus for the first time during college visits, I just had this remarkable feeling, a feeling that I hadn't felt at any other school. From that point on it was an easy decision, and I chose to apply early decision. Colgate is unique in that it excels in so many areas. Whether it is academics, athletics, student lifestyles, or even being ranked the number-nine party school in the country, whatever it is, Colgate is consistently found near the top." Another student says, "Upon visiting, I simply fell in love with Colgate. It was the feeling emitted by the students, the small, close-knit community, and the intense academic experience I sampled in the class I sat in on. Three years later, it's still kind of hard to put into words what Colgate means to me, but the maroon runs through my blood. I love the sense of tradition and the fact that Colgate never leaves one even after he graduates here. We produce strong, intellectual people and we develop leaders. Graduates are social beings who are ready to go anywhere in the world and into any field and excel. It's an exciting place to be, the place I have called home for three years, and I can't wait to be a proud Colgate alum after I graduate from here."

A recent graduate "encourages all those who are interested in the university to take a visit and stop a person at random on their way to one of Colgate's beautiful buildings and ask them why they are at Colgate. I guarantee you the answer will be different for each person, but no matter what, they will feel that common bond to their alma mater and so can you. Colgate is the place for those who want to learn, think, and give back to society."

Who succeeds at Colgate? One student says, "Academically, I feel that anyone who takes time out to care about the classes and the intellectual environment here does well. You have to work hard to do well (B+ or A) and that

means studying a lot and doing reading and research. If you blow off your classes and don't take your work seriously, you will find yourself in a deep hole. I believe the opportunity to learn so much from a gifted faculty and a strong liberal arts curriculum lets one explore new areas. It aids one's intellectual growth and that is something no one forgets. Socially, people from families with college backgrounds will have an easier adjustment because they know what to expect. White students of upper-middle-class backgrounds are the norm here. But students who are ready to get involved and be engaged, people who are eager to be social and work with others will excel. Finally, it is a great asset if one can expand his or her mind and comfort area to accept new people and new ideas." One grad says, "Student leaders who are highly motivated, intelligent, and willing to contribute to the community tend to thrive at Colgate. . . . Students who are willing to learn from other people's differences also do quite well."

Students feel supported and valued on campus: "There is a huge support service here. From students and peers involved in Residential Life and the Link staff (first-year orientation) to the academic and administrative deans, there are a ton of people to seek help from. Every single activity and group is easy to get involved in; I jumped right into student government and student activities by my second semester and was the head of a campus organization as a sophomore. . . . Being social and ready to meet and work with people is a huge asset. Also, being responsible and developing good time-management skills is important to balancing a social and academic life. Also, not being afraid to talk to professors and people about problems or questions will help one understand that there are many students who are uncertain about life in and after college." Another student notes that "the Office of Residential Life was an integral part of my experience. . . . Res Life offered me the chance to excel as a peer counselor and advisor, through resident advisor and head resident roles. The club sport program allowed me to form my own martial arts club, which taught karate to students, faculty, administrators, and local residents."

Another student says, "I love Colgate for what it symbolizes. The deep sense of tradition that comes during Freshman Orientation when the students walk up the Hill for Convocation in the Chapel—and again as seniors when we march back down the Hill and surround Taylor Lake to symbolize the complete circle of Colgate. The look and feel of the buildings. I love the pride that students feel toward this school. You graduate Colgate with a part of the school embedded in you forever. I am proud of the great commitment to academics, the strong liberal arts curriculum, and the rigorous academic experience that molds us into intelligent young adults. The great tradition of athletics here and how well we perform at the Division I level. The active student body where it seems that everyone is involved in something, from student government to outdoor education to the newspaper or the Greek system. And the sense of community that flows from the people here."

While Hamilton—and the college—could be bigger for some students, "the many opportunities for involvement both within the Colgate community and in the village of Hamilton are invaluable experiences that have taught me just as much as I could learn by obtaining any degree here." Another says, "Colgate is really good at providing students with more things to do than could ever be done. Most students wish there was more time in the day so that they did

not have to choose between attending a great lecture, going to dance practice, cheering on the hockey team, and meeting with a professor." She advises students to "get involved in *something*. You don't have to do everything first semester, but join a club, join a team, go to a brown-bag discussion, attend the physics lectures weekly. It is the best way to meet new people, feel an attachment to the university, and get the most out of your short time at Colgate." As with other similarly located campuses, diversity is an issue at Colgate, but in general the college is respected for its efforts: "Colgate is not by any means an overly diverse campus, but I feel that it does a fair job attracting students of all different races. Colgate has unique housing opportunities available for those students who choose to live in a residence hall that is made up of people of the same race. Also, different student organizations exist that enable students of any race to join who are interested in learning about a specific race. But I'm not sure that Colgate does a tremendous job of providing students of different ethnic backgrounds many opportunities to 'express themselves' and their culture on a regular basis. I feel a lot of that is due to the small rural setting we're found in and not so much ignorance on the part of the university."

"The financial aid programs at Colgate have been huge because I *never* would have been able to come here if it weren't for the great aid package I received, and I know Colgate does its best to provide for all incoming students. I hope I become a millionaire someday so I can continue the tradition of giving back to my school to make sure that kids like me have this chance. The study-abroad program was one of the best experiences in my life. As a kid, I never dreamed I would be able to live in London, have class across the street from Big Ben, see the

Eiffel Tower in Paris, and travel to Ireland (a great accomplishment because of my Irish heritage). I was able to immerse myself in the study of British history and culture and I will never forget the people I met or the experiences I had. Study-abroad gave me a new perspective on the world, my American identity, and what is really possible for someone who sets his sights high."

"The beauty of the campus is unparalleled," says a senior. "The small Hamilton village and the campus itself are some of the most serene, peaceful places I have ever lived and my experience of spending my summer here only heightens that fact. This school does a good job of promoting community. The residential life is great, and starting in one's first year, the dorm life produces great friendships. Also, the bond between students and faculty is incredible. I have generated strong relationships and even friendships with some of the faculty here, brilliant men and women. The ability to sit down and talk with them, interact and learn from them have been instrumental in strengthening my intellect. Colgate does a great job of creating strong social life. Be it through the Greek system, or, as in my case, through involvement in student-led activities, there are almost too many places for students to explore and get involved in. It is there that I have become a leader, someone capable of meeting people and making things happen on campus." What do students gain through their education at Colgate? "The greatest thing about this place is that I can do whatever I want with my degree. I feel prepared to conquer new challenges and use my liberal arts background to pursue my interests and, ultimately, do a job I love." A grad with plans to pursue a teaching career (perhaps at Colgate) in psychology wants to maintain academic connections to the college,

as "my advisor and I are in the process of publishing my senior honors research thesis."

What Happens after College

"Colgate students go everywhere after they graduate," says a dean. "Large numbers pursue graduate study or enter corporate training programs, as would be expected. Others travel, work for nonprofits, start their own companies, play professional sports, pursue careers in the arts, and the list goes on. The idea that a liberal education prepares people for a wide variety of careers and postgraduate endeavors is not just admission-brochure copy at Colgate. It is our reality." Another dean argues that Colgate students "are prepared for everything! Our students are sought after by all the major corporations. . . . The recruiters know our students are smart enough to learn anything in the field and socially 'savvy' enough to become leaders/managers." Alumni connections are also there for graduates: "People become very attached and one of the special attributes is the commitment of our alumni. They truly 'bleed maroon' and are ready to step forward to help those in the Colgate family. Our students have many doors opened in the real world by alumni who are eager to help. Colgate fosters a passionate attachment to the college."

Recent statistics for the class of 2007 illustrate the diversity of paths Colgate graduates take. Eighty percent were employed full-time, and almost 20 percent were attending graduate school full-time. The top career and graduate fields included financial services (17 percent); education (11 percent), health-related (9 percent), communications (8 percent), consulting (8 percent), law (7 percent), business-related (6 percent), sales and marketing (6 percent), nonprofit (5 percent), and sports and recreation (5 percent). Top areas of graduate studies included medicine and healthcare (28 percent), natural sciences (19 percent), law (18 percent), social sciences (11 percent), education (7 percent), and humanities (7 percent). A dean points out, "The involvement of Colgate's alumni is unique among liberal arts colleges and universities. Colgate alumni, friends, and parents have endowed seven scholarships for summer learning experiences. These include financial support for students wanting to pursue internships in law firms, public policy, financial firms, non-profit organizations, community service, health care, and other interests. Colgate alumni continue to maintain a relationship with the community long after they have graduated. Many return to campus in January to participate in Colgate's Real World program, a two-day career conference. Seniors and alumni participate in over thirty panels ranging from graduate school to tips on interviewing for jobs and networking. Receptions are also held, giving students a chance to mix and mingle with alumni in formal and informal settings."

Colorado College

14 East Cache La Poudre
Colorado Springs, CO 80903

(719) 389-6344

(800) 542-7214

coloradocollge.edu

admission@coloradocollege.edu

[Number of undergraduates: 1,972
Total number of students: 2,002
Ratio of male to female: 45/55
Tuition and fees for 2008–2009: $35,844
% of students receiving need-based financial aid: 40
% of students graduating within six years: 83
2007 endowment: $523,228,000
Endowment per student: $261,352]

Overall Features

Founded in 1874 as an independent, traditional liberal arts college on the New England college model, Colorado is an old school by Western standards. Over the past quarter century it has advanced dramatically its goal of enlarging its reputation as a top academic institution with a national student representation. One senior faculty member attributes the many positive developments to "a deliberate search by the college for excellence." It is well on its way to achieving national prominence as measured by the quality of education it provides; the credentials and reputation of the faculty; the multitude of new academic, recreational, library, science, arts, and athletic facilities; and the selectivity of the student body, with its diverse geographic backgrounds. Of the students, 75 percent come from outside the state, including all fifty states and thirty-plus other countries. Both the eastern and western states are well represented. The president states that for 125 years Colorado College has prospered by blending its commitment to tradition with its willingness to embrace innovation; that it has implemented its core values through small classes, personal interaction between students and faculty, and academic rigor.

In 1970, the college made its most momentous curricular change in delivering the liberal arts education it still viewed as the heart of the educational experience with its initiation of the Colorado College Plan, familiarly known as the Block Plan (described in detail later). This unique academic structure helps Colorado to attract strong students to its residential campus snuggled in a wide valley with a commanding view of the Rocky Mountains and Pike's Peak. With its magnificent setting in one of the most popular regions of the country for college-age students and a dedicated faculty leading the Block Plan, the college sees itself as possessing the potential to compete with the most selective of the liberal arts schools we include in this book. A senior administrator describes the essence of the college this way: "We offer a major liberal arts education with an altitude!" Given the underlying philosophy that critical thinking, creative expression, and communications skills are best developed by close association between teacher and student, the Block Plan requires teachers to engage actively with students in small classes and seminars and in the field work that is a frequent component of the course work. The ratio of students to faculty is eleven to one and it is a set rule that no class can exceed twenty-five students. The average class holds fifteen students and the majority are conducted in seminar fashion.

Colorado College's location in this distinctive area of the American West also influences the educational environment, with special majors and field work built into many courses and concentrations. The campus is set in Colorado Springs, the second largest city in Colorado, seventy miles south of Denver, the largest. Although there are some cultural activities available in the conservative yet ethnically diverse Colorado Springs, the majority of students spend their free time on the nearby mountains, trails, and waterways, or motor up to Denver where there are more fun outlets on the weekends. Students are more likely to be seen with their inline skates, skis, climbing ropes, and mountain bikes than with the traditional accouterments of eastern college students.

The college leaders are committed to expanding the mix of undergraduates to represent the national demographics. Almost every admitted candidate who demonstrates need is offered an attractive financial aid package, and the college offers a few awards to needy international students. The athletic teams have a clearly stated policy of inclusiveness and any form of discrimination will not be tolerated. Athletics are a major component of life at Colorado College for a majority of students either through very competitive Division III play or intramural teams. The athletic department believes strongly that its programs make a major contribution to the academic and social development of their students. The men's hockey team and women's soccer team compete very successfully at the NCAA Division I level, and seven men's and seven women's teams compete in Division III intercollegiate sports (the college dropped football, softball and water polo for the 2009–10 academic year). Three-quarters of students participate in intramural sports.

There are many outlets on campus for artistic and literary interests for the less athletically inclined. Students describe themselves as a liberal-thinking group with a relaxed living and dress style that can belie the heavy workload required under the Block Plan. The many special interest groups that represent particular gender, political, and ethnic groups found on virtually all campuses today are present, but on the whole they seem to shy away from the aggressive and vitriolic behavior present at the more liberal colleges. This is in keeping with the individualistic nature of the Colorado College community. Studies are demanding but the Block Plan allows for set breaks for traveling, relaxing, community service, and getting out to the mountains. Students are required to live in campus housing for their first three years and are guaranteed housing for the fourth year. As a result, about a third of students live off campus at any time. While there are recognized fraternities and sororities, a very small proportion of students choose to join them. Special interests and theme houses, such as language and arts specialties, are also sponsored by the college.

A senior administrator says of the future, "I believe that we will continue to expand the global opportunities for our students, both in their academic work and by bringing international faculty to the campus. I would hope that our small college will be able to increase its endowment to make sure that the strain of college costs will not prevent the best of the country's students from coming to CC. Our current Achieving Our Vision campaign will strengthen our ability to provide opportunities for students, recruit and retain outstanding faculty, and create a twenty-first-century campus." Notes the dean of admissions, "Since 2001, we have added a 300-bed residential complex with apartments, a new science building, and a dynamic new arts center. Next up will be a new library and a recreational center."

What the College Stands For

The former president of the college supported the intellectual and personal goals as expressed in the mission statement, "To provide an excellent liberal arts and sciences education which challenges well-motivated students of varied social, ethnic, and economic backgrounds and prepares them for positions of professional leadership and civic responsibility in an interdependent world." Underlying this mission are four priorities: enriching student intellect through high expectations for imaginative teaching, learning, and scholarship; promoting understanding of both similarities and differences among people with diverse backgrounds and interests; collegiality and mutual respect; and involvement in and responsibility to one's community. The administration seeks to provide a transformational experience for its students that they will carry with them for their adult lives. The college believes that only by a continuous process of looking at itself critically and creatively will Colorado continue to innovate and meet its goals in providing a first-rate educational and developmental experience. Said a senior administrator, "The college has sought the very best among students, faculty, and staff, and has looked in particular for individuals who believe in the values to which the college is committed. The success of the college is a testament to the extraordinary individuals who have come here. Our faculty is focused on instilling curiosity, wonder, and intellectual excitement in the young women and men who come to study here. This focus, along with small classes and the intensity of the Block Plan, helps students succeed in their academic pursuits and gives them the confidence to explore and discover." The administration is committed to broadening its economic and cultural mix of students and faculty by active recruitment and retention of minorities in order to enhance the environment of the campus and the growth of all students. The willingness of the college to embrace change over time if it betters the way students learn is articulated in these words of the former president in her 125th anniversary speech: "I envision several enhancements that build upon our core values: every student would complete an original creative project, from collaborative research with faculty to documentary filmmaking; every student would have an intercultural experience, whether in Kansas City or Kyoto." A professor's view from the classroom offers additional insight into Colorado's mission and style: "In my view, the only advantage—but it is a huge advantage—that a small liberal arts college has over other types of colleges and universities is our personal approach to education. Small liberal arts colleges like Colorado have the ability to engage and challenge our students as no other type of college or university can. For me, the most tangible sign that I'm accomplishing my goals is when a student of mine has the 'aha' experience and I can visibly see that the student has recognized some new truth or come to some new realization."

Curriculum, Academic Life, and Unique Programs of Study

The Block Plan is the most distinctive academic feature. "The Block Plan requires deep immersion and focus each and every day," says an administrator. "Students must take their head out of the book, look around, and find common ground with peers and professors. It's hard to bluff your way through a three-hour class. Students must be prepared and eager to learn. All of this breeds an openness, a respectful and authentic approach to dealing with others." Students undertake only one course at a time for three and a half weeks with a professor who teaches only that course. Some courses may last for one block while others for two or three, depending on the content of the subject and amount of material to be covered. Students take eight courses or blocks each academic year. This allows both students and faculty to focus all of their attention on the one course. The dean of the college and the faculty says of CC's goals for its students, "We want to reinforce their curiosity and their intellectual confidence by exposing them to the perspectives of multiple disciplines and to the intricacies of one. We want them to tangle with complexity and to learn to manage ambiguity as they study the development of different cultures and societies and as they examine the natural world. In addition, we expect them to sharpen their analytical and critical capacities and to practice the tech-

niques of writing and speaking which allow them to communicate what they know convincingly. In the course of their studies, we hope students discover the benefits of cautious analysis, the pleasures of determined inquiry, and the importance of independent mindedness." The centrality of the Block Plan is clear in every aspect of a CC education: "Colorado College combines a rigorous academic and dynamic intellectual environment with an atmosphere of informality and collaboration. Many comparable liberal arts colleges on the East and West Coasts have a reputation for competitiveness that CC has managed to avoid while maintaining equal or higher standards of intellectual and academic intensity. Students in disparate fields of study feel a sense of camaraderie because of the Block Plan, which requires a daily commitment to course work. When studying on the Block Plan, the focus is on a student's own engagement in class and in other activities and not on how other students are doing." Another unusual element of the curriculum is the way students bid for their desired courses on a point basis. The more points they bid for a favorite class, the more likely they are to win a place, since all classes are kept to a maximum enrollment. Many courses will combine a heavy reading and writing load with field work, since there are no other conflicts in the schedule.

The college describes the flexibility of the Block Plan as one of its major advantages. "Each class is assigned a room, reserved exclusively for its faculty and students, who are free to set their own meeting times and to use the room for informal study or discussions after class. Since competing obligations are few, time can be structured in whatever way is best suited to the material. No bells ring. Nothing arbitrarily intrudes after fifty minutes to cut off discussion. An archaeology class can be held at the site of a dig for one block, followed by the second block for laboratory analysis. A biology class might have a week of classroom orientation, then go to the field for two weeks. An English class can spend one morning reading a Shakespeare play aloud and the next morning discussing it or getting together with an acting class to try a few scenes." The college runs a second campus in the San Luis Valley, 175 miles from the main campus, where students can reside as part of their focus on Southwest studies or other specialized topics. You get the idea. It is common for classes to extend the classroom walls to the Grand Canyon or to Indian reservations or even abroad. With only one subject to cope with, students are free to create independent study or research projects or self-development study such as learning a language, a musical instrument, community service volunteering, or working on their athletic and out-of-door interests. Interesting special programs include facilities that relate to CC's Block Plan opportunities. One is a student-built cabin in the mountains near campus. The other is a facility in the San Luis Valley at the foot of the Sangre de Cristo Mountains near the Great Sand Dunes. There is little questioning of the intensity of the academic work since the content of a traditional full semester is covered in three and a half weeks. Students cannot hide from their teacher and their peers. The assigned work must be completed and contributing to the class discussion is a given. At the end of each block there is a four-and-a-half-day break for students to do as they wish.

The faculty are known to work as hard as their students to make such an intensive learning structure achievable. Colorado College boasts a faculty of 160 full-time professors, forty-eight part-time professors, and another fifty who visit each year to teach a course in their specialty. The Block Plan also makes it feasible to offer a wide range of interdisciplinary concentrations. These include, but are not limited to, environmental studies, international affairs, North American studies, Southwest studies, American ethnic studies, and studies in war and peace. Colorado students must fulfill general education requirements that are rather flexible: two courses in the Western tradition, three in the natural sciences with a lab, and two additional courses in non-Western or multicultural topics. First-year students are required to elect a theme-based seminar that combines course work with outside activities to broaden their exposure and hone their reading and writing skills. One-half of the student body typically takes advantage of off-campus terms for one or more blocks. Students can be found in New York, Paris, Florence, or in any major mountain range doing field work in geology or environmental matters. Approximately 55 percent of all students participate in study-abroad opportunities in every part of the globe on Colorado's own sponsored programs, and a multitude of off-campus programs is sponsored by the Associated Colleges of the Midwest, a group of quality, independent liberal arts colleges. For those already focused on their future plans, there are preprofessional programs in medicine, dentistry, law, physical therapy, education, and veterinary medicine. Reflecting its environs and some of its student body, a major in health fitness and exercise is offered. Cooperative 3/2 degree programs in engineering are offered with Columbia University, Rensselaer Polytechnic Institute, the University of Southern California, and Washington University, and a 3/3 degree program with Columbia.

Major Admissions Criteria

Because the Block Plan is central to the learning experience, the admissions committee puts special emphasis on signs of a candidate's commitment to engaging in a challenging academic environment, one that places great weight on writing and communication skills, on potential for contributing to the small classroom teaching method, a genuine involvement in the liberal arts subjects studied in depth in the Block Plan, and the ability to complete a demanding workload in a timely fashion. The faculty express their major interest in students "who are truly interested in something, students who are willing to work hard at something not because someone else thinks they should, but because it is something they really want to do well." These abilities are best measured by the content and level of the secondary school curriculum. Special attention is paid to students who complete advanced placement, honors, accelerated, or enriched courses. Class rank is one indicator of performance. Over 90 percent of admitted students rank in the top 25 percent of their graduating classes, and more than 70 percent are in the top decile. The application essay and samples of writing ability are given weight, and teacher recommendations and test scores are also taken into consideration. Special talents, extracurricular participation, and international experiences are important predictors of major contributions to the small and intimate community that characterizes Colorado College. Geographic, multicultural, and intellectual special interests are strongly factored into the decision process. Colorado has a substantial scholarship budget that is utilized to recruit outstanding students of all backgrounds. Thirty athletic scholarships are given to topflight men's ice hockey and women's soccer players. Special merit scholarships are available for students who plan to major in the natural sciences. CC's dean of admissions points to these changes in admission: "Since 2002, our applicant pool has increased by 56 percent. Our selectivity has decreased from a 56 percent admit rate to a 26 percent admit rate in the last five years, and our yield rate has increased from 27 to 41 percent. The student body is exceptionally able and talented. Twenty-three percent of the recent incoming class self-identified as American ethnic minorities (vs. 13.5 percent five years ago), 4 percent are international students (vs. 1 percent five years ago), and 2 to 3 percent are dual citizens living abroad (vs. 1 percent five years ago). Most important, faculty report they are engaged scholars who exhibit the more elusive qualities of passion for learning, curiosity, and freshness of mind. In admissions we are looking for a willingness to choose a traditional liberal arts education with a nontraditional delivery system; eagerness to jump in and engage with dynamic peers and professors; an intellectual appetite, a spirit of adventure, and a humble nature; and the traditional qualities—outstanding grades, letters of recommendation, and strong writing."

The Ideal Student

An atmosphere of individualism within a supportive community and a kind of Western, open-plains philosophy informs the environment at Colorado. A longtime professor of English describes what she believes are the qualities of the successful Colorado College student: "A good academic background certainly doesn't hurt, but I believe attitude may be more important. I particularly like students who ask questions, explore new areas, welcome new ideas, cooperate with other students in class, and work hard in extracurricular activities." The direction toward grade-grubbing and competition among students rather than an enjoyment of creative and critical learning is not something she hopes to see develop as the college has become increasingly more selective. Another administrator who works closely with students agrees with what kind of students do well at Colorado College: "Intelligent, self-assured, broad-thinking, extroverted, and possessed of plenty of work ethic and initiative, and a commitment to excellence." It should be clear to a potential candidate that four years at Colorado College will provide a fresh and dynamic experience. It should also be evident that to succeed in the Block Plan approach to learning, one needs to have good study and time-management skills, strong reading and writing skills, and a desire to immerse oneself in a subject and work closely with the faculty and fellow students. The academic program is intense due to the short period of time available to master a subject. Students say that it is especially demanding for science majors to stay on top of their work. The opportunities to combine theory and hands-on learning cannot be beat for those with real intellectual curiosity and the ability to focus. While there is a generally liberal flavor to the political life on campus, there is little in-your-face behavior. There is a mix of conservative and liberally inclined students to make the community work for just about anybody. A "sense of adventure" and a love of the out-of-doors and more physical space to stretch one's legs and mind make Colorado College a terrific place for an active and involved student. Any strong high school student considering the top eastern liberal arts colleges would do themselves a favor by looking into the college that officials refer to as "the only national liberal arts college in our time zone."

Student Perspectives on Their Experience

Students love the intimacy of CC and their contact with the faculty. Says one senior, "I have not come across a professor whose door is not open outside the classroom to help a student. It is amazing how many nonacademic faculty members are willing to get in touch with students through e-mail or in passing or through their department to help them out with both classes and extracurricular activities. Lastly, I feel that a student's peers contribute a lot to their success. Because everyone is on the Block Plan, we can all empathize with each other about a difficult block and support each other during the times of academic stress."

A sophomore speaks for most of her peers: "I chose CC because of its special and convenient schedule (the Block Plan), its beautiful location, its reputation, and the size of classes, which allows for more personalized attention and better chances for individual success." Those who do best are "the people who are committed to their success and do not procrastinate. The ones who take a wide range of courses, and pick their classes according to professors and not according to sexy titles. The people who put all their efforts into learning what the professor is teaching, and trying to get the most out of each class." A senior points to "students who are willing to go, go, go and then totally relax on block breaks. It's an all-or-nothing system—you either work all the time, or, during block breaks, not at all. Students who learn to write well, who don't fall behind in their classwork, who are willing to work hard on a day-to-day basis during a block, do well here." Students like the Block Plan overall and selected CC because of it, but they emphasize the need for potential students to be very aware of the plan's demands and proactive in determining their path through the college. Says a senior, "The Block Plan is a wonderful way to conduct college classes. Especially with a limited class size, students are able to dive into a subject completely—there are stories of students who begin to dream about their class topic." A junior adds, "The Block Plan really allows me to get to know my professors personally. The classes must be small and we see each other every day. I love it."

Activist students at CC are doing some pretty interesting things. A student says, "The school is amazing in its ability to accommodate the brilliant whims someone in its student body may want to dive into. It makes sure that you have the financial and logistical resources to accomplish whatever goal you set in your mind to do. The school aptly prepares you to deal with the high-stakes, intense work that you have to do in the real world. It inherently teaches you to learn to manage your time and work effectively." Another points to CC's location as a real asset: "The school does a great job of connecting with the Rocky Mountain image by providing unique opportunities for CC students. I consider CC the best school in the entire Rocky Mountain region and the college offers ample chances to really enjoy the Colorado lifestyle. There are numerous courses offered in the American Southwest and CC offers many trips to explore the environment, whether they be environmental science classes or kayaking block break trips to Aspen. CC has a prestigious internship program called the State of the Rockies that students get paid to participate in for the summer. State of the Rockies introduces students to issues of critical importance for the Rocky Mountain region; presents

them to well-known politicians, environmentalists, and businesspeople in Colorado; and establishes a strong student base with which to pursue community activist projects. From my knowledge, past projects have varied from immigrant rights in Colorado to the morality of hunting in the region."

Colorado College has increased its diversity awareness in recent years with such programs as the Glass House multicultural residence, the Black Student Union, a First Generation program, and groups for students of many other backgrounds and interests.

One student raves, "I love how my biggest class has been thirty students and smallest has been two." Students who do well at CC are gregarious, active, outdoorsy, and open-minded: "They are ambitious but not in the conventional sense of lining their ducks in a row or checking off items on life's to-do list. Most important, CC students who succeed the most are those who are not afraid of making mistakes."

Others are concerned about how much they are doing on campus: "CC doesn't offer the best atmosphere for a balanced academic and social life. Because the Block Plan is so intensive, students might be up until 3:00 a.m. studying, reading, or writing a paper they were just assigned that day. Then on weekends, people feel like they need to make up for lost time, and party a lot." "A stereotype at CC is that everyone is a rich, liberal, white kid who hates religion," says a student. "But honestly, that's not true. There are more minorities choosing to attend CC, and there is definitely a good portion of students who are on some form of scholarship or financial aid. There does tend to be a stigma against religious students, because they're associated with conservatives and considered close-minded. But while the religious atmosphere here is not exactly welcoming, there are many different religious groups on campus where a religious student can find other students who share the same beliefs." A minority student who feels she does a good job mixing with various groups on campus offers this reflection: "Culturally, there was once a large trustafarian hippie population, but that is slowly becoming a smaller sect that is being replaced by more and more preppy East Coasters. In general, everyone cares about the environment, politics, and the great outdoors, with an exceptionally large ski/snowboard population. The college does its best to bring all of these groups together and create the great dynamic we have."

A junior resident advisor says, "All kinds of diversity exist on our campus, it is just not always visible to the naked eye. The college always seems to be looking for ways to make students with different backgrounds feel welcome, but people always complain. As a head resident I am required by the college to lead at least two activities that focus on diversity issues each semester. This has gotten some healthy conversation going in our house." Some would like more of a voice in governing their affairs and working with college officials on student life, while others wish the college did not have such a "laid-back" attitude. Students appreciate their opportunities for study abroad and community service close to home. They generally feel supported and connected on campus, and as a senior describes, "We have an excellent Writing Center that really helps students improve their writing ability. We also have an excellent counseling center for students struggling with personal issues. Finally, Colorado College has a plethora of student organizations and other opportunities that help students to feel socially accepted and to make friends."

As a whole, CC is special because of the "Block Plan, which allows for travel, research, and concentration in one thing at a time. The school is special because it is like a little bubble in the Rockies; it is like a whole city that is tightly knit as a community." Says another student, "For students who want personal attention from their profs, who want to be challenged by a rigorous academic schedule, and who are interested in learning to appreciate learning, than CC is the right school for you."

What Happens after College

Students major in a wide range of disciplines that will lead them on to graduate school or directly into the workforce. The strength of the sciences, especially biology and geology, prepares many students for entrance to medical school and graduate school in the sciences. Sixty percent of each class enrolls in professional or academic graduate programs within five years of graduation, a figure that is comparable to that for the other top liberal arts colleges. A survey of Colorado alumni who went on to further study indicated that they felt far better prepared in their writing skills than a majority of their fellow graduate students, thanks to the emphasis on writing embedded in the undergraduate curriculum. Many students enter law and business schools after majoring in political science, history, and English. The director of CC's career center says, "Many of our graduates elect to pursue experiential activities after college, including internships, fellowships, personal travel discovery and other gap year experiences such as the Peace Corps, Teach for America, Fulbright/Watson/Coro/El Pomar, and numerous other public-interest/service/leadership programs and fellowships. Our students are passionate about the environment and its protection, are extremely interested in outdoor activities (leisure or otherwise) and are often entrepreneurial in spirit and inclination. They are also extremely interested in global issues, international service, and career activities. In short, while there are identifiable trends in student postbaccalaureate interests and activities, the individuals in question are at least as likely to develop their own ideas, programs, activities, etc., as they are to follow an existing or structured program, career path, or graduate course of study."

A student says of her CC experience, "Colorado College really does prepare its students for the real world. Students leave CC as efficient, self-motivated hard workers, traits that numerous employers value. Words cannot express how thankful and appreciative I am for the job skills CC has enabled me to gain while studying here."

Davidson College

PO Box 7156
Davidson, NC 28035-7156

(704) 894-2230

(800) 768-0380

davidson.edu

admission@davidson.edu

[
Number of undergraduates: 1,667
Total number of students: 1,667
Ratio of male to female: 50/50
Tuition and fees for 2008–2009: $33,479
% of students receiving need-based financial aid: 34
% of students graduating within six years: 92
2007 endowment: $489,461,000
Endowment per student: $293,617
]

Overall Features

Davidson is still affiliated in an informal manner with the Presbyterian Church, which founded the college in 1837. This relationship is evident in the strong emphasis on values, ethics, and the honor code that are part and parcel of the academic and social guidelines prominent on campus. Its mission statement presents a clear picture of what a student and faculty member can expect when he or she becomes a member of this small, intimate community: "The primary purpose of Davidson College is to assist students in developing humane instincts and disciplined and creative minds for lives of leadership and service." The college "emphasizes those studies, disciplines, and activities that are mentally, spiritually, and physically liberating." This ethos is important to understand in considering Davidson as a place in which to study for four years. Given the college's significant commitment to educating its students, Davidson's faculty are evaluated for promotion both on the quality of their teaching and their research, which informs their interaction with students. Davidson has long enjoyed a reputation for close-knit relationships between teachers and students, small classes, and offering opportunities to engage in research projects and independent study with experts in students' fields of interest. The honor code is another defining element. The requirement to report a fellow student for any act of dishonesty is taken seriously by all members of the community. This contributes to the tone of openness and sharing that many individuals comment on in describing their experience. Davidson is a small, handsome campus that is a designated arboretum located in a small town. There are fraternities for men and eating clubs for women on campus, but all are nonresidential. Students therefore get to know virtually all of their classmates, since they all live in campus housing that is rated favorably by students. True to its founding and its location, Davidson is traditional in tone and student body. There is concern for community social action on the part of many students, but not too much interest in political activism, nor is there a great deal of racial, ethnic, or religious mix in the population. The two primary activities on campus are studying and athletics. Both are treated very seriously by a majority of men and women. The academic standards are demanding; faculty expect a high quality of work and active participation in the classroom. Most students plan to continue their education into graduate school. Merit-based academic scholarships are awarded each year to about 20 percent of students to attract students to this serious learning community. Davidson was also one of the first colleges to eliminate loans from all aid packages. For so small a college, the quality of athletics is also of a high standard, with some twenty sports played at the Division I level. Many men and women participate in intercollegiate teams or do club sports. There is ample opportunity for any good high school athlete to join a team at some level of play. A number of athletic scholarships are offered for the very competitive athlete.

What the College Stands For

A dean comments, "Our statement of purpose says, 'The primary purpose of Davidson College is to assist students in developing humane instincts and disciplined and creative minds for lives of leadership and service.' Our educational goals are to expose students to new information, increase their ability to analyze information, do critical research, and to learn

to express themselves clearly. Additional goals are to develop information seeking and problem solving goals while keeping students open to other cultures."

A senior administrator further elaborates Davidson's educational goals: "To provide the best preparation we can for lifelong learning; to prepare our students for lives of leadership and service; to provide a thorough education in the liberal arts and sciences; to teach students to think critically, to write well, to acquire a broad and thorough knowledge in a variety of disciplines." Another notes that the college aims "to produce thoughtful, service-oriented alumni who will be leaders in their communities, who, because of their learning experience at Davidson, are prepared to be critical problem solvers with a thirst for knowledge." The college has increased its ethnic, geographic, and religious diversity over time, admitting women in 1974 and expanding its student population. It has worked to offer social options beyond "the traditional Greek system" and increased opportunities to study abroad. The college's honor code is said to establish "an atmosphere of trust" and to serve as "the bedrock of the learning experience." "In addition to providing an environment where students take self-scheduled exams," says one administrator, "and they don't talk about their exams to other students who haven't taken them yet, the honor code also pervades the whole campus in a more subtle way. Bikes and laptops don't get stolen. Money found on the sidewalk gets returned. The Davidson honor code helps develop in students, faculty, and staff a deeper sense of humanity."

Overall, we hear about the importance at Davidson of the college community and the personal relationships formed on campus:

"There is an uncommon satisfaction with the Davidson experience that is obvious and pervasive in the students and alumni—enthusiasm and gratitude abound—primarily due to the care, personal attention, and quality received in all aspects of their experience here." Another dean says, "Our network of support—academic, administrative, and personal—helps students. From faculty advising to career counseling to personal counseling, we are here to help our students. Also, our students are quite well-motivated, reducing their needs." A senior administrator says, "Davidson has gone from a regional men's college to a nationally ranked coeducational institution. It has remained a liberal arts college but has not constrained the curriculum to narrow studies of Western culture. We have increased our international connections and activities and deepened our science research (given Davidson's history for early experimentation with X-rays, we have a long science history, but the college now sponsors more lab research). There is an increasing emphasis on faculty-student research collaboration that again gives an important twist to the liberal arts theme by making students practitioners and not just receivers." And a senior biology professor adds, "The facilities and equipment are state of the art now. The biology curriculum emphasizes inquiry in our labs instead of cookbook labs. Students are encouraged to think and analyze data more than simply memorize factoids."

In the future, Davidson hopes to increase the diversity of the faculty and the student body, to expand the opportunities for earning academic credit outside the classroom through such programs as service learning and internships, to continue to review its curriculum, and to introduce "in a rational and planned manner more technology into the learning process."

Curriculum, Academic Life, and Unique Programs of Study

The academic requirements are rigorous and traditional in nature, what one administrator refers to as "an unwavering, strong core liberal arts curriculum." There are a number of distribution requirements, including a course in religion, which all students must fulfill. However, students comment on the easy access to their teachers if they need help and the support of their peers. This is not a community known for intensive competition among its students. For the very intellectually interested learner, a two-year humanities program based on the great-books concept is available. Davidson has always been praised for its strong science programs and thus many students have prepared successfully for medical school and other graduate degrees. A faculty member points out the following: "We are the center for the Genome Consortium for Active Teaching, which helps faculty at other schools teach genomics, too. Davidson also has a strong neuroscience program that has served as a model for other campuses for years. Davidson was one of the first schools to participate in the iGEM competition in synthetic biology. Our undergraduates do exceptionally well in this very new field of research. They have published their results and presented their work at regional and national meetings. We host internationally recognized research scientists to present their work and to spend time with our students." Another program of note is the Medical Humanities Program. The life sciences center enhances the curriculum and teaching in this field. Many undergraduates major in history and government and political science as a foundation for attending law and business schools. One-half of all students participate in the many study-abroad programs the college sponsors. These include the Dean Rusk Program in International Studies based on campus and programs abroad in Mwandi, Zambia, India, Germany, France, and elsewhere. Faculty point to the balance between academics and athletics at Davidson, as well as the importance of the honor code throughout the college environment: "Athletes get no special treatment in the classroom and they typically turn in their assignments early if they have away games. The overall GPA of our varsity athletes is slightly higher than the collegewide GPA. Again, the honor code is so pervasive that students know they can trust each other and work in a supportive environment where learning is more important than grades."

The college offers a preorientation program for non-Caucasian students and hosts a number of multicultural and international clubs and organizations. A full-time dean works with these groups. The Black Alumni Network and the Second Family Program, which matches students with local host families, also help students to adjust to the college. Internship possibilities have expanded in Charlotte, a city twenty miles away that has experienced much recent development. The college also stresses its commitment to service programs and student-faculty research opportunities and interaction. Support services, including learning and counseling centers, resident advisors, and faculty advisors also help students to succeed.

Major Admissions Criteria

Davidson is looking for "bright, ambitious, open-minded, curious, honorable, physically and spiritually active, self-motivated students who are self-starters, willing to work hard, to live within a rigorous academic environment with high expectations and equally high results," to use the words of one institutional leader. Important criteria used in evaluating applicants are: "Ability to clearly articulate and present their thoughts in an essay; a strong record of achievement in their academic work; an extracurricular record that displays depth and meaning rather than breadth." Another administrator gives the major admissions criteria as "academic performance, extracurricular involvement, and personal recommendations." Another notes "high school academic success, rigor of high school preparation, and evidence of leadership and/or involvement."

An admissions dean and Davidson alum says, "I think my favorite admissions goal is the one I deal with on a daily basis in the admission office: the goal of creating a college community that's both unbelievably diverse and incredibly cohesive. I've always thought of Davidson as a balanced place—in terms of balance between study and play, between small town and big city, between intimate academics and Division I athletics—but the college's biggest asset is its robust balance of student socioeconomic situation, ethnic or religious background, and political belief structure. The key to that balance, though, is to note that it wouldn't mean anything if the student body didn't fully ascribe to a larger sense of community (as reflected in the Davidson honor code) and to the tradition of listening that has been a part of this institution for more than 170 years. If these students from very different backgrounds didn't talk across groups, across life experiences, Davidson wouldn't be Davidson. Diversity of thought and the ability to truly experience that diversity—those two don't always go together, but they both have a home here."

The Ideal Student

The right student will enjoy an exceptional total experience in this college. Here is one of the finest environments for the more conservative, serious learner who enjoys active participation with others in the classroom, on the playing field, in the dormitory. Any young man or woman who wants to commit to the quality of life and spiritual values, not necessarily religious in nature, that Davidson stands for will enjoy college life and grow in many ways. Be prepared to work hard, make many intimate friends, and find numerous social and athletic outlets for your energy and enthusiasms. Do not expect to find a politicized campus tone, a free-wheeling curriculum, or a highly diverse student body.

"We're looking for students," says an administrator, "who are willing to devote time to extracurricular activities, who have a joie de vivre, and who are decent human beings. The students who seem to do the best here are the ones who recognize the academic and extracurricular opportunities Davidson offers and who take advantage of them." Another dean wants students who are "excited by ideas and the life of the mind," and who have a "willingness to be involved," are "anxious to put ideas into action," and have a "commitment to honor and integrity, leadership qualities, a keen intellect and disciplined work habits." Says a professor of communication, "Davidson students are exceptionally bright, exceptionally highly motivated, and exceptionally nice. Students who need a competitive edge with others—as opposed to themselves—do not tend to be as comfortable here."

Student Perspectives on Their Experience

One graduate who was a Bonner scholar at Davidson had studied abroad in Bangkok. "Davidson sponsored my trip there for a month teaching English and working with a monk on a basic English-Thai audiotape for beginners. All in a day's work I guess . . . it was lots of fun." That says a lot about the kinds of things one can do, even in a small liberal arts college in North Carolina, and many students reflected these kinds of sentiments.

Students choose Davidson for the faculty, the location, the friendliness on campus, and the academic strength of the college. Says one student, "We visited Davidson at the end of our trip, arriving late for the two o'clock tour. What happened next, I think, most aptly characterizes Davidson. A regular student just stopping in to the admissions office introduced herself and offered to catch us up to the tour group, filling us in on what we missed. I thought 'Wow, that's a first.' And the information of the tour about Davidson's excellent teaching staff, academic reputation, and relaxed atmosphere (not cutthroat competition) among students just confirmed 'the feeling' I got when I was at Davidson. I chose Davidson over all the others for many reasons: the teaching staff, the close relationship between students and faculty, the absence of a graduate program. . . . In short, the academic atmosphere was ideal. And finally, just the openness of the student body and administration, the frequent 'Hey, how's it going?', the fact that I can go to my prof's house for a movie and ice cream as part of class; that's what made the difference for me and why I chose Davidson." "I chose Davidson College because of its reputation, its no-loan policy, and its absolute commitment to academics," says one sophomore. "I also admired the sense of community that I felt at the college." A senior adds, "I chose Davidson because I wanted a school with a sense of community and spirit, which is greatly enriched by awesome athletic programs here, as well as great academics, which are taught by caring, accessible faculty."

Students like Davidson's educational style in and out of the classroom. "Davidson does a great job providing students the opportunity to take learning to a new level. By that I mean, Davidson is very much about educating the whole person not just in an academic sense, but as a well-rounded individual. Only so much can be learned from books—it's totally different to visit a historic sight, sit and talk with the homeless or a war veteran, go to Lake Norman and do your own research on the water quality. Davidson students, in general, take advantage of these opportunities by spending time abroad, volunteering in the community, and taking leadership positions on committees. Davidson caters to all of these needs through programs like the Dean Rusk Program, providing students with the opportunity for summer and semester study abroad. The Bonner Scholars Program, of which I'm a part, volunteering and serving in the local community in everything from tutoring to construction, from coaching to free medical-clinic volunteering." A junior praises "teacher availability," explaining, "Teachers are always there for the students. I remember going to the student union on a late-night snack break and being able to stop by my teacher's office and ask for help and also just chat about things that were happening." He praises the library ("Lots of volumes, the interlibrary loan system, a reserve for government documents, and the possibility of getting your hometown newspaper if you can get enough of your friends to fill out the required forms.") and the atmosphere ("Going to school in a small town is a great thing. From the chimes every fifteen minutes to getting your mail at the same post office as the professors and people

in the town, there is a real homelike feel to the campus. Also, being only twenty-five minutes from Charlotte means that if you have to escape the suburbs you can—although most students seem to be content with staying at Davidson.").

The academics are challenging but fair, according to students. "Davidson professors do not believe in grade inflation at all," says one junior. "Students receive what they deserve based on their work during the semester. When you get an A, it means that you worked very hard and that it was exactly what you deserved. I know that when I graduate from Davidson my GPA may not be a 4.0 like friends of mine from high school, but I can tell you that every point on my GPA is something I worked hard for and in my mind will represent everything I learned while I was at Davidson." "Make sure that you want to make a commitment to academics," says one student. "You will be miserable if you do not like to study, because that will be what you do for the majority of your four years here." Another student echoes these sentiments: "Passionate-yet-humble hard workers do well here. You can't come here and succeed if you're looking for a party school, or if you're not excited about learning. Davidson is also all about being the underdog, so if you think you're awesome and love to tell people about it, you're not appreciated." Another student points out the faculty support and attention that accompany academics: "I feel deep admiration and gratitude toward the faculty members for tailoring that curriculum to each and every student who walks through the doors. Coming to Davidson, I was required to take a math course. Math was never a strong suit of mine, and I hadn't received an A in a math course probably since sixth grade. I was scared out of my mind sitting down in Math 110 early in the second semester. Yet on the first day of class, before I had even introduced myself to the professor, Dr. King turned to me (knowing that I wrote about the school's basketball teams for the *Davidsonian*), and right off the bat he explained a concept of statistics using a scenario from the previous weekend's basketball game. The extra effort that Dr. King invested (by attending school sporting events, reading the paper, and making a concerted effort to incorporate that all into a lesson) made all the difference in the world to me and resulted in my receiving an A."

Continuing what seems like a theme among the small colleges, the junior says, "The worst thing about Davidson in the eyes of the general public is our lack of diversity on campus. We're a southern, white, upper-middle-class school. The Davidson administration/admissions office has been making a great effort to remedy the noticeable lack, especially of African Americans on campus." There is not necessarily a great deal of support from students for affirmative action. Yet they recognize the efforts of the college to diversify and appreciate the international and ethnic diversity that does exist on campus, noting that the school is welcoming and friendly to all students. Nonetheless, students "wish there was a greater degree of a social mix" and interaction among students. Another student says, "Davidson has a lot of diversity— ethically, socioeconomically, geographically— and a lot of varied interests throughout the student body, but I don't think it's tapped into as much as it could be."

Who succeeds at Davidson? "Any student (regardless of ethnicity, religious background, and so forth) who is excited about learning, is willing to give 110 percent in the classroom, is willing to be active in both the internal and the external communities, and is willing to take some risks will get the most out of a Davidson education." Again we hear about the social student who is also academically talented and interested. Another student notes, "Everyone who attends Davidson brings their different backgrounds and personal experiences. These experiences, when coupled with the curricula, help to expand on the subject matter within the class. I can recall a complex analysis course in which we had an exchange student from Germany. A great guy, he had already taken courses covering much of what we were learning in this class, yet he elected to retake this course so that he could get a different perspective on what he had already learned by absorbing the discussions in a U.S. college class versus one in Germany. While he felt that he benefited greatly from his experience in our class, I feel as though I gained much more from having him there as well. It is this type of educational environment that Davidson students create with the aid of the professor. A Davidson student will strive to make the most of the educational experience inside and outside the classroom."

"I find that my colleagues/peers are as helpful as my teachers," says one student. "At Davidson, the students are very willing to lend a helping hand. It's not a 'cutthroat, destroy your lab work' kind of a place." Another agrees: "The overall feeling of camaraderie from the individuals, the dorms, even the lawns, aids to the education that one receives from Davidson. I mean, at Davidson students don't compete with each other, there isn't an 'I must do better than you' mentality on campus. Students here don't ask each other what their grades are, nor do they really care. The only person that a student competes with at Davidson is him/herself."

There is some criticism of the fraternities

and eating clubs (sororities) on campus, with some students advocating their removal and others advocating coed houses. However, the opening of the Vail Commons as the open, coed, college eating establishment and the student union, as well as the open-access, no-rush, and no-hazing policies in the houses, appear to have alleviated many student concerns. In addition to numerous social activities on campus and in Charlotte—including the all-school formal—students find other ways to bond with each other across many lines: "I love the community-building that exists on the freshman hall. It is an integral part of life at Davidson, and one that apparently lasts beyond the first year on campus. Flickerball (7-on-7 touch football with unlimited forward passes, a sport played by all freshmen halls and by many upperclassmen as well) is a unique experience that forges lifelong friendships between hall-mates."

Students reflect the administration's emphasis on and appreciation of the honor code, universally praising it in theory and in practice and in terms of its impact on their academic, social, and future development. "I don't think the importance of the honor code can be stressed enough. It provides for an immense sense of community among the faculty, staff, and students, and that sense of community radiates into the town." "The HC governs every aspect of life at Davidson, from the classroom to the playing field, and it's taken very seriously. In a nutshell, it says Davidson students will not lie, cheat, or steal. In my choosing Davidson, the HC was very important because it grants the academic body, ironically enough, a great deal of flexibility. Teachers don't have to be in the exam room, I can schedule my own exams, I can take a test and have confidence that my studying will reflect on my test and no one

else's, and so on. With the HC comes lots of worrying for prospective students . . . it worried me. Will I get kicked out? How strict is too strict? Does this mean I can't work with someone else? It's up to the teachers to what extent they wish to utilize the HC, but the point is, it's there and it works. Very rarely does anyone get expelled from college, but let it be said that there's some gray area that's gone over at orientation. What impresses me is that Davidson seeks people who value integrity in a world where integrity is seen less and less. Again, Davidson educates students in more ways than one. The bottom line is, it's there for the students."

It is worth including another student's detailed description of the honor code in practice, as it may help readers to gain a clearer picture of honor codes at Davidson and some other college campuses. "Nothing has been as powerful an influence on my life as the Davidson honor code. This is something that has become part of who I am as an individual. The honor code is not just a series of rules, it is in the heart and soul of every Davidson student, while they are at school and while they are away (even after graduating). The honor code basically says that students will not lie, cheat, or steal. . . . Coming from a suburb of New York, I felt as though I had to keep an eye on my belongings at all times. But after locking my roommate out after his showers three or four times, we both came to an understanding that leaving the room open was not a dangerous activity. Heck, this last year I opted to purchase a Sony Play-Station, and after other people on the hall learned about it, I had no problems allowing people to go into my room to use it when I wasn't around. That was the trust that I had in my peers and my trust in the honor code.

"Other situations, like academics, are equally affected. For example, my first test (we call them reviews) at Davidson was a Calculus II exam. I was coming from a school where students expected the teacher to remain in the room and make sure no one was cheating. Upon receiving the test at Davidson, the teacher asked us if there were any questions, and since there were none, he told us that he would be back five minutes prior to the end of class to wait for the return of the tests. Then he walked out of the class to do some work in his office. For a minute every student just kind of looked at each other and then we began to take our tests. It is this trust that teachers have in students, knowing that they will not cheat, that allows us to take our tests without teachers looking over our shoulders. It also gives us the opportunity to take no-notes, time-limit, take-home exams. It is not uncommon for a teacher to give an athlete a take-home test to take on his/her bus if their traveling corresponds to a review date in the class. Also, when it comes to finals, students have the ability to self-schedule their exams. What this means is that over the course of the exams periods, students can take any of their exams in any of the exam periods that they choose. This self-scheduling is really what it means to have an honor code—teachers give us flexibility in arranging our exams, and students are trusted to not cheat while taking their tests. . . . Be it having the ability to leave your book bag outside of a classroom, knowing that it will be there when you return or just being able to talk to your teacher as a peer because you each have such a high amount of trust in each other, the honor code is something to live by. . . . On the way to a basketball game, a friend and I walked right over a twenty-dollar bill. Neither of us moved to pick it up, we just kept walking. On our way back we saw a note taped to the ground where the twenty had been; it said 'Twenty Dollars Found Here. If Yours, Please Call x1234.' That's the power of the honor code."

What Happens after College

After graduation, many Davidson students go to work in business organizations. Over time, however, more than 70 percent will earn a graduate degree, and a large proportion will enter the legal and medical fields. An institutional leader says, "Davidson students excel at positions that require creative solutions, leadership skills, and problem-solving abilities. They are good communicators, verbally and in writing. They can be given significant responsibility with little instruction. Although they may not have specific training for a particular job, they're smart enough and resourceful enough to be able to learn how to do a job exceptionally well."

The numbers show a broad array of postgraduate involvements, but Davidson's surveys of graduating seniors and those one year out of college show that about two-thirds of grads go to work after Davidson, with almost 40 percent going into business fields, 19 percent into education, 15 percent into science/medicine, 8 percent into nonprofits, and 5 percent into law or politics. Of the one-fourth of students who go quickly into graduate school, almost one-third head into the arts and sciences, one-quarter into medicine, and 19 percent into law.

Duke University

2138 Campus Drive
Box 90586
Durham, NC 27708

(919) 684-3214

duke.edu

Overall Features

It is only little more than a century ago that a small regional college founded by North Carolina Methodists in the early 1800s was moved to Durham as its permanent home with a generous endowment from the Duke tobacco family. The Duke Endowment foundation's establishment in 1924 enabled this small, undergraduate institution once known as Trinity College to transform itself into Duke University, a leading national university situated on a magnificent residential campus with exceptional facilities to advance teaching and research. While today Duke is fully nondenominational (there are more than two dozen religious-life groups representing all major faiths) on campus, there are vestiges of its religious heritage. The imposing Gothic chapel is the centerpiece of the west campus, there is a graduate divinity school, and the motto established by the founders, *Eruditio et Religio*, is still in force.

The unusual number of academic programs and institutes dedicated to exposing undergraduates to public service and national and international issues reflects the historic mission of the university. Over 40 percent of the student body studies abroad, many doing community outreach activities, particularly through the DukeEngage program, which funds student civic engagement projects throughout the world. The men's and women's separate undergraduate colleges were merged in 1972, and what had been the east campus for women was transformed into the residential campus for all first-year students in 1995, with social life and numerous organized activities centered there. The purpose is to build class unity and a sense of camaraderie.

Today Duke is home to 6,340 undergraduates and 7,100 graduate students. There are two distinct undergraduate schools, the Trinity College of Arts and Sciences and the Pratt

School of Engineering. Duke today competes for outstanding students with the Ivies and other nationally distinctive universities that include Stanford, MIT, Vanderbilt, and Rice. Its rise to national stature in a relatively short time is a tribute to the vision of its educational leaders, its wealth, and in no small measure its outstanding basketball program. What football has been historically to the University of Notre Dame, basketball is to Duke. It is not a mere coincidence that the national media attention that the program has attracted every basketball season has also seen a continual rise in applications from all corners of the nation.

Alumni take great pride in the reputation of the university both for its athletic winning ways and its academics. Duke is more like Stanford, Northwestern, Notre Dame, the Universities of Virginia and North Carolina, Vanderbilt, and Rice than the Ivies or Chicago in its neat balance of academic excellence, serious students, and a major athletic and social environment. Like its peer institutions, Duke believes it offers the best of two worlds in having a medium-small residential college with a research university at the disposal of its students. Although there remains a southern tone to the community, only 15 percent are Carolina residents. Duke reaches out to all corners of the country and internationally as it intentionally builds a world-class student body.

A survey of Duke graduates indicates that they are very satisfied with their education. About 90 percent say that they would strongly recommend Duke to prospective high school seniors. They use adjectives like *wonderful*, *terrific*, and *remarkable* in describing their experience as undergraduates. They feel that Duke provided them with an excellent, well-rounded education that has contributed to their personal development as young adults. The percentage of first-year students who return for the sophomore year is an extraordinary 97 percent, and 97 percent of students graduate in four to six years. These are remarkable statistics that reflect well on the serious nature and quality of the student body and the resources available to them.

The university has articulated its strong commitment to attracting more nontraditional students and breaking out of its predominantly white, middle- and upper-middle class population. Thanks to a generous financial-aid program due to its healthy endowment, Duke can guarantee full financial aid to all accepted students who qualify for financial need. This has resulted in a sizable growth in the number of African American, Asian American, Hispanic, and first-generation students admitted and enrolled each year. More than 45 percent of the most recent first-year classes are students of color. Ten percent are internationals.

What the College Stands For

In the words of the dean and vice provost for undergraduate education, "Duke is a place of 'outrageous ambitions,' never content to rest on its laurels. Students and scholars work to understand and address important world issues. Tradition is combined with a desire to ask questions; and a civic commitment to Durham, North Carolina, and the wider global community informs our work. We want students to be able to think and reason for themselves and to make their academic lives ones of curiosity and exploration. We want our students to be able to express themselves with intelligence and sincerity, while respectfully engaging and empathizing with others of different backgrounds and with differing opinions, as that results in deeper understanding for all."

Duke's stated mission is to provide an outstanding liberal arts education with a strong global and interdisciplinary emphasis as well as student engagement in research under the tutelage of its faculty. The dean proclaims, "We provide resources and numerous opportunities for our students to use their knowledge to make an impact on the world and apply what they learn in the classroom in service to society."

The university has "an unwavering commitment" to affordability and access for deserving students. It has the resources to practice need-blind admissions and meet the full demonstrated financial need of all accepted candidates. A recent financial-aid initiative raised more than $300 million to endow its scholarship program.

Curriculum, Academic Life, and Unique Programs of Study

In keeping with its educational goals, students must meet general education requirements that include not only a distribution of courses across the social sciences, humanities, foreign language, arts, quantitative studies, and natural sciences but also a modes of inquiry course. Intent on seeing that students are receiving the best possible education, both Trinity College and the Pratt School of Engineering assess student learning outcomes on a department-by-department basis, as well as a class-by-class basis. A two-part course evaluation asks students and professors to state independently at the end of each class which broad learning outcomes were met, allowing analysis of how well professor expectations match student perceptions of outcomes. The Pratt curriculum is subject to Accreditation Board for Engineering and Technology regulations. The Trinity curriculum includes broad learning objectives, with each course coded for which objectives it is intended to meet by faculty agreement.

The Focus program for first-year students offers a range of small, cross-discipline, interrelated seminars on a given theme taught by professors. Students who elect to participate in the Focus live together in a first-year dormitory. The program can include travel and community service off campus. The Terry Sanford Institute of Public Policy offers an interdisciplinary that concentrates on political science, history, economics, sociology, and other courses that prepare students planning a career in public service. The university is expanding its offerings in the performing and fine arts, and has completed recently its impressive art museum. There is particular enthusiasm over a new dance concentration. The Nicholas School of the Environment, which offers undergraduates courses and fieldwork in natural and environmental studies, includes a marine laboratory. Duke is also blessed with a 7,000-acre forest in the Piedmont Range, where research and recreation take place.

Major Admissions Criteria

Duke has to be counted among the most selective colleges in the country these days. Close to 18,000 high school seniors apply each year and under 25 percent are accepted. The number of early-decision applicants has increased by 25 percent in the most recent admissions cycle. The middle 50 percent of admitted students scored in the 670 to 770 range on the SAT and from 29 to 34 on the ACT. Subject tests are required with math and science specifically for engineering candidates. The dean of undergraduate admissions states that the university seeks students who will take advantage of the many intellectual opportunities available to them, take on leadership roles through the vast range of organizations on campus, and contribute to the larger society. When considering applicants his committee looks for those students who demonstrate curiosity, ambition, risk taking, and academically excellent preparation. The first and most important factor is the quality of the high school curriculum that shows evidence of stretching oneself and an outstanding performance. Engagement in the community in any meaningful way is also a serious factor. Teacher and counselor recommendations are important in confirming these academic and personal traits. Test scores matter as confirmation of mastery of course work and key skills of writing and thinking analytically and logically.

The senior dean described the three most important criteria in the selection of students this way: "First, a student needs to demonstrate the ability to be comfortable with the academic rigor and opportunities at Duke. Second, a student needs to demonstrate the willingness to fully engage in a challenging and diverse academic and social community. Finally, a student needs to be ready to make a mark, to have an impact, in and outside of the classroom. We learn these things about students from all parts of the application, and the importance of various parts of the application are different for each student."

The Ideal Student

The criteria that the admissions committee uses in selecting students represents, in part, the ideal type of student who will enjoy a successful career on the Duke campus. Smart, hardworking, ambitious, community-oriented, confident, self-initiating, and independent are the requisite qualities. Almost as important is the student's EQ (emotional intelligence) and social skills. Duke is a hyperactive social environment, one that calls for the ability to maneuver one's way through myriad clubs, activities, internships, fraternities and sororities, teams, and residential-life challenges and opportunities.

The ability to balance the demanding workload with the social life is necessary for survival. Sorting out independently which of the many academic and social activities to take advantage of can be a challenge for those who are not risk takers or adventurous. While Duke has stretched its geographic borders to enroll talented students, there is still a large southern contingent and social traditions that promote a fairly conservative environment or mode of pursuing life on campus. The very reclusive or nonsocial individual is not likely to feel as much at home at Duke as on some other equally excellent college campuses. One dean put it this way: "Students who are interested in being partners in their education, who are willing to engage our faculty in critical discourse, who are willing to work to make a difference, who believe that being part of a teams means sometimes leading, sometimes following, and always learning—those are the students who do best at Duke." A senior stated that "the student who does best is the kid that wants a real balance of work and play. This is not to say that Duke is easy—it is not. It is very challenging and there is no point in going to Duke if you do not have serious academic ambitions." This and a passion for Blue Devil basketball will do the job.

Student Perspectives on Their Experience

The perspective of a good many Duke students is captured in the comments of this young woman: "The classroom can be competitive, so the student who does best will have to be able to speak up, defend her opinion, and not be intimidated by her equally intelligent peers. But the person who does best at Duke also cannot take herself too seriously. Duke is very rigorous on weekdays, but we play as hard as we work. Duke is a school for a student who wants a combination of the challenges of an Ivy League school, with many of the cultural and social elements of a state school. It is college in every sense of the word—think *Animal House* meets Harvard."

A senior English major observes, "I don't know about the many other majors, but I know that I have never felt that any professor was not there for me if I needed help. Though Duke does not feel like a small college, there is a feeling that most people know one another, and there is always someone you can turn to for advice and guidance—academic and personal." Student adjectives used to describe the Duke experience are *rigorous, fast-paced, fun, aggressive, unexpected,* and *challenging.*

Another telling observation: "I enjoyed the many Duke-sponsored events for students, which ranged from prominent speakers to Broadway shows to concerts by bands that I listen to often. I enjoyed being a part of my sorority, which added a silly, girly aspect to my academic life, and I enjoyed attending lectures given by professors on subjects that they might not teach in their classes."

What Happens after College

Influenced by the university's emphasis on service to the community, students have made Teach for America the single largest employer of graduates in the last several years. Many graduates enter the world of finance and banking. The overwhelming majority of alumni enroll in graduate school at some point after experimenting in a variety of jobs, volunteerism, and travel. Duke boasts that graduates who apply to law and business schools have a 95 percent acceptance rate and that about 80 percent of those who apply to medical school are accepted.

Emory University

200 B. Jones Center
1380 South Oxford Road N.E.
Atlanta, GA 30322

(404) 727-6036

(800) 727-6036

emory.edu

admiss@learnlink.emory.edu

Number of undergraduates: 5,094
Total number of students: 9,259
Ratio of male to female: 48/52
Tuition and fees for 2008–2009: $36,336
% of students receiving need-based financial aid: 38
% of students graduating within six years: 88
2007 endowment: $5,561,743,000
Endowment per student: $600,685

Overall Features

No college or university in America has changed more dramatically since its beginning than Emory. Founded by the Methodist Church in 1836 as a traditional southern college to prepare young men of social position for the ministry, Emory has risen to the top group of the midsize teaching/research universities, with a student body that includes as many northerners and midwesterners as southerners today. The composition of the student body is fairly diverse as more students of color and those of all religious denominations have been actively encouraged to enroll. At present over one-third of the undergraduate body represents various minority groups in America. Thanks to the infusion of hundreds of millions of dollars in gifts from friends and alumni in recent years, Emory is one of the wealthiest institutions of higher education in the nation. As a result, outstanding faculty have been recruited from many of the other top universities, new programs of study and special research centers have been added, talented students have been offered generous scholarships based on merit as well as need, and the physical plant has become one of the most complete learning and residential centers of all colleges.

Emory fits a particular niche in the world of universities: It is a midsize undergraduate college with major schools of business, law, medicine, and many academic disciplines available at the graduate level. The campus facilities are continually updated or new buildings erected to meet the needs of students. As an example, a $25 million state of-the-art library, new dormitories, a student center, and an athletic center were added in recent years. Emory is currently in the midst of an ambitious ten-year strategic plan for development, called "Where Courageous Inquiry Leads." As the university describes it, "The goals and strategies expressed in the plan will allow Emory to achieve its vision of becoming a destination university, internationally recognized as an inquiry-driven, ethically engaged, and diverse community whose members work collaboratively for positive transformation in the world through courageous leadership in teaching, research, scholarship, health care, and social action." The four strategic goals Emory is concentrating on are: "Emory has a world-class, diverse faculty that establishes and sustains preeminent learning, research, scholarship, and service programs. Emory enrolls the best and brightest undergraduate and graduate students and provides exemplary support for them to achieve success. Emory's social and physical environment enriches the intellectual work and lives of faculty, students, and staff. Emory is recognized as a place where engaged scholars come together in a strong and vital community to confront the human condition and experience and explore twenty-first century frontiers in science and technology." Emory's location on the outskirts of Atlanta and its size make it an attractive campus. Although the university guarantees housing to all undergraduates, only about two-thirds choose to live in the dormitories. Many find appealing the opportunity to live in an apartment in town or share a house with friends. Over one-third of students join a fraternity or sorority, which provides an active social outlet for many students. Others take advantage of the numerous activities the university has created or spend free time in Atlanta. Emory is unique in the South for the fact that competitive sports are low-key, with none of the intensive recruiting and awarding of athletic scholarships that is very much the tradition in the region. There is no varsity football team, for example. There are nine intercollegiate teams for men and nine for women available to those who care to participate. There is a good deal of club and intramural athletic activity for students by contrast. Studies are treated seriously, as the majority of undergraduates have their eye on professional graduate school even as they enter the campus gates for the first time.

What the College Stands For

Emory sees itself clearly as a provider of a liberal education at the undergraduate level. As one administrator put it, Emory focuses on a "broad-based liberal arts education, with a comprehensive core curriculum." It puts a "strong emphasis on internationalism, languages, and computer literacy. Our graduates become leaders across all disciplines. We want to prepare them to live in a diverse world, accommodate change, and to think, communicate, and problem-solve effectively." Both Emory's undergraduate and graduate programs are highly ranked nationally, and the institution sees its rise in stature as beginning with a $120 million gift in Coca-Cola stock from a donor in the late 1970s (due to its Atlanta neighbor and benefactor's prominence, Emory is sometimes referred to as "Coca-Cola University"). "The growth in facilities, faculty, and programs has been meteoric," says one administrator. Contributing to this growth has been "unprecedented financial support, a favorable location in one of the country's most dynamic cities, superb facilities, the faculty and placement record, warm climate, and the growth overall of the Sunbelt."

A dean lays out Emory's education goals: "Continue the search for knowledge as has been demonstrated by students during their formative education; integrate living and learning experiences into the educational process; develop and enhance leadership skills; ascertain that students have a broadbased liberal arts education, thus enabling them to have competencies in multi-disciplines; encourage continued educational pursuits beyond the undergraduate level, enhanced appreciation of the changing dynamics of the world, and enhanced appreciation for individual and group differences." To accomplish these goals, Emory provides "ample resources; numerous libraries; numerous computer labs around campus and in wired residence halls; small student-faculty ratio; the Freshmen Advising and Mentoring at Emory program (FAME); specialized tutorials; individual and group academic tutoring through the Office of Multicultural programs and Services; and well-educated, experienced staff members."

Even with its size, Emory maintains a low teacher-student ratio, a strong residential college environment, and a broad campus life program. "I would like to see the institution continue to stress values, morality, and ethical considerations in a complex world," says an institutional leader. "Emory has already done landmark work in these areas as they pertain to professional education in business, theology, law, public health, and medicine." As opposed to a small-college environment, here we see the interconnections possible in a university, where the graduate programs on campus provide resources and models for the undergraduate college.

One administrator notes Emory's forward-looking perspective: "Emory's best years remain in the future. How many schools can honestly say that? The school is still 'young' as national institutions go. We are still developing, growing, and experimenting. There is not the same aura as a place that is three hundred years old—there are new organizations, majors, programs. Student input is essential and sought."

Curriculum, Academic Life, and Unique Programs of Study

Keeping to its academic tradition, Emory requires completion of a core liberal arts curriculum. The requirements are fairly specific and take up about one-third of every student's program. Beyond that, the range and depth of subjects and majors to choose from is impressive. Students find the grading system rigorous but the access to their teachers very easy. A large cohort majors in chemistry or biology because of their interest in premedical studies and the superb science departments. Others choose psychology, English, history, and business as preparation for graduate school. Students may double-major and pursue B.A./M.A. programs. Emory offers many study-abroad opportunities and a well-developed internship program. The university has unique relationships with the American Cancer Society, the Carter Center, the U.S. Centers for Disease Control and prevention, the Georgia Tech–Emory Collaboration for Regenerative Medicine, and the Emory-Tibet Partnership, among others.

The "philosophy of 'athletics for all,'" says an administrator, "has some 90 percent of students playing intramural sports." This is complemented by the nationally ranked NCAA Division III athletic program. Emory competes in the University Athletic Association, which is modeled after the Ivy League athletic program and includes such universities as the University of Chicago, Washington University, Johns Hopkins, and New York University. The Emory campus is high-tech and offers extensive use of the Internet. Professors and students, for example, can interact on LearnLink, a campus-based network. A key extracurricular program is Emory's national champion intercollegiate debate team.

"Volunteer Emory is one of the largest student volunteer groups in the nation," says one administrator, and students have representation and a voice from the board of trustees level to the Student Government Association. It provides many opportunities for undergraduate research across the curriculum.

Emory maintains a "need-blind" admissions program, providing a great deal of financial aid, as well as a national merit-based scholarship program called Emory Scholars. Emory has, according to one administrator, "extensive diversity. One-third or more of the entering class each year represents ethnic minorities. The *Journal of Blacks in Higher Education* rated Emory as the number-one school among the *U.S. News* top twenty-five for the number of black undergrads and faculty." Emory provides many clubs, sororities, and fraternities catering to students' varying interests and backgrounds, and has the additional attraction of the diversity of the Atlanta area itself.

Major Admissions Criteria

Emory looks for "active students from high school, who are not only bright but student leaders in some way. Students comfortable with diversity and who seek ethnic, geographic, and religious diversity in their college experience. Students who enjoy being challenged and who are motivated to 'push the envelope' academically." Its major admissions criteria, according to an admissions office representative, are "curriculum chosen; outside time commitments

beyond class; letters of recommendation from counselors, teachers, others."

The admissions office details its admissions approach as follows: "The admission committee will pay closest attention to an applicant's high school course of study and grades. . . . Within the context of the applicant's school, we will expect that the student has taken a solid load of the more challenging courses available. Most competitive students will have a B+/A– average or better within a rigorous course of study. SAT and/or ACT scores are very important but are not the deciding factor. Strong grades in rigorous courses may cause the committee to overlook below-average standardized test scores, but high board scores will never make up for an applicant's weak course selection or grades. . . . The admission committee looks closely at how a student has spent his or her time beyond the classroom. We look for leadership and/or commitment in extracurricular activities. We seek active students who will contribute to our dynamic community, students who will bring to our campus many different backgrounds, experiences, interests, opinions, and talents. We pay close attention to the contact an applicant has had with the Office of Admission during the application process. It is important that applicants have done their research on Emory, whether through ordering a Video Visit, talking with a representative at a college fair, attending an information session in your city or at your school, or visiting our campus. We also expect a competitive candidate to articulate why Emory is a particularly good match for them. . . . Applicants should choose recommenders who know them well, who know their academic strengths, and who can tell us about their character. . . . We read essays and short answer responses closely. These writing samples are an applicant's best opportunity to communicate who they are, what they enjoy, what issues interest them, and what arouses their curiosity. There is no evaluative interview during the Emory application process. In the writing samples, applicants should include any information they would like the admission committee to review when their applications are considered."

The Ideal Student

Emory has great appeal for the serious student who is goal-oriented, even if he or she is not quite certain what that eventual goal might be, who wants to experience a diverse campus life near a large metropolitan setting, who is willing to study hard, and who does not care greatly about a high level of campus spirit or spectator sports. Emory has great resources in all regards that are readily available to the mature, intellectually curious individual who will appreciate the solid foundation in learning the university provides. Opportunities to work at the elbow of a renowned teacher, writer, or scholar can be obtained but will call for individual initiative. As the admissions office puts it, "Students come from far and wide to live and learn within a diverse community of talented artists, accomplished researchers, inspired scholars and skilled educators. This diversity makes Emory a destination for students seeking the benefits of a first-rate liberal arts education paired with the resources and facilities of a top-tier research university. The curriculum is challenging, giving students the skills needed to pursue any career or succeed in graduate study. The nurturing and supportive faculty are readily available for advice and guidance. Emory's campus life thrives on constant activity, and students are encouraged to get involved, share opinions, and flourish in their new envi-

ronment." Students who like to take advantage of opportunities (and many do, such as the 40 percent who study abroad), will find many choices to diversify their education at Emory.

Student Perspectives on Their Experience

Emory students choose the university for its reputation and its focus on undergraduate education, but, also important, for its graduate and research resources as well as its urban location. "Emory was quite an enticing school," says one senior, "due to its wide variety of majors, activities, and resources to benefit from. As a student who has interests in the fields of medicine, anthropology, and business, I am always able to meet my needs and explore numerous options near the campus for gaining experience."

The university is said to do well at "introducing students to diverse cultures. Many students are introduced to cultures that they otherwise would have been unfamiliar with. Also, Emory has so many student organizations to choose from that it allows everyone to find their niche. The school does its best to provide for students with disabilities through the Office of Disability Services. They even provide note-taking services to help students who have a problem taking notes in class." The university "provides students with the ability to determine their own path. There are numerous options from which students may select the way they wish to get involved and grow."

"The students who do best," says a senior, "are those who immerse themselves in the culture of Emory. I truly believe that the students who do well in school are not those who always have the highest GPAs. Rather, the ones who are truly successful are those who have achieved a balance between scholastic achievement, personal development, organization involvement, and social growth. When you finish college, you should have grown as a person, not just academically." One drawback of Emory's diversity and multiplicity of options is its fragmentation, and thus poor communication across campus, according to one senior, "The multiple paths to self-determination and self-definition also serve to undermine the connectedness of different groups. Once people branch out they find much in common with others, but most are hesitant to do so." Thus, those who do best at Emory are "people who are willing to seek out all of the opportunities the school offers. Personality does not matter as much (whether you are an extrovert or shy). If a person is fundamentally concerned with learning and expanding their horizons, they will succeed."

Students at Emory are helped not only by their own drive, but also by the university's peer culture and support programs. "Personal motivation is the key factor that leads to success. Yet there are some services that help students to lead successful academic lives. The Office of Multicultural Programs and Services has a free tutorial program for challenging classes, and many students choose to tutor students on their own. There is also positive peer pressure to do well academically, and groups such as sororities and fraternities compete for scholastic awards." The honor code is felt to be important, as are the Hughes Science Initiative program for freshman minorities and women in science and the ESEP program (Elementary Science in Education Partners program), which allows students to teach science to elementary school students in the Atlanta public schools. "This gives students with expertise in science the opportunity to share it with children throughout the school system." A senior praises the "wonderful

professors who care about students' development, helpful administrators, great facilities, an active student body, food, technology resources, and an overriding philosophy (shared by professors, deans, staff, alumni, and students) that a place of learning is only as strong as the commitment of all members of the community to that goal."

"Emory is full of professors and administrators who care about the students. This makes life more full and enriched and has made Emory a school that I am proud to call my own," says a senior. Another recalls "opportunities for student leadership in campus life; the openness of professors to instructing in all aspects of life; academic collegiality between all— these allowed me to develop intellectually and become more articulate and better at organizing group efforts."

Emory seems to succeed in its mission of providing a supportive, small-college educational program within its research university environment. "Emory is able to maintain a very personal and concerned atmosphere despite being a top research university. You truly feel a part of a community of learning instead of feeling you are being 'processed' through an academic system. The Emory family of alumni, trustees, faculty, staff, administrators, and students is a warm community that is willing to embrace each student in holistic development if the student is willing to reach out and contribute to the community."

What Happens after College

At some time after graduation, says one Emory administrator of the university's graduates, "about 70 percent enter graduate and professional schools. Emory has a tremendous reputation in medical and law school admission.

The other 30 percent seek employment through our Career Center. Emory is heavily recruited by most Fortune 500 companies, as well as education, government, and other agencies." Emory reports a medical college admission rate of about 50 percent, and a law school admission rate of just over 80 percent. Of the class of 2008, about 80 percent participated in community service or volunteer work, 45 percent in intramural athletics, and 45 percent in study abroad. More than a third had an off-campus internship or worked on a research project with faculty. A third planned to attend graduate or professional school immediately following graduation, half of whom were headed to law or medical school.

Georgetown University

103 White-Gravenor
Box 571002
Washington, DC 20057-1002

(202) 687-3600

georgetown.edu

quadmiss@georgetown.edu

Number of undergraduates: 6,692
Total number of students: 12,022
Ratio of male to female: 48/52
Tuition and fees for 2008–2009: $37,947
% of students receiving need-based financial aid: 55
% of students graduating within six years: 93
2007 endowment: $1,059,343,000
Endowment per student: $88,117

Overall Features

Georgetown holds a unique position within the world of American higher education. It is one of the oldest universities established in the country, founded in 1789, and the first within the Jesuit Catholic tradition. Its founding purpose was to serve both Catholics and non-Catholics who would gain a rich intellectual education with an emphasis on intellectual inquiry, preparation for service to the public, and religious and cultural diversity within the Jesuit tradition. Virtually all other Catholic universities were created decades later with the primary mission of serving recently arrived Catholic immigrants from other countries.

Georgetown's founding mission was much more in line with the early colonial colleges like Harvard, Yale, Princeton, Dartmouth, Brown, and William & Mary. Its student body, curriculum, and training were designed for a different population than the working-class families who were to seek a more practical, affordable, local education in succeeding generations. Georgetown's history is reflected today in a number of ways: the curriculum of its four undergraduate schools, the relatively diverse cultural and religious student and faculty population, the socioeconomic composition of the student body, the emphasis on national and international studies and languages, the pervading tone of moral issues in and out of the classroom, and encouragement of service to the community.

Many students considering Georgetown initially do not even know that it is a Catholic-sponsored university. The university is composed of four distinct schools: the liberal arts Georgetown College, the McDonough School of Business, the Edmund A. Walsh School of Foreign Service, and the School of Nursing and Health Studies. There is a unique division of Languages and Linguistics that offers majors in nine different languages. At the time of applying to the university, a student has to indicate in which of the four schools he or she plans to study. Georgetown's location has enormous appeal to students nationally and from all corners of the world. Washington is viewed as a cosmopolitan city with limitless offerings of activities socially, artistically, politically, and athletically. The availability of internships that can be combined with one's studies is one of the particular appeals of attending college in Washington.

Georgetown's most successful athletic team, the basketball Hoyas, has added to its appeal in recent years. Other than lacrosse, all other athletic programs seem to function at a less competitive level than a student would experience in the major Division I programs. While it has always attracted many outstanding high school students to its campus, the university has become especially popular in the last decade as more students want to have a collegiate experience in the traditional vein within a stone's throw of an interesting city. Georgetown and Washington combine nicely to provide both.

Competition for admission has risen dramatically as a result of this popularity and has made Georgetown one of the most selective colleges in the nation. As testimony to its selectivity combined with the satisfaction with their experience on campus, 97 percent of first-year students return for their second year and 94 percent graduate within a six-year span. There is no Greek social system on campus and the administration has come down hard in recent years on campus drinking and rowdy behavior. Consequently, most social life revolves around the pubs on the main avenue nearby the campus. Clubs and volunteer organizations are a major involvement for many undergraduates. For all its emphasis on diversity and respect for other religions and cultures, Georgetown's

Catholic traditions and beliefs are evident in the majority population of Catholic students, the religious services held on campus, the Jesuit professors (twenty-five hold teaching positions, fifty-eight members of the Jesuit Community live on campus in residence halls, and others serve in pastoral or administrative roles), and the prohibition on advocating publicly for certain issues such as abortion and birth control.

In the main, students are serious about their education and their future, while also determined to enjoy college life. A majority plans to go on to professional careers, especially in law, politics, medicine, and public service. They take advantage of internships to gain experience and make connections, study abroad (almost 40 percent of all undergraduates), and the outstanding faculty who are attracted to the university for reasons similar to those of its undergraduates. With its outstanding reputation in the social sciences, international relations, languages, business, and premedical studies, the university has focused more recently on enlarging its performing arts programs. This is reflected in the new center for the performing arts, which hosts an expanded theater program.

What the College Stands For

Diversity in all its forms—cultural, religious, international, racial, and geographic—is a constant theme expressed by the university's leaders. As a consequence, students represent 130 different nations, and over a quarter are of color. The university promotes "the Jesuit educational mission of caring for mind, body, and spirit" as the underpinnings of its teaching and programs. Students of all religious persuasions are invited to enroll and to worship through the extensive campus ministries in their own faith. The core curriculum underscores the mission of educating students not only in knowledge and skills but also in moral and ethical considerations as a key component of their experience and preparation for future leadership in their communities. Well-known religious figures are brought to the campus to lecture, and religious services and retreats are numerous for all denominations.

Tolerance and respect for other forms of belief and cultural traditions are emphasized through the many speaker events, interfaith organizations, and subjects taught. Most important is the tradition of intellectual pursuit that is at the heart of the Jesuit community. Protestant, Jewish, and Muslim ministries are supported actively on campus through religious leaders and many social and celebratory events. The Center for Social Justice Research, Teaching, and Service directs the community outreach programs, which include forty student volunteer groups. The university encourages and abets student engagement in the community in keeping with its Jesuit tradition of advocating moral principles and social service.

Curriculum, Academic Life, and Unique Programs of Study

The core curriculum requires all undergraduates to take two courses in philosophy and theology. The philosophy course entails the study of general philosophical principles or ethics. The theology requirement includes either a course called the Problem of God or the Introduction to Biblical Literature plus one theology elective. A number of first-year seminars on interdisciplinary topics are available. These stress reading, writing, and active participation in the seminars. Each of the four specialized

schools has its own set of required courses that lay a foundation in its discipline. Undergraduates can concentrate their studies on one of a wide range of academic concentrations—from business and economics to international, regional, and ethnic studies; from history and classical studies to languages and linguistics; from law, government, and politics to science and technology; from theology and religious studies to the visual and performing arts. To be familiar and comfortable with the overall program, the prospective student should review carefully both the curricular content and core requirements of each of the schools under consideration. For instance, the popular School of Foreign Service requires, in addition to the basic two courses in philosophy and theology, two courses in humanities and writing, two in government, three in history, four in economics, proficiency in a foreign language, and a course called Map of the World. Language studies abound in the college, with offerings that range from Arabic to Turkish.

Major Admissions Criteria

With its record-setting number of applications—over 16,000 for an entering class of 1,600 in the most recent year—the admissions committee admits about 3,400 candidates. Keeping to its stated mission, academically talented young men and women who represent a wide range of interests, talents, and socioeconomic, cultural, racial, and religious backgrounds are intentionally offered admission. Unlike a majority of its peer universities, the admissions committees are composed of faculty, student representatives, deans from each of the four undergraduate schools, and the professional admissions staff. The major emphasis is put on the candidates' high school curriculum and performance. The level of the courses taken (accelerated, honors, Advanced Placement, independent study) is viewed as an indication of intellectual curiosity, motivation, and aptitude. Test scores on the SAT or ACT, and also SAT subject tests, are taken into consideration. The middle-range SAT scores of accepted students is 1,300 to 1,500 on the combined critical writing and mathematics sections. The ACT middle 50 percent range is 29 to 32. Volunteerism, leadership, and athletic talent matter as well. The committees pay attention to the input from teachers, guidance counselors, coaches, and employers, where appropriate. Georgetown regards the interview with an alumni representative of the Admissions Program committee as important to the application. The university maintains a policy of need-blind admissions, considering all candidates irrespective of their ability to fund their education. It guarantees to meet the needs of every accepted candidate. Georgetown has an early-action admissions program that admits approximately the same percentage of applicants as the regular applicant pool.

The Ideal Student

The student who consciously seeks a social and educational adventure very likely to differ from his or her high school and home community will be pleased to be at Georgetown. The mix of people brought to the university, the wide array of academic offerings, the bustle of campus life, and the city that bumps up to the campus make for an exhilarating and oft-times challenging adjustment. A desire to get out of one's comfort zone can be essential. A personal belief system in some higher being or order can also matter. One would find it uncomfortable to be a renegade in this environment unless it re-lates to issues of social injustice. Self-direction, self-initiative, and an interest in working pretty hard to do well in one's studies are also essential. The level of social and emotional maturity can define success or disappointment. An interest in current issues of a political, economic, social, or religious nature will lead to much interaction in the dorm, the beer hall, and the classroom. The Georgetown community is not the best college environment for a retiring or reclusive student. Students on campus are likely to say that it helps to like semi-dressy or preppy outfits as well.

Student Perspectives on Their Experience

Few students seem disappointed in their time at Georgetown. Those who have the characteristics mentioned above express their enthusiasm for their fellow students, faculty, and programs. All state their happiness with the location and opportunities Washington offers them. One senior woman reflects on her first year as "a trial by fire and errors until you get used to the people and the academic expectations of the faculty. Once you manage this transition, you will have an extraordinarily fun time while learning." A junior said, "I chose Georgetown over other top schools partly because of the internship opportunities. I have not been disappointed. I know for sure that I am going to head to law school because of the interning I have done for two law firms while here." A student from Greece described his involvement with other international students: "I wanted to experience American culture and gain a business education here. I was concerned about a cosmopolitan environment since I have lived my life in Athens and Paris. I have found all that I could ask for here." Students generally feel challenged by and excited about their university, but do not seem to focus on it as a particularly close-knit or nurturing environment.

What Happens after College

In line with the particular interests and aspirations that bring students to Georgetown, one-third of graduates enroll in graduate schools immediately while the great majority enter directly into the work world. Many of the latter will eventually undertake graduate studies in a professional field of interest. According to university officials a great many of its graduates choose service-oriented careers. A survey of a recent graduated class indicated that approximately 10 percent chose jobs in the public sector including nonprofit institutions, volunteer service, education, government, the Peace Corps, and the Jesuit Volunteer Corps. Teach for America was the largest employer of the most recent graduated class. Georgetown ranks very high among its peer institutions in the number of its graduates who serve in the Peace Corps. However, careers in law, business, medicine, and education continue to be the major attractions of a majority of graduates. Thanks to a multimillion dollar gift from a major bank, Georgetown maintains an active Career Education Center that serves students in career counseling, internships, preparation for the job search, and preparation for applying to graduate schools. It also oversees an active alumni career-services network that enables students to seek advice and potential job opportunities with Georgetown alumni. The esprit de corps that develops during their undergraduate experience continues as alumni, which has fostered an active network of Hoyas helping recent graduates to gain employment in the public and private sectors.

Grinnell College

1103 Park St.
Grinnell, IA 50112-1690

(641) 269-3600

(800) 247-0113

grinnell.edu

askgringrinnell.edu

> Number of undergraduates: 1,623
> Total number of students: 1,623
> Ratio of male to female: 45/55
> Tuition and fees for 2008–2009: $35,428
> % of students receiving need-based financial aid: 60
> % of students graduating within six years: 86
> 2007 endowment: $1,718,313,000
> Endowment per student: $1,058,726

Overall Features

Grinnell is something of an oddity due to its location: It is a small liberal arts college of national intellectual reputation located in a small town deep within America's breadbasket. Its nearest academic neighbors are large, state land-grant universities that serve the majority of in-state residents. Grinnell was founded in 1846 during the infancy of the settling of the Midwest on the model of the traditional New England independent liberal arts colleges. Renowned for its outstanding teaching and the quality of the student body, the college has developed over its long history a loyal following of generous alumni. This has helped to create the largest endowment in the nation among small colleges. The ability to attract outstanding students regardless of financial need together with top faculty who want to focus their career on teaching makes Grinnell a place where one of the truly top college educations can be had. Grinnell is one of a dwindling number of insti-

tutions that guarantees full financial support to all needy students. Each year the college gives out 300 merit scholarships alone. This suggests rather strongly the talent pool a serious learner will be engaged with for his or her four years. The environment on campus is open, intimate, intellectually liberal, and serious in tone. Students rave about the access they have to their professors and the opportunities to engage in research projects or independent study with them.

Grinnell has long enjoyed a first-level science curriculum that appeals to serious students who want to prepare for medical school or careers in teaching and research. Its graduates are perennially among the most successful in completing medical degrees and doctorates in the sciences. At the same time, many students are attracted to the humanities and writing programs that are also considered top-of-the-line. There are no fraternities or sororities on campus, and with a student body of only 1,600, this is an especially close community of learners and teachers. A multimillion-dollar student social center provides for many student activities and brings many social and cultural activities to campus to compensate for the very limited off-campus offerings. Athletics are informal and treated casually by students. Here again Grinnell is an anomaly in the land of big-time intercollegiate football and basketball competition. Overall, Grinnell compares favorably with the very best of the residential liberal arts colleges in America.

An administrator says, "We are now squarely placed among the nation's best liberal arts colleges. Facilities are brand new and designed for the new ways that students learn and live in the twenty-first century. The academic program is opening up across the boundaries that used to separate the traditional liberal arts disciplines. Now departments like physics, French, and economics are supplemented by offerings in film studies, global development, and neuroscience."

What the College Stands For

According to the dean of the college, "For over 160 years, Grinnell has prepared alumni to launch creative ventures that will transform the future of their society. Well-known Grinnellians of the past and present include abolitionists, New Dealers, inventors, diplomats, teachers, poets, and popular entertainers."

Grinnell describes itself as "an undergraduate, four-year, coeducational residential college that seeks to develop in students analytical and imaginative thinking in the liberal arts. The college exists to serve students directly and society indirectly. The college's ultimate goal is to educate citizens and leaders for our republic and the world beyond our borders. To this end,

Grinnell graduates should be equipped to pursue successful careers, satisfying personal lives, effective community service, and intellectually satisfying and physically active leisure."

"Social responsibility and serving the community remains a value of our students. . . . Independence of thought and rigorous academics remain the basis for a Grinnell education," says one administrator. "Grinnell faculty hold learning and teaching absolutely highest in their priorities. Students build very close relationships with professors. Grinnell is a place where a student works closely with faculty to set a course of study that will broaden perspectives and serve a student's future goals."

The dean of the college articulates these six "enduring strengths" of the college:

1. Grinnell College is deeply committed to working, collectively and individually, for the common good.
2. Grinnell College inherits a powerful history of positively transforming the future through bold innovation.
3. Grinnell College pioneers new methods of teaching, both in our campus life and academic programs.
4. Grinnell College minimizes general-education requirements to give each student guidance and responsibility for designing a unique liberal arts program suited to individual interests, needs, and goals.
5. Grinnell College maintains a need-blind admission policy.
6. Grinnell College has forged strong connections over decades with special off-campus sites, such as our academic exchange with Nanjing University and Grinnell College programs established in

London; Washington, D.C.; and Namibia. Most Grinnell students study abroad for a full semester, not just a few weeks.

He adds, "Grinnell College is both comfortably rooted and shaped by our place (as illustrated by the Center for Prairie Studies, historic Ricker House, our Conard Environmental Research Area, and many points of engagement in our local community) while globally diverse in our view, programs, people, and interests."

Grinnell's endowment has opened up many opportunities for the college and allowed it to carry forward its egalitarian heritage. As one dean notes, "The explosive growth of the college's endowment has enabled the college to concentrate on ways to achieve excellence, rather than on ways to balance the budget. This has resulted in both confidence and creativity among faculty, staff, and students in shaping the college's future. It has also allowed the college to remain need-blind in admissions. . . . Our need-blind admissions policy and financial security have resulted in our having a very egalitarian atmosphere. All social, cultural, and academic events are free to everyone in the campus community. A significant amount of money has been set aside to support summer internships and research projects for students, so that these opportunities are not available only to those students who can afford them. Computers abound, so that students who cannot afford to purchase one have ample access." Another administrator says, "The financial accessibility of a Grinnell education reinforces the college's core values of equality of opportunity, respect for individuals, social justice, and community diversity."

Faculty, and their relationships as teachers, advisors, and mentors to students, are the keystone of Grinnell's education. As one dean puts it, "A caring, involved group of faculty are committed to teaching and student learning. Faculty members engage students in research in every academic division of the college. In addition to teaching, they support independent studies and guide readings, sponsor internships, and lead study-abroad programs. Grinnell has a strong sense of community. The academic atmosphere, though rigorous, is noncompetitive. Students do not compare grades. They are expected to work collaboratively in groups. They read and correct each other's work. This sense of community is enhanced by the absence of a Greek system on campus. There are few 'in' groups and 'out' groups among students at Grinnell."

Grinnell is a college that knows itself and what it stands for, and it knows the kinds of students who feel drawn to this small community. As one dean puts it, "I think that most underestimate what an advantage it is for us at Grinnell that students who come here come because they have selected Grinnell with intention and deliberation. Although the surrounding prairie is beautiful in its own way, students don't come for the atmosphere; they certainly don't come to ski; they don't come for the city life. They come because they find here a group of students and faculty who are stimulating to them, an atmosphere where they feel simultaneously at home and deeply challenged." As Grinnell looks to the future, with its strong endowment and growing national reputation, it is "positioning itself to be a leader in the liberal arts," according to one administrator. It is building on campus and solidifying its extracurricular and cocurricular programs. Notes the dean of the college, "We now have the same percentage of international students that we have of students from the state of Iowa (about 12 percent of each), while the majority of our students come from still other backgrounds and traditions, representing domestic as well as global diversity. The face of Grinnell looks very different compared with a few decades ago, but its heart—intellectual curiosity and eagerness to engage with new ideas—is in the same place."

Curriculum, Academic Life, and Unique Programs of Study

Grinnell places a premium on writing and thinking skills, which are thus emphasized across the open curriculum that was instituted in 1971. There is only one specific academic course required of all students: the first-year tutorial. A student's tutorial professor is also his/her advisor. Faculty members design and teach over thirty varied topics, all of which emphasize intensive writing and discussion throughout the term. The college is also experimenting with a "senior capstone" elective program for upper-level students. Grinnell's great financial resources also have enabled the college to build a new science center, a learning/study library, and a fine and performing arts complex. These brand-new facilities mean students can gain a top education in the sciences, the arts, and the humanities. There is an impressive array of off-campus sponsored programs that students can choose from, especially the study-abroad centers. This appeals to a majority of Grinnellians both to satisfy their intellectual interests and to take a break from the small, rural campus. Grinnell has created a number of cooperative arrangements with larger universities' professional graduate schools that allow a student to complete the undergraduate and professional degree in engineering, business, law, medicine, and education over a shorter time than normal. Interactive learning is the engine that

drives the Grinnell education. Students are expected to engage actively with their peers and professors in the classroom and to take a good deal of responsibility and initiative for learning in their field of interest. "The students are 'self-governed,'" says the dean of the college, "meaning that they participate on important committees, set their own policies in the residence halls, and even share responsibility for the evaluation and appointment of their own professors. The intellectual community at Grinnell—faculty and students—lies at the heart of the college, and this centrality ensures that everything at Grinnell supports teaching and learning as our most important work."

Two key offices that represent Grinnell's service orientation are the recently established Office of Social Commitment and the Community Service Center. The college pursues a policy of student self-governance "which is now second nature to the campus, and which colors policymaking in most areas of college life." A dean notes that "the open curriculum and the concept of self-governance work in tandem to create a distinctive atmosphere at Grinnell. On the academic side, the open curriculum presumes significant conversation between students and their advisors in crafting each student's academic program. This sort of freedom, flexibility, and engagement is mirrored on the student-life side of the college, where self-governance is at work. Self-governance presumes that students have the ability and the will to govern themselves, to agree on rules of residential living, and to respect the rights and property of others in the community. Residence-life coordinators do not have responsibility for discipline and punishment, but are charged with community building." On-campus student resources include updated computing facilities and academic support from "learning labs" in writing, science and mathematics, and reading. Grinnell has a peer tutoring program and a Science Project program to help improve retention of students from groups that are underrepresented in the sciences. Grinnell's Office of Multicultural Affairs is available for multiethnic and international students, offering a multicultural student orientation and three mentoring programs. International students are matched with community host families.

Major Admissions Criteria

Grinnell seeks "individuals who are motivated, independent, intelligent, creative, and interested in being part of a vital and rigorous learning community," says one administrator. A senior dean looks for "students who have a thirst and passion for learning." Another says, "The college seeks high achievers who are genuinely committed to a liberal arts education and to a sense of community that encourages everyone's potential." According to a senior administrator, Grinnell looks for these qualities in its applicants: "A sense of curiosity and wonder, a mind that cultivates both imagination and logic, and a heart that feels passion for justice. If you catch us employing the traditional measures of grades, course choices, test scores, personal essays, and lists of extracurricular activities to try to figure out whether you have those things, please believe in the dream that underpins our efforts."

The Ideal Student

As several Grinnell undergraduates have commented, the decision of whether to choose this college for four years is a no-brainer. Grinnell offers no apologies for its size, location, and academic tone. It offers an exceptional set of human and physical resources for the intellectually curious and motivated learner who will embrace the advantages of a very intimate community. Those seeking a large university environment and style of education with all of the trappings that go with it will not find Grinnell appealing. There are no big-league sports teams to cheer for or intensive small college teams to participate in. Those who cannot imagine themselves living in a small-town, country setting are likely to suffer from claustrophobia quickly. A young man or woman who has thrived in high school on strong relationships with committed teachers who are passionate about their discipline and who take great pleasure in discussion and debate on academic and social issues will feel at home at Grinnell. Says one student, "At other colleges I got the impression that you were left alone and allowed to do your own thing, but at Grinnell students would take an active role in being interested in whatever it was you were doing." A sophomore notes, "People at Grinnell are smart but not pretentious, involved but not superficially, and are worldly and globally conscious."

"The students who thrive at Grinnell," says a dean, "are those who are willing to be actively engaged in their education. Because students and advisors work together to build a student's four-year academic plan, students need to bring ideas, opinions, and interests to the table. They certainly do not need to be decided on a major when they begin. Most aren't. But they do need to be ready to bounce around a lot of ideas and to go after the things about which they feel passionately. In addition, Grinnell has proved to be a very welcoming place for students who may have been marginalized in high school because it was not 'cool' to be smart or different. Every fall it is exciting to watch students who, often for the first time, find others like themselves. Grinnell is known for its egalitarian, nonjudgmental atmosphere. Finally, Grinnell is a particularly good choice for students who have a commitment to social activism and service. The college encourages an ethic of service, and opportunities abound." Another administrator says, "Grinnell College has a tradition of respecting individual diversity. Campus life is characterized by thoughtful discussion, attempts to understand people whose experiences and choices are different from one's own, and receptivity to new ideas. Another strong tradition, dating back to the founding of the college, is the determination of generations of Grinnell graduates to put their ideals into action by addressing the central social and political questions of their time."

Says an administrator, "Academically strong, creative, independent thinkers will thrive in this intense environment for learning. This is a great college for gifted, adventurous students who seek to translate their ideals into action and to use their education to make a difference in the world."

Student Perspectives on Their Experience

Students choose Grinnell because of its professors, location, size, financial-aid generosity (even for international students), and philosophy. "I wanted to go to school in a rural location," says one junior. "I was sick of cities. Also, I wanted to go to a school with very strong academics that would allow me to participate in cross-country and track. Grinnell is incredible at professor-student relations. You can approach any faculty member as their equal and most take time to talk to you about any issue. Also, the philosophy of self-governance followed by the college is a good 'real-life' reflective experi-

ence for all." Another junior says, "Grinnell's faculty is top-notch. Not only are they incredibly knowledgeable, their first priority is teaching. Their commitment to teaching makes the faculty Grinnell's most valuable asset."

"Grinnell enables students to start clubs, fund activities, and get involved," says a senior student. "And since it is a small campus, you can develop close groups of friends that you see on a regular basis." Another says, "One would never guess there's so much to do in rural Iowa. The problem is more often that there is too much to do rather than not enough."

"Grinnell does a great job of catering to student interests," says another senior. "Grinnell's open curriculum allows for a true liberal arts education. You can take diverse courses without worrying about fulfilling arbitrary requirements. Almost all classes are challenging and interesting." A sophomore says, "The reading, writing, and math labs are all staffed by competent individuals. The Career Development Office also does a lot of outreach to the Grinnell College community. Student Affairs is also great."

One student says Grinnell is "accessible, small, passionate, unassuming, astute." Another uses these words: *challenging, community oriented, personal, fun,* and *different.* Says another student, "What makes Grinnell even more special is the fact that people can embrace their passions in a place where people are generally open-minded and socially conscious."

A senior notes, "Grinnell attracts the type of student that works very hard for their success, in many cases unnecessarily." Says another student, "Grinnellians can be a strange group, but I have met some of the greatest, most brilliant people I think I will ever meet. We tend to be more on the awkward side, but we're of the endearing type."

"Self-governance," says another student, "means that administrators and rules are not very ubiquitous on campus." Says one junior, "The number of writing assignments and the amount of reading makes it impossible for a student not to improve their ability to read and write. The discussion-based classes, along with the reading and writing assignments, force students to think and express their ideas both orally and in writing."

"Students who are able to keep their academics in perspective do best," says one student. "Those who understand when it is time to buckle down with work, but also when the work does not need to be done. Highly motivated people do better, as do those who do not need a city to be entertained." A junior says, "Grinnell caters to the widest variety of students. I don't think a 'nerd' or a 'cool' person really exists because everyone has their own group of friends and can do well. Academically, the school is amazing. I've learned a lot!" A type of student who does well at Grinnell is "an open-minded liberal student dedicated to academics and to making a real contribution to campus life." Another junior adds that "the school tends to reward those who take initiative. Students who are active academically or extracurricularly gain the most from attending Grinnell."

Students feel supported and connected. "There is a huge academic support network to which one can go for assistance in reading, writing, math, science, or academics in general. Also, your roommates, floormates, and classmates are always willing to lend a hand. Grinnell students are the best support network."

The campus is friendly at Grinnell, but diversity could be improved. Says a junior, "The college needs to improve upon its services for multicultural students. Campus life for them is improving." Some students report a decent amount of international and overall diversity on campus and a lack of a focus on socioeconomic distinctions among the students. Nevertheless, one student argues that "while Grinnell has committed itself to diversity, the vast majority of students still tend to be white and middle class. Grinnell, despite its efforts, cannot seem to attract those of a lower class or racial/ethnic minority. While students are open-minded, open-mindedness cannot substitute for actual diversity."

For those worried about Grinnell's location, students who enjoy the college see it as a plus. "There is no problem finding enough to do in the middle of Iowa. Nothing beats a midwinter Iowa sunset." However, "the isolation of Grinnell is great for some, but suffocating for others. It's really difficult for some to get used to a very small town and small school in the middle of Iowa." Many students thus take advantage of Grinnell's strong on-campus and off-campus internship opportunities. As one notes, "The summer student-faculty research project helped me learn a lot and get some hands-on experience in working with equipment related to my field."

Grinnell is special because of "the people," says a junior. "There are people who are the most intelligent I have ever met, and who also love to have fun. It is academically rigorous, but not competitive. I have gotten to continue dancing, be a part of student government, and be on numerous committees. I don't think this experience could happen anywhere else."

Finally, a senior makes this argument for the special qualities of Grinnell: "Let me make one thing clear: Grinnell is *not* a Hidden Ivy. It is nothing like any of the Ivy League schools I visited and overnighted at. We are our own place, with our own successes and drawbacks. If you think you might like a laid-back, fun, and academically rigorous school, come to Grinnell. Know, however, that it is Grinnell and it is absolutely unique."

What Happens after College

"We measure the success of Grinnell by the lives its alumni lead," says one administrator. "Grinnell prides itself on graduating students who become leaders in business—like Richard McGinn, '68, former CEO of Lucent; leaders in music—Herbie Hancock, '60, jazz pianist; or leaders in their communities. Grinnell students come here with academic talent, independence, and creativity. They leave to lead successful lives and to give back." Grinnell students take many different routes upon graduation, including graduate academic and professional programs, and a diverse array of career paths. "Our first priority," says an administrator, "is to work with the students until they are able to clearly see the path they want to be on. Our Career Development Office also offers a wide range of services to alumni. . . . The commitment to servicing our graduates extends decades beyond graduation."

A senior professor says, "One student who was in my Humanities 101 class is now a highly regarded physician and member of the U.S. Olympic cycling team. Another student, who was in my Virginia Woolf seminar, is an off-Broadway actress who also acts in films and regularly appears on a current television series. . . . The college gives Joseph F. Wall Ser-

vice Awards to alumni who propose the most interesting project that will serve to improve society." According to the administration, "Forty-four percent of our graduates from the classes of 1970 through 2000 have informed the college that they have completed at least one advanced degree since graduating from Grinnell. Of these, 59 percent obtained advanced degrees in the arts and sciences, 11 percent in business, 12 percent in medicine and health professions, and 18 percent in law. For its size, Grinnell College produces an exceptionally large number of alumni who go on to earn Ph.D.'s; in this regard, Grinnell even ranks tenth among all U.S. institutions. On average, more than ten members of each graduating class receive prestigious national fellowships and scholarships (such as Fulbright and Goldwater scholarships, NSF graduate fellowships, and Watson fellowships). Hundreds of Grinnell College alumni have served in the Peace Corps since its establishment in 1961. When we survey all of our alumni, 20 percent of respondents tell us they have made their careers in the business world (banking, insurance, management, or marketing). Eighteen percent of respondents tell us they have careers in primary, secondary, or higher education. Thirteen percent work in the health professions as doctors, nurses, counselors, or other positions. Careers in government and public service, law, writing and publishing, and library or museum work are also represented among a significant number of our alumni."

Hamilton College

198 College Hill Road
Clinton, NY 13323-1293

(800) 843-2655

(315) 859-4421

hamilton.edu

admission@hamilton.edu

[
Number of undergraduates: 1,834
Total number of students: 1,834
Ratio of male to female: 44/56
Tuition and fees for 2008–2009: $38,600
% of students receiving need-based financial aid: 41
% of students graduating within six years: 91
2007 endowment: $701,670
Endowment per student: $382,589
]

Overall Features

A college with lots of tradition that was founded in upstate New York in the early 1800s to educate young men of that region in the liberal arts, Hamilton has undergone several significant changes in its recent past. As Hamilton has gained an increasing national reputation for excellence, its student population is becoming more diverse. During the 1960s Hamilton, an all-male undergraduate college, assisted in the creation of the nearby Kirkland College for women. Kirkland was designed to offer a more progressive style of education with a strong bent in the arts and humanities. Its campus was built in a modernistic style across the road and through the woods from the handsome, traditionally designed Hamilton campus. The two colleges shared many programs and social events but there was clearly a difference in the nature of the student populations and the modes of teaching. Over time it became self-evident that the most sensible plan was to merge the two schools and offer one curriculum taught by one faculty and managed by one administration.

Today, Hamilton College reflects the merger to some extent in tone and curriculum, but it remains a moderately traditional academic and social environment. The historic emphasis on presenting students with a liberal arts education directed by top-quality professors who are there to teach small classes and seminars remains true. A more recent change of import was the decision to make all fraternities and sororities nonresidential as part of a residential-life plan created by the college's trustees and administrators. The goal was to diminish the social role of the Greeks in campus life and their purported anti-intellectual force. All students now must reside in dormitories on either the original Kirkland facility or the Hamilton campus. Hamilton, like all of the small residential colleges of note, wants to put the emphasis on a more engaged, sharing community of young men and women, especially as it endeavors to enroll more students of color and other diverse backgrounds. The dormitories have been developed into clusters for social and academic activities. The administration is aware of the need to develop more social opportunities in this small-town, rural setting. Students have always praised the college for the friendliness of the faculty and the quality of its teaching. A significant portion of each graduating class will go on to graduate school, often having been influenced by a particular teacher/mentor.

The president of the college says, "The most significant change in recent years is the adoption by the faculty in 2000 of an open curriculum, but the fundamental character and purpose of a Hamilton education have not changed." Other new programs of note include centers for writing, oral communication, and public policy. The dean of admissions describes changes on campus further: "In the last five years alone, we have opened a state-of-the-art science center; a fabulous fitness center with flat-panel TVs, a juice bar, and the best climbing wall in the northeast; an Outdoor Leadership Center; two completely renovated residence halls; a new admission and financial aid center; and the first half of an extraordinary social science center opened fall 2008 (the other half will open in fall 2009). Additionally, plans are in the works for a new student, center, theater, art gallery, and studio-arts space . . . all with planned completion dates in the next five years."

What the College Stands For

"I want our graduates to leave this campus with an insatiable intellectual curiosity," says one top administrator, "with an unequaled ability to communicate thoughtfully, eloquently and compellingly; with the confidence, flexibility, and adaptiveness to meet unanticipated challenges; and a willingness to contribute actively to the societies in which they live." President Stewart points to these key areas that define Hamilton: "The college focuses on four areas which, in combination, help define the essence of a Hamilton education: The open curriculum; an emphasis on writing, speaking, and research, part of the college's commitment to individualized instruction; an enduring community, a network of faculty and alumni for a lifetime; and tangible outcomes, preparation for leadership in one's career and one's community. Hamilton is a community in the truest sense of the word." Hamilton College prides itself on the development of an "intellectual spark and social conscience" in their students, and measures its success in part by the significant involvement of graduates in the college. The key to Hamilton's success and approach lies with the faculty and with the college's attentiveness to students. An administrator says, "I think our distinguishing feature is that everyone at Hamilton always puts students' needs first. Our faculty are all exceptionally talented teaching scholars, many with national reputations in their fields, but the real stars of this campus are the undergraduates. We have an extraordinarily able division of student life, whose deans, health and career counselors, and student activities professionals are devoted to making every student's experience as rich and satisfying as possible. Perhaps the most important reason students succeed here is because of other students. This is a supportive and caring community where competition is healthy, productive, and mutually gratifying." "Hamilton's open curriculum attracts students who are confident, self-assured, motivated, and willing to accept the freedom and responsibility offered by this type of academic program," says President Stewart. The dean of the faculty notes, "This is a place, set in an environment of stunning natural beauty, where people from all over come together to pursue educational goals and ambitions, and where alumni, years afterward, recount an intensity of experience that has fueled a lifetime of intellectual and personal development."

Hamilton is seeking to expand its diversity, "not only socioeconomically and racially but increasingly ideologically, geographically, and culturally," says one administrator. The college uses a generous financial-aid budget to help achieve this goal and support a number of multicultural organizations on campus. "Hamilton has a strong commitment to diversity in all its forms," says the college president. "The percentage of the entering class comprised of racial and ethnic minorities is 19 percent, the highest in the college's history, and an increase from 13 percent just five years ago." The admissions dean continues, "Hamilton has students from 49 states and dozens of countries. Five percent of students are international, and 7 percent more are dual citizens from the United States and another country. Roughly half of our students are on financial aid and in any given year 7 to 9 percent of our students are the first in their family to attend college." One administrator defines Hamilton's most important goal, critical thinking, this way: "This includes how to define an issue, to examine it from multiple perspectives, and to assess the merits of the various viewpoints." And outside the academic sphere, Hamilton students "are encouraged to participate in a variety of activities that provide them with exposure to different groups of individuals, as well as with the chance to make lasting contributions to the lives of others in the community."

Hamilton's faculty are committed to their students, and the residential community helps facilitate their engagement. "The fact that faculty are in close proximity to the students makes the formation of personal relationships between faculty and students possible," says a campus dean. Another administrator adds, "Hamilton students succeed largely because they themselves are a driven group of people, but the culture of this place is so focused on each student's success that they cannot help but do well. Obviously, faculty members play a key role here. Students cannot hide on this campus; it is a very connected place. As a result, a swimmer who is having difficulty in chemistry is surprised to learn that the coach and the professor not only know each other, but they have talked and have agreed to a plan to get the student back on track." The dean of students comments, "Hamilton College students engage with faculty in an intellectually rigorous liberal arts curriculum that promotes the development of strong critical-thinking, writing, and speaking skills."

Curriculum, Academic Life, and Unique Programs of Study

Hamilton takes a strong stand on the skills it believes are essential to learning. Three writing-intensive courses are required for graduation in addition to the completion of an intensive independent research project in the student's major concentration. Departments such as physics, geology, history, English, and political science are particularly well regarded. There is a commitment to maintaining small classes that are led by established professors for all four years. There are no teaching assistants at Hamilton. Numerous off-campus programs are made available, from studying abroad to interning in

Washington, D.C. Students are able to complete a combined undergraduate-graduate degree course of study in law, medicine, and engineering with a number of prestigious professional graduate schools. For its size, the college offers a strong variety of arts courses.

The dean of admissions describes Hamilton's innovative curriculum: "Unlike most liberal arts colleges that require students to fulfill distribution requirements, we cater to a more independent student who wants to stretch and grow and challenge herself, but not because she has to, rather because she wants to. . . . We also believe this curriculum raises the level of engagement in the classroom. Every single student in every class is there because they choose to be, not to simply fulfill some requirement. The success of this curriculum hinges on good faculty advising, another component of the Hamilton experience that we take very seriously. Finally, the one thing we aren't flexible about is writing; all of our students need to take three writing-intensive courses during their four years at Hamilton."

One institutional leader points to three practices he sees as essential in a Hamilton education: "The first is a historic and ongoing commitment to the singular importance of writing and speaking substantively, cogently, and eloquently. We pursue this goal across our curriculum, through small classes where no one can be a passive learner. Our second distinctive practice, one which also crosses curricular and departmental boundaries, is our emphasis on a capstone senior experience for every student. Each Hamilton student is expected to write a thesis or pursue an intensive senior program that serves as the culminating experience of his/her four years. And the third practice, deeply ingrained in our college's heritage, is the expectation that students will always publicly demonstrate what they are learning through presentations, poster sessions, and performances." Another administrator comments that "while we no longer have a four-year public-speaking requirement, that ethos remains a part of a Hamilton education. The opportunity to present and then defend one's ideas before one's peers is now more broadly incorporated in our curriculum."

Some other key programs include the Higher Education Opportunity Program, a six-week summer enrichment program to boost achievement and retention of minority students; and the promotion of undergraduate research opportunities that join the practical and the theoretical, such as the summer research program allowing undergrads to work on projects with faculty. Says an administrator, "I have seen in my seven years on campus an emphasis on increasing opportunities for serious undergraduate research at Hamilton—and not just in the natural sciences. Students studying the impact of welfare reform on the local county government (and then presenting the results of that research to elected officials); students studying the impact of a proposed industrial park in a neighboring community (and, again, presenting it to local officials); a student studying the evangelical movement in Latin America (and then having a book published as a result of that research); and a student writing a novel as a senior fellowship project—all offer evidence of a pedagogical approach to undergraduate education that is increasingly focused on original research." Also of note is the Arthur Levitt Public Affairs Center, an undergraduate public policy think tank. The Levitt Scholar Program allows students to deliver their senior projects as oral presentations to high school seniors. Says an administrator, "For the Hamilton student, the experience is invaluable in developing confidence and understanding of one's subject matter, and it demonstrates again the college's commitment to oral and written communication. It also reinforces the notion that if you 'perform' an idea you own it."

Major Admissions Criteria

Hamilton is looking for "active learners, contributors to the intellectual, cultural, and athletic life of the college, whose passion about life and learning enriches the experience of everyone around them," says one administrator. In terms of important admissions criteria, "I believe that everything begins with academic excellence. We are looking for outstanding students whose performance in secondary school confirms their promise as lifelong learners. Through some combination of rank in class, standardized test scores, and counselor recommendations, I want to be convinced that a student's admission and matriculation will strengthen Hamilton's commitment to the pursuit of excellence. What gifts does an individual bring? The answers to that question drive our admissions decisions."

A dean says, "Besides outstanding academic records, we also expect our students to have engaged to a significant degree in extracurricular activities, including sports, community service, and cultural events. One predictor of success at Hamilton is a student's openness to new experiences. A second predictor is motivation to seek assistance, which all students, regardless of academic caliber, will need as they progress through the Hamilton curriculum." Another administrator says that "to the extent that it can be measured and revealed in teacher and guidance counselor recommendations, indicators that students will contribute and enrich our campus are very important. A student's high school record and extracurricular activi-

ties are also considered highly." A faculty member lists the three top criteria as:

1. Demonstrated work ethic.
2. Motivated to do something positive with his/her life.
3. Intellectually curious about the world.

A senior administrator says, "Hamilton looks first at the high school transcript. We expect students to take the most challenging academic program available to them and to perform well in that curriculum. We also look for students who demonstrate leadership qualities and are involved in extracurricular and volunteer activities." The dean of admissions looks for "smart students who don't take themselves too seriously. I've been told by several guidance counselors that our students don't seem to have the same arrogance that is prevalent at many of our peer institutions." The most important admissions criteria are "rigor in the high school curriculum, strong performance within that rigor, and demonstrated intellect or ability to learn as measured through testing. We are SAT optional but not test optional . . . giving students a menu of test options to choose from to prove to us their testing ability."

The Ideal Student

Students who are happy and fulfilled at Hamilton like the friendly, traditional, and fairly homogeneous population and the strong relationships easily developed with faculty through the small classroom. They have little need for the amenities and attractions of urban life. A majority of students anticipate continuing their education in graduate school, especially in the professional fields. There are ample opportunities to become engaged through varsity sports teams, clubs and organizations, and student government. The collegian who enjoys a highly diverse community and a student body that focuses on political causes and social action will not find a large body of takers at Hamilton. An enjoyment of the out-of-doors helps greatly in making it through the long, cold, upstate New York winters. Anyone who expects stores and restaurants to be open all hours of the night will not be satisfied with the small town of Clinton.

Intellectually and socially, "The students who thrive here are those young men and women who are willing to take risks, who care deeply about giving something of themselves to the campus community, and who are unafraid of change and growth." A faculty member adds that "hardworking, curious, bright individuals are what we want and they do very well." An administrator says, "Those who do best here are students who want to be involved—in their classes and in the life of the college community." Hamilton is not a place for students who want to isolate themselves or get lost in the crowd. The president of the college notes, "Hamilton students are liberally educated problem-solvers, able to communicate their ideas clearly and persuasively, both in written and oral form. Students come to Hamilton to find their voice and make it heard. . . . We are in the ninth year of a longitudinal assessment of student learning in a liberal arts setting. The study clearly shows, among other things, that student writing improves throughout a student's four years at Hamilton." Says the dean of the faculty, "In all possible instances, our goal is for students to be engaged in original research, both in collaboration with their professors and with their fellow students." The admissions dean says of current Hamilton students, "I think you will find them to be smart, accepting, diverse (in all ways, but especially attitude), goofy, fun, athletic, laid-back, funky, crunchy, geeky, intellectual, and independent."

Student Perspectives on Their Experience

Hamilton students are clear in their enthusiasm about the college and their reasons for selecting it. Says one senior, "I chose Hamilton . . . for three reasons. One: Academic reputation and programs (at the time, I was looking for a school that offered creative writing). Two: Dedicated faculty (I was blown away after I visited classes at Hamilton, one of which was an English seminar with *four* students and a professor—what attention!). Three: Frankly, I enjoyed myself at Hamilton more than at other schools. People were unpretentious, easygoing, and seemed to know how to enjoy themselves!" A classmate says, "Hamilton made me feel wanted from day one. The admission folks went out of their way to set up meetings with professors and coaches. When the time to decide finally did arrive, Hamilton made it clear how much they wanted me by offering a generous aid package. They never fell short throughout the whole admission process."

And clearly, the faculty-student connecting that goes on at the college is paramount for students. "Hamilton's greatest attribute, to my mind, is dedicated faculty. The college offers amazing opportunities to work closely with highly regarded and respected professors, whether in a very small senior seminar or class discussions continued over lunch or a celebratory dinner at a prof's house at the end of the semester. I never received anything less than spectacular attention and dedication from faculty members—it seems they truly enjoy teaching, and especially at a college that allows and *encourages* such close interaction between students and faculty. I consider many faculty members close friends, especially those in my department (English). . . . When I return as a seventy-year-old alum, I may not remember the name of this or that building, but I sure as hell will remember that life-changing conversation I had on its front steps with my advisor, mentor, and friend Professor Briggs. . . ." Another senior says that students are really helped by "professors, administrators, and resident advisors (I feel like once Hamilton lets you in, it does its best to support you and help you succeed)." "The people" make Hamilton special. "I know my professors and most of the administrators and many staff members on a personal basis. It really is like a big family community." "The advising system has really been terrific," says one student. "The college's commitment to funding undergraduate research and unpaid internships has also been invaluable, and I've benefited from both." A senior uses these words to describe Hamilton: *close-knit, demanding, driven, fun-loving, outspoken.* Students who do well at the college "aren't wallflowers and we thoroughly enjoy hearing ourselves talk, but at the same time we know how to listen. It's not uncommon to be discussing recent *Wall Street Journal* articles and world news topics at dinner. Hamilton students tend to be driven, focused, and goal oriented. At the same time, however, we know how to close the textbook and play a game of soccer." Another senior says, "Everyone on campus loves meeting new people and having as many friends as possible. Our community is strengthened by the fact that we are isolated and end up spending all of our free time on campus getting to know people and hanging out with our friends. The ideal Hamilton student is someone who wants to be a part of the community. You don't come to Hamilton to just sit in the library 24/7. All of our

different involvements bring us into contact with more people, thus tying us to the school even more." And, "Hamilton makes it easy to find out what you are passionate about. The school not only helps students find their academic purpose and interest, but more often than not, lifelong passions."

One graduate testifies to this happening in his first year at the college: "Academically, Hamilton is extremely strong in the writing department. Teachers emphasize clear and effective writing in their classes. It was difficult to meet these high standards as a freshman, but over the course of the year I noticed my writing improve significantly. Hamilton also offers access to a writing center that many students take advantage of. Its main purpose is to help improve student writing through one-on-one discussion. There are so many different clubs and organizations on campus that it is easy to find a new and interesting activity to try or club to join. When I arrived on campus I was intending to row for the crew team. Instead I ended up trying activities I had never done before. I joined an a cappella group, having little musical experience (but enough to get in the group, I guess!). I also gave acting a shot by joining the student-run theater organization. Both experiences were an integral part of my freshman year. I never thought it would be so easy to join a club and jump right into something I'd never done before. Students are able to be part of so many activities due to Hamilton's small size." Says one student of the new curriculum, "The fact that Hamilton has no core classes really makes your college experience unique. You are never bogged down in a 40 percent required course load, and the classes you take are because you have some genuine interest or curiosity." She continues, "Hamilton does a great job allowing their students to run the show. As a student here you really feel like you can get things done. If there's something Hamilton doesn't offer (which happens rarely), you can feel free to start a group or bring up your concern with [the] administration."

Other concerns expressed by students include a desire to be more involved in campus decision making, to increase all forms of diversity (although socioeconomic diversity, encouraged by Hamilton's financial aid program, is praised), and to have more communication between various campus groups. Says a senior, "Hamilton could do a better job bringing students of more diverse backgrounds." Yet another student argues, "Nearly a quarter of students fit into a diversity category of one form or another."

"Alcohol tends to be a problem at Hamilton," says a student, "especially during the cold winters." This is despite the removal of residential Greek houses. "The administration does a great job providing other options for students during the weekends, but unfortunately about half the population decides to drink themselves into oblivion instead. That being said, Hamilton does provide great entertainment options during the weekends for those who don't wish to participate in various drinking games."

The physical environment of the campus continues to be beautiful: "Hamilton's campus is very special and distinguished. Hamilton is *gorgeous*, especially in the fall, and its grounds are spacious and well kept—we even have our own huge nature preserve, ranging from hiking/biking trails to a rare flora preserve to several college-owned reservoirs."

Who does best at Hamilton? Students seem to echo the thoughts of the administration on this count. "Students who create and take advantage of opportunities do best at Hamilton. There are plenty of people who sit back and give

only 50 percent during their time here, but the ones who do the best, get the most out of the college, and contribute the most are the ones who give 100 percent—or even 150 percent! Self-motivation helps people succeed, although students are sometimes pampered a bit too much, in my opinion . . . And the courage to forge one's own distinct path is always a plus!" A recent graduate says, "Hamilton students are good at motivating each other to keep improving and doing their best. However, there is generally *not* cutthroat competition so there is a much more relaxed atmosphere." Another adds, "Students with energy, a love for learning, high academic endurance, and who are willing to take risks will do best at Hamilton. The expectations are high. Professors set high standards and expect that they will be met. In a small classroom environment there is no hiding at Hamilton.

Students strongly back Hamilton's study-abroad opportunities, which expose them to new experiences and return them to campus "to realize just what a privilege it is to be at Hamilton, a place where professors know the faces and names of all their students, a place where one can finish one's senior thesis and celebrate with a drink in the Little Pub afterward with faculty and friends . . . Study abroad also helps students broaden their perspective and bring *back* to Hamilton lessons learned elsewhere." Another admired program is Adirondack Adventure, an eight-day-long, preorientation wilderness program for first-year students. Meant to ease the transition to college life, Adirondack Adventure often turns out to be a student's best week. The bonds of friendship formed on the wilderness trips often last all four years and more! The program lets students meet people, share anxieties and dreams, make friends with students *and* faculty (who often lead backpacking or canoeing trips themselves), and learn to put their trust in the community that will become their home for the next four years. "And best of all, Adirondack Adventure is fun! Even when, the year that I served as assistant director of the program, it rained constantly for three days while students were out in the mountains and lakes of Adirondack Park, they still came back *beaming* from their trips—with even more stories and bonds than if the weather had been gorgeous!"

For students interested in community service, HAVOC (Hamilton Association for Volunteering, Outreach and Charity) provides a good outlet. "This idea of service to the school and surrounding community," says a sophomore, "seems to be getting stronger with each incoming class."

Finally, students leave Hamilton ready for the challenges ahead of them and appreciative of their learning experience. As one recent graduate points out, "You will leave Hamilton with confidence and with the ability to express yourself and share your passions with others. Hamilton will make you work hard for this end result, but as I am already finding, it is well worth it."

What Happens after College

Hamilton graduates go on to a diverse array of careers and graduate programs, with over 60 percent in graduate school within five years. The college says, "Our students are prepared for everything" and notes law, diplomacy, health professions, teaching, advertising, and business as key areas of interest for alums. "We also claim an award-winning pastry chef, an internationally acclaimed ceramist, a former president of the Metropolitan Museum of Art, and numerous members of the foreign service with ambassadorial rank."

One dean notes that "perhaps the most important among the attributes we instill in students is the flexibility to respond to changing circumstances and interests as they travel along their own unique professional paths." Another administrator adds, "Over and over again, alumni tell us how Hamilton's three core values—writing, speaking, and critical thinking—have helped them outperform their professional peers." Hamilton's undergraduate research opportunities across many fields help students gain entry into top graduate programs and provide access to top jobs.

Alumni commitment and success are strong: "Hamilton is a small place with a big reach. In our peer group, we are among the colleges with the fewest number of alumni; in some cases, we have 50 percent fewer alumni. But they make up for it. They are passionate (in some cases, rabid) about this college. Both in dollars and volunteer time, they make their affection known. Typically, about 55 percent of all alumni contribute to the Annual Fund—an important measure of support, and one that ranks Hamilton among the top ten of all colleges in alumni support. The volunteer support is also amazing—from alumni willing to meet with seniors looking for jobs to admissions volunteers to fund-raisers, we have thousands of people wanting to be helpful. . . . This is a very connected place." A survey of the class of 2007 indicated that almost 80 percent were employed full-time, and 19 percent were attending graduate school full-time. Top career fields include financial services (17 percent); education (11 percent); health related (9 percent); communications (8 percent); consulting (8 percent); law (7 percent); business related (6 percent); sales and marketing (6 percent); nonprofit (5 percent); and sports and recreation (5 percent). Top areas of graduate studies include medicine and healthcare (28 percent); natural sciences (19 percent); law (18 percent); social sciences (11 percent); education (7 percent); and humanities (7 percent).

Haverford College

370 Lancaster Avenue
Haverford, PA 19041-1392

(610) 896-1350

haverford.edu

admission@haverford.edu

[
Number of undergraduates: 1,169
Total number of students: 1,169
Ratio of male to female: 46/54
Tuition and fees for 2008–2009: $37,525
% of students receiving need-based financial aid: 48
% of students graduating within six years: 94
2007 endowment: $539,589,000
Endowment per student: $461,580
]

Overall Features

In most of its qualities, Haverford resembles the greater number of small residential liberal arts colleges described in this book. The curricular emphasis is clearly dedicated to exposing students to a wide range of courses in the arts and sciences, with an underlying goal of enhancing their critical-thinking and writing skills. A high premium is placed upon interaction between faculty and students. Classes are kept remarkably small and, consequently, the faculty expects a good deal of classroom preparedness and participation. A distinct advantage for Haverford students is that three-quarters of the faculty live on campus and make themselves available to students at all times. Haverfordians are intelligent, intellectually committed, and liberal-minded. They are headed to professional careers, as are their peers at the other Hidden Ivies, but frequently with time out for travel, community service, or interning. The population of under-

graduates is small, and all of them live on campus, fostering lots of interaction.

The ways in which the Haverford community takes on its unique personality have to do with its founding in 1833 as the first Quaker college in America and its special relationship with its neighbor, Bryn Mawr. While there are few birthright Quakers in the community today, the philosophy that underlies the Society of Friends is still much in evidence. There is a spirit of acceptance of all beliefs and racial and ethnic backgrounds; resolution of conflict over campus issues is arrived at by consensus and goodwill; a commitment to social action to assist the less fortunate is articulated in all areas of campus life; and the acceptance of the honor code as the active principle guiding academic and social life is acknowledged by all students.

Haverford has a generous financial-aid program, which has led to a diverse population that includes over 25 percent minority representation and about 5 percent internationals. Under the guiding force of the honor code, students take all examinations unproctored. They may even choose their own schedule and place to complete their exams. There is a required seminar on issues of social justice, and students gather together each term to review the honor code and school policies. If an individual violates the code in any way, he or she is required to apologize before the entire student body in addition to suffering the consequences of the student judicial committee. Consequently one can imagine the importance community plays on this campus. While students are concerned about social and political issues in the larger world, they tend to be moderate in their expression and actions. To serve others is more of the Quaker theme than political activism and pressure to change others' thinking. Another distinctive feature of the college is its unique and valuable relationship with Bryn Mawr. Since it has such a small enrollment, it is a great advantage for Haverford students to be able to take any course offered at Bryn Mawr toward their diploma. These are two separate colleges with their own individuality, and the relationship greatly enlarges the learning opportunities, not to mention the social advantages, for students. Those who wish to do so can even live on the other's campus and eat in the other's dining halls. A combination of the small size and the academic/social attitude of the college translates into a low-key athletic program that encourages participation by all who are interested. There are a fair number of intercollegiate teams, the most popular being soccer and lacrosse. There is no football program. There are no fraternities or sororities, so the residential focus is in the dormitories. Students tend to participate in the many extracurricular organizations on campus. Philadelphia is within easy reach by commuter train to allow time away from this most intimate of campuses when desired.

Haverford's campus constitutes a nationally recognized arboretum with more than four hundred kinds of flora, as well as gardens and woods. Recent additions to the campus include the Integrated Natural Sciences Center, and a major athletic center. As the college puts it, "Haverford students are entrusted with freedom and responsibility from their first days on campus, an act of trust aimed at developing the whole person and creating a lasting bond between individual and institution. Students leave Haverford with an increased sense of independence, curiosity, and confidence, as well as lasting bonds with friends and faculty. Whether 'Fords go on to advance their educations, enter the workforce, travel the world, or better the world in one of any number of ways, they leave grounded and well equipped for rich, dynamic lives."

What the College Stands For

Haverford sees itself as having several unique attributes, among them: "*A critical balance*: Haverford has maintained what a former president described as 'the balance between intellectual rigor and moral judgment, of knowledge and its application to society.' *Quaker heritage*: Its egalitarian, inclusive community where individuals are important, all views are tolerated, and there is a tradition of trust and responsibility among the students who attend. *A sense of community*: Through its Quaker heritage and its 100-year-old honor code, a climate of respect and concern for others is central to Haverford's ethos. *A conscious effort to remain small*: Haverford has purposely remained small, but it has also remained connected to the world at large and to the individuals who comprise the college community. *Fundamental commitment to a broad liberal arts education*: Haverford has remained committed to educating young people in how to think critically, to understand fundamental theory and concepts, to weigh values, and to form independent judgment."

Says the college, "Haverford provides its students with a liberal arts education in the broadest sense, to develop their capacities for thinking critically and comparatively. Students are expected to go beyond the mastery of facts and evolve into creative and innovative thinkers, experimenters or scholars. The college's rigorous academic program rests on the assumption that students who attend Haverford will become fully engaged in and contribute to the intellectual (and social) life of the campus. As a result of these goals, Haverford has developed an array of major programs that combine much of the disciplinary training of graduate programs with the kind of direct, imaginative and exploratory engagement appropriate to undergraduate education."

Haverford has developed and matured since the introduction of its coeducation thirty years ago and since it has fostered a national reputation. As one administrator notes, "The curriculum has been broadened significantly, the applicant pool and student body are now national, the faculty is larger and offers many more areas of academic inquiry. The social and athletic programs have increased significantly, the administration is highly professional, and the finances of the institution are strong."

The honor code is central to Haverford's mission and identity, and applicants are required to write an essay about how the honor code will relate to their educational experi-ence. Says an administrator, "Haverford is the oldest institution of higher learning with Quaker roots in North America. While the college is nonsectarian, the intellectual and social life is very much influenced by the Friends' emphasis on spiritual insight, intellectual honesty, and leadership. Many of the college's more important programs and practices stem from its Quaker heritage. More of a philosophy than a code of conduct, the honor code means students are expected to develop a strong sense of individual responsibility as well as intellectual integrity, honesty, and genuine concern for others."

Curriculum, Academic Life, and Unique Programs of Study

True to its emphasis on intellectual exploration and independence, Haverford allows students to choose the majority of their curricula over four years. Thus, there is a great deal of flexibility and opportunity in the academic program, and the college sees over time an "increasing trend toward interdisciplinary, integrated learning." There is a language requirement regardless of one's major field of study. Haverford seniors complete a thesis, which allows them to concentrate on a special area of interest as a capstone project. Although the majority of students concentrate in the social sciences and humanities, the college provides an excellent life-sciences curriculum that many future doctors elect. Here is another of the quality liberal arts colleges with an outstanding record of preparing students for admission to medical schools. A majority of classes in all subjects place a good deal of emphasis on reading, writing, and presenting in a logical, systematic, and critical style. Faculty place much of the responsibility for mastering the subject on the student. And "the individual attention of faculty, administrators, coaches, advisors, and other students is key to students' succeeding at Haverford," says an administrator. Another adds, "The college is somewhat unusual in that the ties between student life and curricular areas are strong. The dean's office, for example, is structured to support both academic and personal advising. Because the deans work closely with faculty around individual student advising, there are few, if any, cases where students do not receive needed help. Besides the formal avenues for counseling students, there is a writing center and a math question center, there is career and preprofessional advising, and there are several programs for students of color and a network of student advisors who are trained to help classmates with a host of student life issues. An Office of Academic Resources helps to coordinate support for all students."

Some key aspects in the development of Haverford's academic programs are mentioned by one institutional leader: "Cooperation with Bryn Mawr College has matured to the point where students can enroll in individual courses, take complete majors, and reside at either institution. As the faculty has grown, so too has the breadth of the curriculum. There are many more instances of collaboration among faculty across various disciplines within the college and with faculty at neighboring institutions like Bryn Mawr and Swarthmore. Extensive renovations to a number of academic and residential facilities, construction of a campus center (and an integrated natural science facility) have occurred in the past decade, as has the installation of a number of technological improvements: an automated library system (a virtual library combining the collections at Haverford, Swarthmore, and Bryn Mawr Colleges) and a networked campus with instruc-

tional technologies in our labs and classrooms." Other facets of Haverford's personality and academic life again stem from its Quaker roots. "Consensus is the primary decision-making process used in all levels of governance among students, faculty, the administration, and board of managers," says an administrator. "Given the Quaker tradition against hierarchies, there is a minimum amount of titles and rank at the college. Few students refer to faculty members by their doctoral degree. As part of their requirements for graduation, all students must fulfill the social justice requirement by completing one of a number of courses available across the curriculum in which students examine issues of prejudice, inequality, and injustice."

While there is not a heavy atmosphere of competition for grades or graduate school admission, many Haverford students do plan to continue their education beyond the bachelor's degree some time in the future.

Major Admissions Criteria

Haverford's admissions office is clear about the three criteria they consider most important in the admissions process: "*Academic achievement and performance*: These include the strength of a student's secondary curriculum as well as grades. *Testing:* Standard measurements like SAT or ACT; however, less emphasis is put on the importance of testing results in those cases where students have excelled in strong high school programs. *Contributions outside the classroom*: community service, other extracurricular activities, students who demonstrate well-rounded lives." The admissions office says, "The admission process at Haverford is conducted as a comprehensive review in which each application is treated personally and individually, with extraordinary care and attention to detail. We provide you with the opportunity to convey the broadest sense possible of who you are, what you have achieved during your secondary school experience, and how you will both contribute to and grow from a Haverford education. Our primary consideration in the evaluation process is academic excellence." The college strongly recommends an interview, especially for those living within 150 miles of Haverford.

The Ideal Student

Haverford is a most attractive college for the student who thrives within a small, friendly community and enjoys an active learning and intellectual experience in small class discussions and research projects. People who care about others on and off the campus will find a home at Haverford. "Highly intellectual students who are interested in contributing to a small community are the kind who would do best and be happiest at Haverford," says one administrator. By contrast, anyone who wants to hide out in large lecture classes and take one or two exams a term to demonstrate mastery of the subject will not find Haverford College appealing. This is not the right college for the traditionalist who favors an atmosphere of conformity and uniformity, nor is it right for those who are rabid athletic fans who follow nationally ranked teams. While there is easy access to a major urban center, the majority of Haverford students develop their social life within the small campus community.

According to one dean, "Being part of the community and giving something to others could involve participation in student government, the newspaper, or community volunteer work, for example. Because of the emphasis on student self-governance socially and academically, there is an expectation that everyone will become involved in some way or another. Close to 70 percent of the student body does volunteer work in some formal program. Seventy-five or more student organizations are run only by students. A very high percentage of students play sports." Others comment that "when students refer to the social life at Haverford, they mean something more than going to parties—it's a sense of conscience." And "participation is just as important in academics where classes are small and student involvement is expected as well."

Haverford sees itself as a diverse and "unfailingly friendly place. Respect for the individual is implicit in all campus interactions and is one of the key tenets of the social honor code." Almost 30 percent of Haverford's students are persons of color. The college notes, "First-year students are made to feel welcomed and part of the community at the outset. A general orientation program called 'Customs' is designed and implemented by teams of students who serve as resource people and mentors for groups of students throughout their first year. Besides the mentors who live on the same dormitory floors with freshmen, there is a cadre of students trained to address social and academic issues who conduct weekly peer awareness workshops throughout the first semester on issues such as diversity. Prior to the start of their first year, students of color are invited to a program designed to help them identify and handle the challenges they will face at Haverford as persons of color. A winter institute is organized over winter break for all students and focuses on a specific issue of social justice. An Office of Multicultural Affairs is primarily responsible for supporting students of color academically, socially, and emotionally as they progress toward their degrees. The office is also charged with developing and coordinating cross-cultural programming on campus. The college also has a number of 'interest-based' centers, lounges, and houses set aside for diverse campus organizations."

Student Perspectives on Their Experience

Students choose Haverford because of its academic programs, its location near Philadelphia, its philosophy of education, and its student body. A senior from Delaware tells of her decision-making process, contrasting Haverford's environment to that of her state university: "When I was deciding which college to attend, my final choice came down to U of DE (which has a pre-vet program in which I was interested) and Haverford. I decided to attend Haverford for many reasons. While I felt confident that I wanted to pursue veterinary medicine, I also wanted the benefit of a liberal arts education. I ultimately decided that I would spend the rest of my life working with animals, but I would not get the opportunities that Haverford offered later in my career. . . . I also like the small-school, tight-knit atmosphere at Haverford. After spending time on campus, I loved the feel of the college. Over the years, this feeling has become stronger and stronger. I honestly think of the Haverford community as a big family. While, of course, not everybody in the family gets along all of the time, there is a closeness here that I know I would not have found at a large state university. Not only am I close with the other students, but the community extends to include faculty and staff members as well. The professors go out of their way to get to know you. Dinners at professors' houses are a frequent occurrence and you often see students eating at the co-op with professors or staff members. You get to know your class-

mates and the professors very well. I have taken many of the large classes offered on campus (Intro Psych, organic chemistry, Bio 200—the introductory bio course—Intro to the Study of Religion), but these classes are small compared to their counterparts at other schools. Once you get past the intro-level classes, the class size falls to an intimate level. This year, one of my biology courses had only eight people in it. I definitely would not have had this experience at a larger school."

Who does best at Haverford? A senior mentions that "you need to have an internal drive that causes you to want to learn, to work hard, and to be the best person you can. Haverford is an academically challenging college. You have to work long hours studying hard, fitting in your extracurricular activities as well as the social aspects of college. You need to be the type of person who lives the honor code, both academically and socially." Students are helped to succeed by their own initiative and drive. "With the honor code so prevalent in Haverford life there is very little outward competition to do well. No one tells you their grades if they don't want to, so as a student you have no idea how others are doing. In order to succeed you have be motivated by yourself, not by competition with others. Profs really care about you and know your name. You can talk to deans very easily. We have mostly self-scheduled exams; it is a trusting place, and conducive to learning and doing well."

The honor code is seen as a very important component of life at Haverford, in and out of the classroom. "This code is by far the most important part of this place, and I've noticed it affects my actions outside of the Haverford community as well." Another senior points out, "There are two main parts to the honor code. First, there is the academic honor code. This is what most people think of when they think of an honor code—no cheating, no plagiarizing, etc. We also have a social honor code. We are to respect each other and be respectful of each other. If we are having a problem with a student, faculty, or staff member, we should meet with them to talk things out. The social honor code is probably one of the most distinctive parts of Haverford. We are very committed to it. The honor code is very idealistic, and it is important to understand that not everybody follows it all of the time (for instance, somebody may be loud late at night without really thinking about the fact that they could be being disrespectful to somebody), but everybody tries to follow it. All decisions are made by consensus. Haverford is not just a school, but just as important, it is a community. I honestly feel at home while I am on campus. I think of my classmates and professors as part of my extended family. This closeness influences every part of our lives. Closeness leads to respect and this respect causes people to be respectful of each other. You do not lie to, steal from, or in any way hurt somebody whom you respect."

This is illustrated in the academic environment of the community as well. "The community aspects of the college help students to succeed. This is felt throughout the college. The professors are wonderful. They are always willing to meet with students and help them work through any difficulties they are having in class. My professors have always gone out of their way to help me. The community aspects of the college also lead to wonderful student-student relationships. We work together to reach a deeper understanding of subjects. Because of our commitment to the honor code, students are able to balance group studying with the individual work that is necessary. I feel comfortable going up to almost any person in my class to ask a question and I would take the time to answer the questions of any of my classmates. Professors take advantage of this group atmosphere—I have had an exam for which I was supposed to meet with a group to discuss all the questions before we submitted individual answers."

In terms of the college's downside, says a senior, "This place can feel very homogeneous. If you don't think or act in a very liberal way, it is easy for people to look down on you. People here are also exceedingly busy. Everyone is stretched too thin. Sports, singing, school . . . Everyone gives one hundred percent at everything and it exhausts you and can stress you." Diversity overall at the college is seen as moderate, but the college is praised for its efforts to overhaul its programs in this area and to help students of different backgrounds adjust. Another senior comments about Haverford students that "they are generally kind, thoughtful, intellectual, and interesting people who have had some amazing life experiences and come from surprisingly different backgrounds, or at least have their own well-thought-out and strong opinions on all kinds of topics. I think that those who see Haverford as homogeneous aren't looking very carefully at those around them." And a junior points to such programs as Customs and Summer Tri-Co as helping to create "safe spaces" to address topics of concern. "Respectful, intelligent discourse is one of the most essential and wonderful parts of the 'community' atmosphere that is cultivated at Haverford." Another senior notes that "Haverford does not always show you how the 'real world' works. Sometimes people say that we are in our own 'Haver-world.' Out in the 'real world' people are not always nice, not always honest or respectful. People do not always trust you, nor do they always deserve your trust. I love Haverford and I would never give up the wonderful atmosphere that is present here, but you must remember that this is not the 'real world.'"

"It's tough," says a senior, speaking of the academic program. "It takes a lot of effort to really excel. You can have gobs of fun at college, but if you go here, also be prepared to work. It's safe. We are in a little suburb next to Philly, and although we are near a big city, we are really a tiny community, where everyone knows everyone else (this can also be overwhelming when you notice you know *everyone* on campus). I feel I should also mention Bryn Mawr, in order to get an accurate picture of Haverford. This institution, two miles away, provides a whole other college of classes to take. It is like being at a much bigger university because you can take classes there. It also, though, really distorts the male/female relationship on campus, which can be a bit odd."

Finally, Haverford students admire and respect the attention they get from faculty. As one junior puts it, "Haverford's major strength is the interaction that occurs between students and professors. Because it is a small college, Haverford's faculty members have the opportunity to really get to know their students, and vice versa. The professors I have had the privilege to know and learn from truly seem to have the students' best interests at heart because they seem to have such a strong passion for what they do. . . . Haverford is such a terrific school because with the 'external' competitiveness removed, students and faculty can appreciate learning in and of itself."

What Happens after College

One administrator commented that "a degree from Haverford has always meant the same thing: an excellent educational foundation that carries with it a notion of social responsibility and lifelong learning. As a result, our graduates are very interesting people, often with many dimensions to their lives, and they are very involved in their communities and their professions." Says another, "Haverford College graduates are prepared to go into a variety of fields after graduation."

A ten-year survey of graduates in their first year after Haverford indicated that 17 percent on average have headed to graduate school, while 60 percent were employed, with others traveling, on a fellowship, or unknown status. Top graduate school fields in that first year have been arts and sciences (8 percent), medicine (5 percent), and law (3 percent). Education (13 percent) and business (11 percent) have consistently been the top employment fields, with science (7 percent), human and community service (6 percent), communications (5 percent), and law (5 percent) representing other popular career areas.

The Johns Hopkins University

3400 North Charles Street
Mason Hall
Baltimore, MD 21218

(410) 516-8171

jhu.edu

gotojhu@jhu.edu

Number of undergraduates: 4,578
Total number of students: 6,244
Ratio of male to female: 52/48
Tuition and fees for 2008–2009: $38,200
% of students receiving need-based financial aid: 45
% of students graduating within six years: 91
2007 endowment: $2,800,377,000
Endowment per student: $448,490

Overall Features

Hopkins occupies a unique historic position in American higher education as the first university founded as a research graduate institution for the purpose of encouraging original research and scholarship that would advance human knowledge and science. Since its inception in 1876 it has held firm to this mission even as it has evolved into a substantial undergraduate residential college for 4,400 talented and motivated young men and women. Research activities abound among the prominent faculty to the extent that Hopkins receives more federal funding for research projects than any other university in the country. This tradition explains the serious academic tone on the campus. More so than at other leading research universities, Hopkins undergraduates have unlimited opportunities to participate in scholarly projects with their professors. Some 70 percent of students will participate in at least one research project during their four years in the college.

Students come to Hopkins to take advantage of the outstanding academic faculty and resources at hand. Hopkins is a relatively small research university, which enables it to keep the great majority of classes small. There are few large, lecture-based classes. Social life and athletics are not priorities for those who elect to attend the university, though there are a few very competitive sports programs. Hopkins is a serious place for serious students from all corners of the globe. It attracts a talented and diverse group of students from all fifty states and seventy-one international countries. Hopkins maintains a generous financial-aid program in order to attract a diverse group of talented students; one half of all undergraduates receive some form of aid.

While it is true that a third of students enter Hopkins as intended premedical concentrators and another third enroll in the Whiting School of Engineering, the remainder have forty-nine majors and forty minors to choose from. Faculty and programs in international relations, English and writing, the social sciences, and music are considered outstanding. Students in the humanities and arts rave about their departments and faculty, and refute the image of Hopkins as only a "premed and engineering factory." The renowned Peabody Institute school of music is a division of Hopkins and while it remains a distinct performing arts program, Hopkins students can avail themselves of courses and a concentration through the Zanvyl Krieger School of Arts and Sciences.

The founding purpose of the university also is reflected in the campus setting. It is only in recent years that the trustees and administration have concentrated on developing a more comprehensive campus environment by building new dormitories, a student commons, and an elaborate exercise center, while also enlarging the athletic and social programs. The very attractive Homewood campus for undergraduates is located north of downtown Baltimore. Students do speak of the encroachment of the city in terms of crime. The university recognizes this by maintaining an elaborate security system.

Virtually as famous as its academic reputation, intercollegiate lacrosse is to Hopkins as football and basketball are to a number of large universities. Hopkins competes in Division I in lacrosse only, and has won national titles a number of times. It is the primary activity that draws students together when they are not studying. A wide range of traditional athletic teams for men and women are supported and compete in the Centennial Division III league. There are off-campus fraternities that attract about 20 percent of students. A great many students participate in social service programs, which are readily at hand in the Baltimore community.

What the College Stands For

The concept of excellence in academic life, from research to teaching to learning, is at the heart of the Hopkins enterprise. As the underpinning of all that the university does, the official mission "of the Johns Hopkins University is to educate its students and cultivate their capacity for lifelong learning, to foster independent and original research, and to bring the benefits of discovery to the world." The belief that research and scholarship go hand in hand with excellence in teaching is embedded in the university's philosophy. Undergraduate as well as graduate students are encouraged, in many disciplines, to engage in scholarship activities through independent projects and research collaboration with their teachers. The breadth of the curriculum across the sciences, technology, international relations, and the humanities and social sciences reflects the vision of gaining knowledge and sharing it beyond the ivy walls of the campus for the betterment of society.

In keeping with its intellectual roots, Hopkins is a culturally and intellectually open and diverse community of teachers and learners. A humanities major states that "what Hopkins does best is both encouraging and supporting ingenuity in all fields, not just the sciences. Hopkins students, past and present, are constantly breaking new ground and making significant contributions to their areas of interest. If the school did not value and support this creative energy, all of the inventions and new information that our students help to develop would not be possible."

Curriculum, Academic Life, and Unique Programs of Study

An undergraduate can choose a major field of study from an extraordinary range of disciplines, from anthropology to writing. Languages, sciences, social sciences, mathematics, arts, music, film, and cultural studies are available to students to delve into. Engineering students also have a number of specialized concentrations to choose from which include aerospace engineering, biomolecular engineering, biomechanics, chemical engineering, computer science, electrical engineering, environmental engineering, and entrepreneurial and management engineering. Faculty advising plays an important role in helping students choose their field of concentration and their courses. There is no core curriculum. There are distribution requirements to ensure that students gain exposure to the arts and sciences and humanities.

The presence of the Peabody music conservatory offers special opportunities to study music at an advanced level within a liberal arts curriculum. Many students take lessons on an instrument from members of the conservatory. Several double-degree programs for engineering and liberal arts majors are offered. The grading system is demanding, with many courses graded on a curve. There is little talk of "gentlemen's Cs" on the campus. To keep the academic pressure on students at a lower boil, first-year students receive "satisfactory" or "unsatisfactory" notations on their official transcripts and grades are not factored into students' overall GPA.

Opportunities to delve into a subject or project of strong interest abound. Faculty encourage students to get involved in their research work wherever possible. Asked what the university does best, a senior replies, "Hopkins is great at offering students a ton of opportunities—academic and social. The administration realizes that Hopkins students are driven, impassioned, and capable of accomplishing a lot, so they open up opportunities for us to utilize these strengths. These include a plethora of five-year B.A./M.A. programs offered throughout the arts and sciences and research opportunities and grants for independent studies and research. My professors have all been great resources for me, helping to drive my academic interests outside of the classroom, as well as offering advice and guidance for internships, research, and further studies."

Major Admissions Criteria

Admissions to Hopkins is very selective. In the most recent year, 14,800 students applied and 3,600 were accepted, a 24 percent acceptance rate. The caliber of the high school curriculum and grades are the primary factors considered in reviewing candidates. The admissions committee looks for those students who have challenged themselves by taking advanced-level courses, including in those subjects that relate to their future field of study. Test scores of admitted candidates are at the high end of the range; the middle fiftieth percentiles for admitted students on the SAT are 660 to 670 on critical reading, 690 to 780 on math, and 670 to 760 on writing. The range on the ACT is 30 to 34. Given its serious academic environment, the admissions committee gives less weight to extracurricular activities unless the candidate meets their academic criteria. Personal interviews are offered on campus or with alumni as an avenue for learning more about the university and present undergraduates' experience. In line with its peer institutions, Hopkins undergraduates have a 93 percent graduation rate over a six-year period, a significant indicator of the highly selective admissions process and student satisfaction with their experience.

The Ideal Student

Hopkins is not the college for the academically faint of heart. Seriousness of purpose, and a passion for learning or doing well in one's studies as a preparation for graduate school and the professions, matter a great deal. Students state that the image of Hopkins as a cutthroat academic environment is simply not the case. What is true is that students work hard to meet the demands and expectations of the faculty as well as their own. There is less of the "work hard, play hard" culture that attaches to a number of other top colleges. Hopkins and the surrounding city can be an interesting and fulfilling place to explore and experiment intellectually, artistically, and culturally for the curious and independent individual. Confidence in one's abilities will go a long way to ensuring a positive four-year experience.

In summarizing her reasons for choosing Hopkins, a senior woman majoring in international studies spoke for a great many of her fellow students: "I chose Hopkins for many reasons. I wanted an atmosphere where I would be surrounded by people much smarter than me. I wanted to be able to learn from my peers, as well as be challenged by the academic environment. Everyone I had met when I visited the university seemed very passionate about their field of study, about the activities with which

they were involved, and about the university in general. I thought that would be a very exciting environment to be in for four years. The people who do best are those who are ambitious and passionate. You have to be self-driven and passionate about your field of study in order to succeed in such an academically rigorous environment. But more so, the students that are happiest here are the ones that get involved with activities, that utilize all the resources." A chemical and biomolecular engineering major who chose Hopkins because he wanted a small university that was strong in research and faculty interaction observed that "people who are active and organized do best at Hopkins. There are so many clubs, events, research opportunities, fraternities, etc., that people can get involved in many meaningful ways. However, there needs to be a balance between academics, which are stressful, and extracurricular activities, or else your grades will suffer. There are many internships, grants, and scholarships that are posted every day. There is strong alumni support with grants and networking. There are many tools to cope with the demanding academics. Being proactive and reaching out for all these resources and opportunities pays off." A humanities major believes that the ideal Hopkins student is "anyone who is highly motivated, hardworking, passionate about learning, and not scared to make mistakes and think outside of the box."

Student Perspectives on Their Experience

A senior who had transferred after her first year at a highly respected university had this perspective on Hopkins: "Many reasons factored into my decision to come here. Hopkins has more of an emphasis on intellectualism and self-motivation while my former school is more a 'learning what you are taught' environment. While no one was there just to get a degree, education seemed overwhelmingly a means to an end—that end being a job—whereas at Hopkins many students seem to be genuinely interested in their chosen subjects." In reflecting on the campus community this student observed that "there is a social disconnect. Hopkins can sometimes feel less like a cohesive college community and more like a *place* where athletes, nerds, Greeks, and various student groups, among others, just happen to learn the same things. You will find few campuswide traditions and events. It isn't that they aren't offered, but rather that they suffer from low turnout and lack of interest." Regarding the diversity of the community, one minority woman feels that "there are scores of different cultural groups at Hopkins. The population is exceptionally culturally, ethnically, and racially diverse. The university itself does a fine job of being multiculturally accessible and friendly." Another student of color's perception is that "Hopkins is diverse in every sense—ethnically, racially, socioeconomically. Everyone on campus is different and from my experience, the atmosphere is very accepting and tolerant of everybody's differences. Students embrace the diversity on campus also in terms of interests, hobbies, views, values, and ideas." One young man who is an international studies major reflects the feelings of a great many of his fellow students: "I cannot emphasize this point enough: Hopkins is not just for premed and science students. Obviously we have the top programs for those subjects in the country, but just because we're good at those subjects doesn't mean we're bad at others. I came to Hopkins for the strength of our humanities and social science programs, particularly our number-one international studies major. You will find fantastic programs in everything from our top-ranked writing-seminars program to our bioethics program, a hidden jewel of the campus."

What Happens after College

In recognition of its talented and ambitious student body, Hopkins maintains an Office of Pre-Professional Programs and Advising. The office takes an active role during students' undergraduate years in advising on course planning, internships, and research opportunities. Students comment favorably on the support they have gotten from the staff. Alumni as well as undergraduates are encouraged to utilize its counseling services. Our survey of several dozen undergraduates reflects the intention of the majority of students to continue their education at the graduate level. Medical and public health, international relations, law, business, education, engineering, and technology are the most popular fields of professional study. Hopkins's reputation for academic rigor and a highly motivated student body, and the opportunities for undergraduate research with its prominent faculty, lead to a high rate of acceptance into top graduate school programs.

Kenyon College

Ransom Hall
106 College-Park Street
Gambier, OH 43022

(740) 427-5776

(800) 848-2468

kenyon.edu

admissions@kenyon.edu

[

Number of undergraduates: 1,653
Total number of students: 1,653
Ratio of male to female: 46/54
Tuition and fees for 2008–2009: $40,240
% of students receiving need-based financial aid: 46
% of students graduating within six years: 88
2007 endowment: $192,934,000
Endowment per student: $116,717

]

Overall Features

Kenyon is the quintessential residential liberal arts college that dedicates its faculty, resources, and facilities entirely to the education of undergraduates. Located on an attractive campus in a rural area of Ohio, Kenyon provides a powerful sense of a community where students and faculty work closely together in exploring the broad range of arts and sciences. There are no graduate students, and all faculty are committed to teaching and mentoring. With only 1,600 students, all of whom reside in campus housing, social and academic life is centered on the people and the programs within the immediate community. Kenyon is a very selective college with demanding academic standards. In virtually every facet, it resembles its peer New England liberal arts colleges.

The articulated mission and goals of Kenyon's administration and faculty are one and the same for its students. Regardless of their academic and personal interests, virtually all stu-dents rave about the quality of the faculty and their engagement with them in and out of the classroom. Said one senior major who is engaged in the advanced theater and writing programs, "Kenyon makes academics personal. Class is not a chore for the professors as they want to help students learn. I go to office hours for my professors at least once every two weeks. We don't even have to talk about class. I cherish the conversations about contemporary theater I had with my biology professor. And if you want to have lunch with one of the faculty they always love to do so. They are interested in hearing what you have to say for yourself and not just what you take from their lectures."

Many bright, motivated students choose Kenyon for its outstanding reputation overall, and particularly its English and writing, theater, languages, political science, art history, and premed departments and faculty. Students majoring in almost all fields speak to the quality of their teachers and the combination of high expectations they put upon them and the support they provide. A biology major from a suburban high school encapsulated the essential features of the college that are shared by most students: "Kenyon is blessed with many really wonderful people more than anything else. Particularly I have found that the professors, including the biology department (I am a bio major), are so intelligent and enjoyable to work with. You develop real relationships with them beyond the classroom and they become great resources. The community is also really great. While it can get old being in a small town with 1,600 people and not getting away often, it rarely feels exhausting because the people are so friendly and enjoyable, and all are united in their goal of receiving a great education while really enjoying themselves and building bonds with one another."

Although its enrollment is small and its campus is in the midst of Ohio farmland, Kenyon is not a regional institution. All fifty states are represented with substantial numbers of students coming from New England, the mid-Atlantic, the upper Midwest and the West Coast. There is a reasonable number of international students from twenty-seven foreign nations enrolled as well. A young man from Serbia said that he enrolled in Kenyon "for its academic prestige; 'studying in the company of friends'; and the warm, friendly, and welcoming community; and plenty of one-to-one student-professor communication."

There are many Division III athletic teams for men and women. Kenyon is a founding member of the North Coast Athletic Conference, which is composed of a number of academically selective small colleges. The men's and women's swimming and diving teams have been exceptionally outstanding over the years in national competition. Students are excited to have their brand new athletic center, which includes facilities for all indoor sports as well as a fitness center.

The nearest major city is Columbus, the state capital, where students will go occasionally for entertainment and a breather from the campus bubble. As one would anticipate, half of every junior class studies abroad in order to broaden their education and to experience a world larger than Gambier, Ohio. The college sponsors its own programs in England, Honduras, and Italy. The Office of International Education encourages students to gain an international experience and guides them to some 150 approved programs beyond its own offerings. There is such a plethora of campus social, interest, cultural, and Greek and athletic organizations available that students comment on the need to find a workable balance between the academic workload and extracurricular commitments. The college has established a Good Samaritan policy as an effort to have students look after one another in times of difficulty related to alcohol and drug abuse. Students who report a peer who is in trouble will not be penalized in any way. This is emblematic of the Kenyon campus community.

What the College Stands For

Kenyon states unabashedly that it is an outstanding liberal arts college "where academic excellence goes hand in hand with a strong sense of community." While the curriculum is grounded in the traditional liberal arts and sciences, the attitude of the faculty is anything but traditional in their teaching and interaction with their students. Open discussion, debate, innovative thinking, independent research, and interdisciplinary studies are strongly encouraged. In the words of the administration, "We set high academic standards and look for talented students who love learning. Small classes, dedicated teachers, and friendly give-and-take set the tone. Kenyon welcomes curiosity, creativity, intellectual ambition, and an openness to new ideas. We see learning as a challenging, deeply rewarding, and profoundly important activity, to be shared in a spirit of collaboration." Further, "Our greatest strength is our faculty, outstanding scholars who place the highest value on teaching. Close interaction with students is the rule here: Professors become mentors and friends. Requirements are flexible enough to allow for a good deal of exploration. Other notable strengths include our distinguished literary tradition, many opportunities for research in the sciences, and programs connecting students to our rural surroundings. The Kenyon experience fosters connections of all kinds—to classmates and teachers and friends, to the life of the mind, to global perspectives, to our own unique traditions and history, and to a place of inspiration."

Curriculum, Academic Life, and Unique Programs of Study

True to its liberal arts tradition, Kenyon offers eighteen major fields of study and thirteen interdisciplinary programs. There is no core curriculum but there is a language proficiency requirement and one quantitative reasoning course. All seniors must conclude their studies in their major with comprehensive exams or an original project or an artistic performance to demonstrate mastery of their field. There are numerous special programs, which include working on the highly regarded *Kenyon Review* literary journal, a summer science scholars program for independent research with a faculty member, environmental science studies that utilize the college's five-hundred-acre na-

ture preserve, and the rural life center for the study of the farm communities that surround the campus.

Major Admissions Criteria

Admissions is very selective, yet highly personal. 4,626 students applied for admission in the most recent year. 1,352 were accepted, and 458 enrolled. The mid-50 percent SAT range on critical reading was 630 to 730 and 600 to 690 on the math. The mid-50 percent range on the ACT was 28 to 32. A great deal of attention is paid to the quality of the high school curriculum and performance in the core college prep courses. All applicants rank in the top half of their class and all those accepted have a GPA above 3.0. Personal talents and commitments in their school and community, together with the potential to take seriously their education at Kenyon, are also important factors. The content and quality of the personal application factors in the admissions review. Interviews on campus or with alumni in the home area are encouraged.

The Ideal Student

Kenyon students speak articulately about the type of individual who has a successful experience. "Anyone who loves what they do will do well here. You have to have a passion for something and Kenyon fosters these passions." A junior encourages prospective students "to spend the night and go to classes. Make sure that this academic and social life is for you. A number of freshmen transfer out because they realize that they really want to go to school in a city and want more off-campus options." A sophomore from a small-town high school advises that "students need a strong academic background before they enter Kenyon. What helps them to succeed here is to be well-rounded in their interests before they arrive and know the things one should know coming from a good high school. They should have a good literature, writing, history and science/math foundation to cope with their studies here." Another campus leader observes that "the people I know who are happiest here are those who are highly motivated. You can really make the most of your time here if you are willing to work hard in your classes, identify interests and pursue them by getting involved, and be willing to make Kenyon your home. It doesn't always start as a perfect fit, but people who are open to the experience and really want to fit here will find a way to really enjoy it and profit from their time here. The Kenyon student body is best described as highly motivated, yet not outwardly competitive and always willing to have a good time. We live life to the max here." Says another junior, "Kenyon is best suited for students who want to figure out for themselves who they are. It is not a good place if you only want to blend into the crowd."

Student Perspectives on Their Experience

The great majority of students feel that the college administration and faculty fulfill their promise of presenting an outstanding array of academic opportunities and first-rate teaching in an intimate, interactive environment. From a campus athlete and leader: "The faculty at Kenyon really helps students succeed. I have never been surrounded by more willing people in my life. They are always seeing to it that we students have the best of what we need to learn and that our personal goals are met." Another campus leader encourages students "to take the initiative and get involved. There are a large number of student organizations in proportion to the number of students on campus. This means there are lots of opportunities for the individual to take on leadership positions.

Diversity in the student body receives mixed messages. One student's perception is that "Kenyon offers a litany of on-campus resources geared toward embracing diversity. These organizations include Hillel, the Black Student Union, the Crozier Center for Women, the Asian Awareness Club, and the Unity House for homosexual students." Another student has a different take on this subject: "Diversity is a challenge for Kenyon due to its size and location. I think that the college has done a good job of recruiting athletes from different cultures and foreign countries, which adds a lot of variety. However, racial diversity is still pretty limited, though it has gotten better since I have been here. The socioeconomic diversity is least evident." Happy times, happy community, close relationships with peers and teachers, lots of avenues for involvement in campus life, and lots of academic work pretty well sums up the perspective of Kenyon students.

What Happens after College

Based partially on her own daughter's experience at Kenyon, the dean of admissions and financial aid describes the career and graduate school advising "as a robust and active parent and alumni career network . . . there are many ways that Kenyon sets up graduates to become successful. But perhaps our graduates' greatest asset is their vision, the ability to believe that they can do whatever they imagine with their lives." Grounded in the learning tools and expansive perspective their liberal arts education has given them, Kenyon graduates are successful in entrance to major graduate academic and professional schools. Many take time to explore their potential career interests before committing to further studies. Law, business, and medical schools are popular as is working in English and writing fields. A number enter the Peace Corps and Teach for America. A recent enthusiastic graduate has found that "in the outside world, just dropping the name *Kenyon* will get you opportunities you never thought of. In my summer internships and travels abroad, I have been surprised by the number and kind of people who graduated from or know someone who graduated from the college. Everyone always seems to love Kenyon students, and so doors open unexpectedly."

Lafayette College

118 Markle Hall
Easton, PA 18042

(610) 330-5100

lafayette.edu

admissions@lafayette.edu

Number of undergraduates: 2,352
Total number of students: 2,352
Ratio of male to female: 55/45
Tuition and fees for 2008–2009: $36,090
% of students receiving need-based financial aid: 52
(Class of 2012)
% of students graduating within six years: 89
2007 endowment: $734,421,000
Endowment per student: $312,253

Overall Features

Founded in the 1820s to educate young men in a combination of the liberal arts and a variety of sciences and practical arts, Lafayette has evolved over time into a coeducational undergraduate college with a wide range of resources and educational programs on an attractive, self-contained campus. Lafayette's students, faculty, and senior administrators are in agreement that theirs is a very special college in a number of ways, that they are distinct from the majority of their sister liberal arts institutions, and that they are on a roll today in terms of their enriched academic programs, selective student body, new facilities, and a theme of global education for all its students.

In the past decade the college has experienced a 42 percent increase in applications, a 19-point improvement in the admissions rate, a 57-point increase in the average SAT scores, a 25-point increase in the percent of students graduating in the top decile of their high school class, and a 69 percent increase in minority enrollment. With a total student body of 2,400 men and women in equal numbers, Lafayette maintains an outstanding engineering program within a traditional arts and sciences curriculum, which only a handful of strictly undergraduate liberal arts colleges provide. The college is proud of the broad range of disciplines taught by the faculty and the opportunities for exposure to the sciences and liberal arts by all students.

With the support of a loyal alumni body and energetic leadership, the college has moved forward on a number of fronts in recent years. Situated on a 100-acre hillside campus overlooking the town of Easton, Lafayette impresses all visitors with its attractive grounds and buildings. While it is a nonsectarian college today, there are two chapels on campus that support the religious interests of all the major religious groups represented in the student body. A number of new residences; classrooms; a center for the arts; and a college center, which houses the dining hall and the many clubs and social activities on campus, have been built in recent years along with a major sports center that serves both its twenty-three Division I varsity men's and women's teams and active intramural and fitness programs. Lafayette competes in the Patriot League in the major sports. One of the oldest and most fiercely contended athletic rivalries is the annual football match with Lehigh University.

A "Plan for Lafayette's" future calls for building on the college's traditional strength of outstanding teaching and mentoring by faculty; for diversifying the student body and increasing access to a Lafayette education; and for strengthening the study-abroad, life sciences, and arts programs. The goal is to hire thirty new faculty, thus reducing the student-to-teacher ratio to 9 to 1. These are ambitious goals that underpin the confidence in Lafayette's present and future success in attracting bright and ambitious students and maintaining the support of its loyal and proud alumni. The college's endowment on a per-student basis places it in the top 2 percent of all private colleges in the nation. Lafayette has made a major commitment to expanding its socioeconomic, racial, and geographic diversity by implementing a financial-aid initiative to help all qualified students enroll at Lafayette regardless of their ability to meet the costs. All students with total family incomes of below $50,000 have their demonstrated need met entirely with the inclusion of a $1,500 workstudy award and grants. The loan component of the aid package will be limited to $2,500 per year for students whose total family incomes are between $50,000 and $100,000.

What the College Stands For

The college's official mission statement is referred to by senior administrators as the guide to what the faculty seeks to accomplish for its students: "In an environment that fosters the free exchange of ideas, Lafayette College seeks to nurture the inquiring mind and to integrate intellectual, social and personal growth. The college strives to develop students' skills of critical thinking, verbal communication, and quantitative reasoning; it encourages students to examine the traditions of their own culture and those of others and their capacity for creative endeavor. . . ." To this end, the college fosters a long list of experiential learning opportunities that include internships, study-abroad options, and one-on-one research programs with faculty.

Internationalizing the curriculum through global studies and study-abroad experiences is increasingly emphasized in the course offerings. The stated goal of the college for some time has been to diversify the student body, the faculty, and the social and cultural offerings to change its historic composition of a homogeneous population of white, eastern-centered, economically comfortable students. The college supports 250 interest-based organizations on campus in order to move away from a social scene that has centered around the Greek houses and dorm drinking parties. Today students come from thirty-eight states and over fifty foreign countries. About 15 percent of recent classes are students of color. Lafayette cherishes its tradition of hands-on teaching by a talented faculty. States one senior dean, "Ultimately it is the caliber and commitment of the faculty, all of whom hold the highest degree in their fields and all of whom are committed to teaching undergraduates and to the success of each student, that helps our students in the most significant ways. The fact that Lafayette is exclusively focused on undergraduates, that the vast majority of classes enroll less than twenty students, and that the teacher-to-student ratio is under 11 to 1, enables the faculty to provide each student with a good deal of personal attention and mentorship." What has remained a constant strength at Lafayette has been this commitment to teaching and influencing the lives of its undergraduates. One longtime dean observed, "It is impossible to have a conversation with a Lafayette graduate without mentioning the impact of one or more our faculty members had on the life of that individual as a student. The amount of time our faculty devote to individual students is staggering." A major goal of the college is to expand its honors and independent study programs in order to have more students experience scholarly work and be more involved with the faculty in their own research. In keeping with its emphasis on community sharing and responsibility to others, Lafayette has established the Good Samaritan policy, which allows, without penalty to either party, underage students to call campus security on behalf of a fellow student who is intoxicated to the point where they need medical attention.

Curriculum, Academic Life, and Unique Programs of Study

Lafayette grants the bachelor of arts degree in thirty-one fields of study, and the bachelor of science in nine fields of science and four of engineering (chemical, civil, electrical and computer, and mechanical). Interdisciplinary majors are offered in African studies, American studies, biochemistry, international affairs with engineering, mathematics and economics, neuroscience, and Russian and East European studies. There is a common course of study that includes a first-year seminar and a Values in Science/Technology (VAST) seminar. The seminars study some of the major current technological issues from a multidisciplinary perspective. The seminars are writing intensive and challenge students to tackle the ethical issues surrounding technological advancements and gain greater appreciation for cultures and traditions other than their own. Liberal arts students must satisfy a foreign language requirement through formal classes or study abroad in a non-English language locale.

A five-year, two-degree program is also available in certain fields. In keeping with the educational mission of the college to have all students "examine the traditions of their own culture and those of others," a foreign-culture requirement is in place. About 80 percent of the students complete at least one internship, and over one-half of all students will study abroad on their way to their diploma. Over 150 students each year participate in the EXCEL program, which enables undergraduates to help faculty members with their advanced research, and another 200 students will do independent-study projects with a professor. Engineering concentrators are also encouraged to study abroad through cooperative programs that grant them credit toward their Lafayette degree. A proactive Office of Intercultural Development guides and encourages students toward a wide variety of international study options. The college is expanding its three-week, interdisciplinary, intensive courses abroad to entice even greater numbers of students to study abroad. The Academic Resource Center provides a full range of support services that range from study-skill workshops, time management, test anxiety, and

peer tutoring to disability services and resources for students with learning and physical
disabilities and attention disorders. The Military Science program sponsors ROTC, which
prepares students for duty as commissioned officers in the U.S. Army Reserves, or National
Guard. ROTC scholarships that cover full tuition and fees are available to enrolled students
who commit to active service in the army upon
graduation.

Major Admissions Criteria

In the most recent year, Lafayette attracted
6,364 applicants, accepted 2,224 or 35 percent, and enrolled a first-year class of 594 students. Sixty-five percent of the students were
ranked in the top decile of their high school
class, the median GPA was 3.53, and the middle 50 percent had SAT section scores that
ranged from 580 to 670. The specific criteria
weighed into the decision process are the quality of the high school program (the number and
level of academic courses taken), writing ability, and standardized test scores. Either the
SAT or ACT with writing is required. According to the admissions office, counselor, teacher,
and other references, together with interview
reports when available, help the committee appreciate special qualities of mind and character
in many cases. A separate evaluation is made of
the quality of students' involvement in organized extracurricular and athletic activities as
well as in jobs held by the student. The dean of
admissions says that one reason the college enjoys a retention rate of 95 percent and a six-year
graduation rate of almost 90 percent is the care
that goes into selecting students who are a good
match. The admissions staff looks for candidates who have demonstrated intellectual en-

gagement, excitement about learning, academic
ability, a willingness to take risks, discipline,
industry, and a willingness to contribute to the
life of the community. The admissions process
is highly personalized, according to the dean:
"We are watchful for students who demonstrate
an unusual strength of character, such as those
who have overcome severe hardship. We are
also cognizant of the shape of the entire class;
as we value diversity, we try to put together the
most interesting class we can from among many
well-qualified applicants." In line with the college's long-term strategic plan, a recruitment
plan has been implemented that is designed to
increase applications from American students
of color, low-income, and first-generation-to-
college students from geographic outreach areas, prospective humanities majors, and students
interested in the arts. This plan is linked to the
healthy financial-aid policy and the new academic programs that recently have been put in
place.

The Ideal Student

Self-motivation and confidence, a strong work
ethic, a desire to engage with the faculty, an
interest in joining clubs or athletics, and good
organizational skills are essential to success at
Lafayette. Lafayette favors the traditional, outgoing, academically motivated student who enjoys working hard and engaging in community
activities and a social life. The students who do
well at Lafayette are those who are able to jug-
gle multiple tasks and challenges. They are
often pursuing academic interests well beyond
the classroom, often across several different
disciplines, and extracurricular and/or athletic
talents and passions at the same time. One
student commented that "students need to ac-

tively seek out all of the innumerable resources the college provides; my advice would be that if a student is looking for something in particular, to not be afraid to question, to explore, and reason through all options. Lafayette can help you on your way to anything, but you have to take the first steps of motivation. If you want something from your education, there is nothing here that will stop you, so long as you go for it." Another upperclassman echoes a similar theme: "Lafayette really works under the idea that you get out of the experience what you put into it. The people who seem to do the best here—and who seem to have the best time doing it—are those who are actively involved in pursuits inside and outside the classroom. They are the ones who are in class every day, and then do sports, music, theater, or whatever else they are into on the weekends. It is very easy to meet a great group of very driven, incredibly intelligent, and diverse individuals this way, namely just by going a little outside of your comfort zone and saying hi to someone."

Student Perspectives on Their Experience

Undergraduates appear to agree that their college is academically demanding, is a stimulating and active environment, and is filled with outstanding and committed teachers. Says one junior, "Homework dictates quite proportionately how happy students are. We work a lot (some more than others) and sometimes homework loads can become a burden." One senior observes that "our professors are very accessible and offer any help we need; the college has the resources to provide me with all that I could possibly need; if there is an activity, a program, or a class that I am interested in, the faculty and deans will make it available to me

and my classmates." What makes Lafayette special to most students "is the connection between our professors and students. Lafayette professors are always there for their students, and they are always interested in what their students are doing. Not only does this close connection make students more excited for classes and discussions with professors and other students, it also opens all sorts of doors: large portions of students do research with their professors." Says another upperclassman, "The college creates an atmosphere that is the perfect balance between academic prestige and demands on the students and a fun and social life. I feel that my classmates and professors are my friends." Students feel safe on the campus, as noted by one young woman: "I don't have much concern about student life. The campus is very safe, with few incidents of crime. Whenever something does happen, students are notified through e-mail and by mail, and we even have a text message service for emergencies now." This same student declares that "the college is as good as it sounds. There is no place I would rather be."

What Happens after College

Close to a third of students enroll in graduate or professional schools immediately after graduation. Two-thirds are employed in full-time jobs in a wide variety of professional fields within six months of graduation. A large number pursue graduate studies after several years of work experience. Another 5 percent pursue other paths, including applying to graduate school, part-time employment or temporary employment, internships or volunteer work, travel, and freelance work. The outlets for both the B.A. and B.S. graduates are wide and varied, with many major companies and organizations represented. Engineering graduates find positions in a number of major technical and public organizations. On the basis of her own experience, one senior believes that "it is the alumni network that gives Lafayette its greatest edge over other similar schools—the college keeps track of their alumni, who are more than willing to provide connections for externships, internships, and jobs for Lafayette people."

Lehigh University

27 Memorial Drive West
Bethlehem, PA 18015-3094

(610) 758-3100

lehigh.edu

admissions@lehigh.edu

Number of undergraduates: 4,856
Total number of students: 6,974
Ratio of male to female: 57/43
Tuition and fees for 2008–2009: $37,550
% of students receiving need-based financial aid: 43
% of students graduating within six years: 85
2007 endowment: $1,085,639,000
Endowment per student: $115,669

Overall Features

Founded in 1865 as a private, nondenominational college by a leading industrialist and businessman to educate young men in the fields of science and technology and to carry out cutting-edge research in these fields, which would have practical applications in the expanding American enterprise, Lehigh today continues to carry out its traditional mission with an unusually broad curriculum and first-rate faculty that includes the arts and sciences of a liberal arts college. Blessed with excellent facilities for learning and research, and a highly selective student body of 4,750 undergraduates, the faculty is committed to providing a great many fields of study for its students, hands-on teaching in the classroom, and opportunities for undergraduates to carry out research projects with them. All of the teaching faculties hold either the Ph.D. or the highest degree in their field. Lehigh is a unique combi-

nation of a relatively small university that emphasizes small classes and interaction between students and teachers, together with a good deal of original research and advanced level courses in a number of disciplines that align more with larger universities.

Lehigh is comprised of three undergraduate colleges and one graduate school: the College of Arts and Sciences, the P. C. Rossin College of Engineering, the College of Business and Economics, and the graduate College of Education. Over ninety academic major concentrations are offered. There are a good number of cross-disciplinary majors that make it possible to study in several of the colleges simultaneously. A symbolic indicator of the changes that have occurred in recent years is the university's mascot. For most of its history, Lehigh was proud of its Engineer mascot that would appear at athletic events and rallies in support of its teams. The transformation of the engineer into a mountain hawk in recent years reflects the effort of the trustees and administration to broadcast the expansion of the university on a number of fronts, including the admission of women and the major expansion of the arts and sciences and the performing arts studies that complement the highly regarded engineering and technology and business programs.

Today 38 percent of undergraduates are enrolled in the College of Arts and Sciences, 29 percent in the business college, and 30 percent are in the engineering college. The university prides itself on the many special programs that combine different but interrelated fields, something that smaller or less academically rich colleges can provide. In contrast to the majority of four-year undergraduate colleges across the nation today, the ratio of men to women is heavily weighted toward men: 59 percent to 41 percent. This is due to the strong reputation in the sciences, math, computer science, technology, engineering, and business that Lehigh enjoys. Historically, Lehigh has attracted a majority of its students from the neighboring states of New Jersey, New York, Delaware, and Maryland, in addition to Pennsylvania. As a result of a strategic outreach effort in recent years, its composition includes students from 47 states and 44 foreign countries. It also has broadened the socioeconomic and racial diversity of recent entering classes, as well as encouraging talented women to apply.

Lehigh is situated on a 1,500-acre campus overlooking the Lehigh Valley, replete with imposing Collegiate Gothic buildings as well as modernistic facilities for its advanced scientific and technical studies. The Zoellner Arts Center is a state-of-the-art facility that attracts a broad array of professional musical and theater artists from nearby New York, Philadelphia, and Washington. Students participate in the many artistic activities available through classes and student-run organizations. The business college also enjoys its own facility that was designed with classrooms that enable the faculty to teach the case-study methodology, lecture halls to host speakers from the world of business and industry, and the most recent technology utilized in the global business community. The only complaint students are likely to voice about their campus is the steep hill on which their dorms and classroom buildings are situated and the hikes up and down required to maneuver daily life.

Beyond the traditional emphasis on demanding academics and professors' engagement with their students are the popularity of Greek life and athletics that involve a large proportion of undergraduates. Forty percent of men and women belong to fraternities and sororities, and many of them live in the Greek houses. Social life on the weekends tends to center around the parties sponsored by the Greeks. A large number of students participate in twenty-five Division I varsity teams. Students enjoy turning out to support the teams, particularly for one of the oldest intercollegiate competitions in all of college supports, the annual football match with their neighbor, Lafayette College. The athletic facilities both for varsity athletes and intramural participants and fitness fans are excellent, in the opinion of undergraduates. For those students who are not interested in the Greek system or athletics, there are many social organizations that center around specific interests.

Another longstanding tradition among the student body is the heavy alcohol consumption that is perceived as a badge of honor. Students are expected to work hard, very hard, at their studies during the week, and then blow off the stress on the weekends. Observes a sophomore, "Everyone here works hard and plays hard; if you come to Lehigh you know you are going to get the full college experience." An engineering major commenting on the special sense of place that students feel about their school captures the essence of Lehigh: "The university's traditions really make Lehigh a special place. Young alumni, older alumni, and current students can all share in a common experience, and while the world changes, Lehigh adjusts to the change, but also keeps its core values and traditions alive." Lehigh presents a serious learning environment with unlimited academic opportunities, a chance to experience campus-centered college life in all its aspects, and to prepare for a successful professional career, this being the ultimate goal of all Mountain Hawk students.

What the College Stands For

Lehigh's mission for its students and the institution as a whole is straightforward: to provide a rigorous liberal and scientific education for leadership roles and practical service to the community. The combination of intensive teaching engagement with students, collaboration among the different faculties and the interdisciplinary studies that benefit students, and opportunities to learn through research are the means for achieving this mission. Internships in a wide range of fields, community service programs, and leadership development through the multitude of student-run organizations are generously supported by the university to deepen undergraduates' personal, emotional, and intellectual development.

Curriculum, Academic Life, and Unique Programs of Study

Arts and Sciences students can choose degree programs from among more than fifty disci-

plines in the college's eighteen academic departments, and in the interdisciplinary programs with the college or with the business and engineering schools. Students can also design independent majors with the guidance of the faculty. Although the curricular requirements are relatively flexible, students must complete several distributional and foundations courses. These include, in the first year, two semesters of English that stress both composition and literature; a college seminar with reading, writing, and discussions of topical subjects in a small group; and a one-credit course called Choices and Decisions that is led by the students' advisor and reviews transition issues of college life and the individual's personal and academic responsibilities. In order to graduate, students must complete courses in mathematics, arts and humanities, natural science, and social science. In the junior year students must complete an intensive writing course.

The College of Business and Economics has a core curriculum that all students must complete. These include first-year English writing and composition, calculus, statistical methods, and a number of economics and introductory-business-principles subjects. The next three years build on these foundations with higher-level courses in finance, accounting, business principles, organization, and marketing. The last two years of study leave room for more electives and cross-registration with the other colleges.

The College of Engineering offers eleven distinct specializations, and core requirements apply to all of them. Common foundation courses include advanced mathematics, chemistry, and physics; biology is a requirement for bioengineering majors. Cooperative training programs and study abroad for full credit toward the diploma are available to engineering students.

Unique programs include a free fifth year of

study toward a master's degree; a combined five-year B.A./M.A. program with the graduate College of Education for students who want to prepare for a career in education with full accreditation to teach; a seven-year B.A./M.D. medical program with the Drexel University College of Medicine; a five-year B.A./B.S. engineering dual-degree program; fifty recognized study-abroad programs; and eight-month paid cooperatives in major companies and nonprofit organizations with full credit toward the Lehigh diploma.

Major Admissions Criteria

In the most recent year, 12,155 students applied, 3,882 (32 percent) were accepted, and 1,116 enrolled. The middle 50 percent range on the SAT critical reading was 580 to 670, and the math was 640 to 720. The university has a generous financial-aid program and meets the financial need of all admitted students. It also awards 400 merit-based scholarships and 37 athletic scholarships. An early-decision admissions program is in place which attracts a large number of students.

The most important factors in the admissions review are the level and quality of the high school curriculum, and academic performance as reflected in the overall GPA and class rank (if available). Either SAT or ACT admissions tests are required. Teacher and counselor recommendations, personal interviews with an admissions officer or alumnus interviewer, the quality of the personal application, and special interests and talents are additional major factors taken into account.

The Ideal Student

In the words of one undergraduate, "The student who does best at Lehigh is the person who strives to better him/herself. No one at Lehigh is going to hold your hand and walk you through the process. Those who are ambitious and willing to take the first step and seek out the many resources when needed are the students who succeed." Others stress the need to develop good work habits, time-management skills, and the ability to separate study and academic life from social time. A marketing and communications major believes "the people who do best at Lehigh are those who are actively involved in organizations around campus, and know how to manage their time wisely." Still another student says that "students who put themselves out there and get involved on campus will do better than those who focus solely on studying." Undergraduates emphasize the importance of being well prepared to cope with the academic expectations of the faculty and the level of the curriculum, whatever their field of study. They encourage high school students to take advanced-level courses in their junior and senior years in the subjects that relate to their intended field of study.

Student Perspectives on Their Experience

A junior science major feels that what is special about Lehigh is its balance: "In talking with my friends from high school who attend other colleges, they do not seem to experience the balance we do at Lehigh. Every student on the campus not only strives for academic excellence but also is involved in campus activities and leadership opportunities while also developing a social life. Also, Lehigh has a very strong alumni base. I am very confident that when I enter the working world, I will be able to use a connection with an alumnus to help guide my path." States an upperclasswoman, "The students are so passionate about the school and its mission, and that transfers into the next parts of our lives. Lehigh alums are always eager to help out fellow Lehighers, and in this ever-changing world, it is nice to know that there is a group of people out there who will always be willing to lend a hand. The students are incredible, the faculty is highly qualified and eager to teach, and the college has more to offer than I ever thought possible." Some students find the popular fraternity and sorority scene socially exclusive and irrelevant to the main purpose of the college experience, while others find it an invaluable part of their development. One young woman's view is shared by many of her friends: "I have gotten a great deal out of my involvement in Greek life. I have been given various leadership opportunities through my sorority that I would not have had otherwise. I have developed my communication, organizational, and motivational skills. Moreover, I am always impressed by the efforts of the Greek chapters to improve the Lehigh and surrounding communities. I find that Greek and non-Greek Lehigh students alike really step up to improve the area in which we live, to give back to the university and community that has given us so much." Many students find the location a positive feature. While Bethlehem does not have much to endear itself to students, the reasonable proximity to Philadelphia and New York makes it easy for those who want to get away occasionally from the campus bubble.

What Happens after College

Lehigh graduates perennially fare well in acceptance to top graduate schools and employment. The rigorous academic standards and particularly the foundation in the sciences, math, computer sciences, and business and economics provide key skills that appeal to graduate schools and employers. Lehigh's highly active alumni network also proves helpful to young graduates as well.

In the most recent graduating class, 30 percent entered graduate school directly; 64 percent were employed by 290 business, technical and nonprofit organizations; and 5 percent entered the military. A large number of alumni enroll in professional and other graduate schools at a later date. Lehigh maintains an active professional career and graduate school placement office that students state they find helpful.

Macalester College

1600 Grand Avenue
St. Paul, MN 55105-1899

(651) 696-6357

(800) 231-7974

macalester.edu

admissions@macalester.edu

[
Number of undergraduates: 1,912
Total number of students: 1,912
Ratio of male to female: 44/56
Tuition and fees for 2008–2009: $36,504
% of students receiving need-based financial aid: 67
% of students graduating within six years: 87
2007 endowment: $675,987,000
Endowment per student: $353,549
]

Overall Features

Macalester is decidedly more than the sum of its parts. A small liberal arts college of some 1,900 men and women, it has a deep and abiding commitment to a global perspective in its curriculum, student body, and special programs that is unmatched by the majority of its peer institutions. Inherent in its approach to education is a very culturally diverse student body and wide-ranging academic programs that include international experiences and internships for a majority of its students. Global sustainability, which includes environmental, social, economic, and human concerns, is a thread that runs through the academic and out of classroom offerings.

Macalester was founded in 1874 by the Presbyterian Church to bring higher education to the settlers of the newly recognized state of Minnesota. Since then, it has grown from a well-regarded regional college into a highly respected college that draws its students from every state in the country and more than eighty-five foreign countries. Today the college is identified with a cluster of highly selective undergraduate liberal arts colleges with a liberal bent. These include its neighbor, Carleton; Colorado, Grinnell, Oberlin, and Vassar Colleges, and Wesleyan University. Macalester is one of only a handful of selective undergraduate-only liberal arts colleges located in a metropolitan setting, a particularly appealing one, which provides unlimited opportunities for community service, internships, cultural, and social outlets.

Students perceive its environment favorably; as one student puts it, "Macalester benefits from a combination of big-city resources and everything that is great about a small liberal arts college—small classes, a sense of community, professors deeply committed to teaching. The setting matches the attitude of the college in that socially aware students prosper at Macalester." Observes another student, "People are always looking for better and new ways to utilize the amazing city, whether it is school-sponsored trips or easy ways to get to and from campus for a concert. Most students come here because of the Twin Cities, so actually having them play a role in our time at Macalester is important to us."

Beyond its appealing location and small size, Macalester's strong sense of community, commitment to building a diverse student culture, and generous financial assistance program attract talented students from a wide range of economic, ethnic, and religious backgrounds. Twenty percent of the students are of color and 12 percent are international. A majority of students share in the opinion of this junior: "Macalester is truly a uniquely cosmopolitan place in the midst of the 'Minnesota nice' midwestern environment. People wave and say hello to people they have never met, simply assuming that the other person would be great to know. I personally find the lack of arrogant elitism highly refreshing." The campus is set in a pleasant, quiet section of Minneapolis with the vast University of Minnesota main campus within walking distance. Activities, music, and food geared to a student population abound in the area.

Macalester students are known to be smart, active, and engaged, and concerned about social, political, and ecological issues. By the time they graduate, over 90 percent of Mac students will have engaged in local community work through course studies, applied research, work-study, or civic leadership. More than 250 students each year earn academic credits for internship experiences in the Twin Cities. Macalester's students do not experience the social and intellectual bubble effect common to many of the small liberal arts colleges located in small towns and rural communities. More than 50 percent of all students pursue studies across the entire range of academic disciplines abroad for at least one semester and in many cases for a full year. The college's financial-aid policy ensures that every student who desires to go abroad will have the opportunity.

Macalester has become much more selective in its admissions and more diverse in all respects over the past fifteen years. There is a wide range of new academic programs that reflect the college's commitment to preparing its graduates for the world they will encounter, including Chinese and Arabic, and new concentrations in such areas as global and community health, and human rights and humanitarianism.

There is no Greek system on campus. Students tend to be joiners, with a great many social, civic, and athletic organizations available to them in addition to the community service and internships that keep them very occupied. There are ten varsity men's and women's teams that compete in NCAA Division III and the Minnesota Intercollegiate Athletic Conference.

What the College Stands For

Macalester's administrative leaders articulate clearly its mission and ambitions for the institution and the students it educates. In the words of the college's president, "As our mission statement indicates, our central goal is to educate students at the highest level of academic rigor while also instilling in them an understanding of the importance of internationalism, multiculturalism, and service to society." The chair of the extensive international studies program wants students "to achieve a deeply informed transnational citizenship, and multidisciplinary abilities to understand and then act in global situations." Civic leadership at the local, national, and international levels underlie all the academic and campus programs, which are continually designed and reviewed to ascertain if they are meeting this ultimate purpose. The college places a premium on sustained attention to student learning by "a high-quality faculty who bring active engagement in scholarly work to their teaching and mentoring." The president believes that "faculty members are engaged with their community in a profound way. New courses and majors have been developed, and new activities and organizations for students have appeared. Everything changes, but the core values (academic excellence, internationalism, diversity, and civic engagement) remain the same."

Curriculum, Academic Life, and Unique Programs of Study

A very flexible set of distribution requirements are in place. These include two courses in the social sciences, one in quantitative reasoning, one in a writing seminar or workshop, two in natural sciences and math, two in fine arts and humanities, and two related to cultural diversity. In the senior year, students complete a capstone experience that can be an independent project, a performance, a portfolio of artworks, or original research. International studies is one of the most popular majors. Others include language studies, social sciences, and life sciences for those planning a career in the health fields. A senior

science concentrator advises entering students "to take one of the wide variety of first-year seminars and to try a bunch of different fields of study during your time here. One of the great benefits of Macalester is its many strong departments. Where and when else can one take Portuguese and American Environmental History just for fun?" Through the help of its internship office, in the past school year 303 students completed academic internships working with 216 community partners, sponsored through twenty-five different Macalester College departments. The college's Institute for Global Citizenship assists students in creating international study programs and research projects.

Major Admissions Criteria

In the most recent year, 4,967 students applied, 2015 were accepted, and 485 enrolled. The SAT range was: critical reading 620 to 720, math 630 to 710. The ACT range was 27 to 32.

There is an early-decision admissions program. Personal interviews are available and are encouraged. Both need-based financial aid and merit-based awards are offered.

In the words of the dean of admissions, academic excellence as demonstrated in a rigorous high school curriculum with strong grades is the most essential quality sought in applicants. Admissions test scores count as well but can be less important than indications of a commitment to contributing both to the school community and the larger community. The admissions committee looks for students who do not simply follow trends; students who are not looking for traditional college cliques to establish their identity, students who are intellectually curious and socially adventurous; and students who like being different, slightly edgy.

The Ideal Student

The faculty believe that the students who succeed at Macalester are self-disciplined, ambitious, open-minded, and willing to push their personal boundaries beyond their particular social, cultural, economic values and priorities. A desire to learn from others in a community composed of many different kinds of peers is important. Since all faculty teach in small class settings and encourage their students to engage in the classroom and take on independent work, Macalester is not a place where one can hide in the back of the classroom or fail to complete required work. Advises a senior prelaw major, "Be ready for anything and be open to all types of experiences. The only type of person who would not end up working well at Mac is someone who is closed off and happily content with life as it is. You must be willing to grow—a lot." A senior psychology major believes that "Macalester is an ideal school for students interested in getting involved in their community, exploring ideas of multiculturalism and internationalism, and working in an academically rigorous but collaborative environment." States another student, "A passion for an academic area or some aspect of the community or larger world gets students engaged and happily fulfilled."

Student Perspectives on Their Experience

Students use these words to describe their college and their peers: *open, progressive, international, engaged, eclectic, beautiful, stimulating, rigorous, diverse, activist-oriented, intelligent, friendly,* and *dedicated.* One senior believes that "what Macalester does best is build community. This is, in part, a function of the size of the college, but more so, students come here because of the present students and the atmosphere the college leaders have created for us." Professors are viewed as very friendly, accessible, and sincerely caring about their students succeeding. One student who transferred from a large research university at the end of her first year has this perspective on Macalester: "I came to the realization that I wanted to spend my formative college years in an environment that would challenge me both inside and outside of the classroom. I wanted to meet individuals of vastly different backgrounds and perspectives from my own. My subsequent search brought me to Macalester. The benefits of a small college and a personalized education came as icing on the cake."

What Happens after College

Within five years, the majority of graduates go on to graduate school in a broad variety of fields, the most popular being international affairs, environmental studies, public health, medicine, architecture, education, economics, languages, law, and business. Popular destinations are the University of Cambridge, Chicago, Columbia, Georgetown, Harvard, Johns Hopkins, the London School of Economics and Political Science, Princeton, and Yale. Many recent graduates have gone directly into community-action programs that include Teach for America, the Peace Corps, AmeriCorps VISTA, Human Rights Watch, and the United Nations. Others work at the business organizations and nonprofits where they served in internships as undergraduates.

A sample of the class of 2008 graduates

finds one pursuing a master's degree in finance at the London School of Economics, one in a Mississippi elementary school through Teach for America, one studying at the National Theater Conservatory in Denver, and another working in China in a health-advocacy organization. In recent years graduates have furthered their studies through three Rhodes scholarships, twenty-six Fulbright scholarships, twenty National Science Foundation fellowships, two Mellon fellowships, twelve Watson fellowships, and two Truman and three Goldwater scholarships.

Middlebury College

The Emma Willard House
Middlebury, VT 05753-6002

(802) 443-3000

middlebury.edu

admissions@middlebury.edu

Number of undergraduates: 2,422
Total number of students: 2,422
Ratio of male to female: 50/50
Comprehensive fee for 2008–2009: $49,210
(includes room and board, tuition, and fees)
% of students receiving need-based financial aid: 45
% of students graduating within six years: 93
2007 endowment: $936,354,000
Endowment per student: $386,603

Overall Features

Middlebury has long enjoyed a reputation as a strong academic college in combination with a very upbeat social life and physically attractive environment. Its Vermont location and beautiful campus have made it an appealing place for generations of socially outgoing, athletic, intelligent young men and women who find such a combination of salubrious elements at the top of their wish list. Similar to its peer institutions, the small residential liberal arts colleges, Middlebury concentrates on delivering a strong foundation in the critical skills of thinking analytically and creatively, clearly expressing ideas orally and in writing, and appreciating the great ideas and issues of the past and present through exposure to the arts and sciences. The vehicle for achieving these goals has always been through the presence of top faculty who teach and inspire students in the small-class situation. This has been the college's mission since its founding in 1800. An honor code

is well entrenched in the academic life of the community. Exams are unproctored; students must take responsibility for their own behavior and report their peers for infractions of the code. A student-led judicial council hears cases of violations and renders judgment.

Middlebury is on a steady ascendancy in the selectivity of the student body, the caliber and breadth of its programs of study, and the physical resources for enhancing the academic, social, and physical well-being of its undergraduates. As impressive evidence of meeting its goals, the college's trustees have raised significant endowment and capital funds that have helped to build and renovate many campus structures. Most residence halls have been built or refurbished within the last two decades. A large arts center was built in the early 1990s, which includes an art museum. A newer academic building, Mc-Cardell Bicentennial Hall, is an award-winning environmentally friendly center for natural and social sciences. The most recent major campus addition is a multifaceted library and technology center.

Middlebury's administration has stated that the goal is to attract the very best of the high school talent pool and to compete directly with its closest rivals—Dartmouth, Williams, and Amherst. In order to help expand its academic and extracurricular offerings, as well as to bring greater multicultural diversity to the campus, Middlebury has increased its enrollment at a gradual pace to some 2,400 undergraduates. More faculty were hired to maintain the traditional access by students to their instructors and the small-size classrooms. Other signs of change have been the removal of the Greek system, replacing it with social houses based on students' interests, and the development of a commons system that brings students and faculty together in five clusters to share interests and ideas. The hope is that the social fragmentation and antiacademic attitude of the old fraternity system will be replaced by a more intellectually positive ethos on campus. As the college states, "Middlebury's unique residential system exemplifies the college's conviction that an excellent liberal arts education takes place around the clock—as easily over dinner as in the classroom. The residence halls are grouped into 'living-learning communities,' called commons. The commons combine the academic, social, and residential components of college and foster close and abiding relationships, not only among the student residents, but also among the faculty and staff who are part of their commons. Students are part of the same commons for all four years, and live in dorms associated with their commons for their first two years at Middlebury." Many students are attracted to Middlebury both for its extensive off-campus programs of study and for its athletic and recreational options. Two-thirds of

all students choose to study abroad during their four years. The college has long excelled in intercollegiate ski competition and Division III ice hockey, winning many championships in both sports recently. There are 31 varsity and many club sports for the serious to casual male and female athlete and some 28 percent compete in varsity sports. The Middlebury Ski Bowl is one of the best collegiate facilities in the East, and the gymnasium and golf course get positive reviews from students. The vast majority of students take advantage of the location to ski and do other outdoor activities.

What the College Stands For

One institutional leader sets out Middlebury's goals: "We seek to educate students broadly in the liberal arts in the expectation that they will be leaders in whatever they may choose to do. We seek well-rounded students who will contribute to the college during their four years here, and contribute to society thereafter, and who will, not just through narrow technical expertise but through a range of skills developed here, lead lives of value to themselves and to society at large."

Middlebury sees itself as having a particular identity and continuity that is shared by students, faculty, the administration, and graduates. "In its essential, human features, the college has changed very little, and that is part of its appeal," says an administrator. "This is still a place where able, well-rounded students can come and be part of an active community that encourages them to broaden the limited reach of their own understanding. Our special claims continue to rest upon size and scale, the encompassability of this community, and the special type of education that results in such a place. This has less to do with enrollment numbers than with a commitment to preserve the essential spirit of the place. Nor is the nurturing of that spirit dependent upon permanent institutions or structures: Fraternities had their place, now they don't; we are decentralizing our student deans' offices and now have a system of continuing residential communities with dining (the commons system). These may seem radical departures. In fact, they are done in order to preserve those things that make Middlebury—and Middlebury students and alumni—recognizable to one another over the years. So in one sense we have changed a great deal; in another we have changed hardly at all."

As Middlebury grows, the college will seek to maintain and continue to develop its residential liberal arts community and identity, an aspect of itself that one institutional leader sees as crucial: "Keep in mind that, though we claim to be small and intimate, we are larger by far than most of the high schools attended by most of our students. To them we seem large, bureaucratic, impersonal. We need to do all we can to combat those tendencies. The creation of commons communities offers the best antidote. Moreover, the compelling claims of Middlebury will continue to rest upon the added value that comes from human interaction. Technology may broaden the definition of what it means to be a student. . . . If we are to continue to be valued, [we] must offer education on a human scale, with human contact. This will be increasingly prized in the future."

Some of the things that make Middlebury special, according to one administrator, are "our location in Vermont. Our emphasis upon 'peaks of excellence,' which truly do define us and our special strengths: language, literature, international studies, environmental studies, our residential system, and opportunities for 'applied experiences.' The size and composi-

tion of our community, with its own brand of diversity. The small-town rural setting, the large number of international students, and Middlebury's outreach through Bread Loaf [the highly regarded summer English school and writers' conference] and the language schools. Not much doubt where you are when you're here."

Another dean notes that what helps students to succeed at Middlebury are "relationships with mentors of all kinds: faculty, staff, coaches, alumni, and townspeople. These relationships are facilitated by a size and scale that allow for multiple daily contacts with students both in and outside of the classroom. Everyone understands and supports the need for these relationships." Says the college, "The college's commitment to environmentally responsible and sustainable practices is apparent at every turn—from the college's commitment to become carbon neutral by 2016; to the natural landscape design of the campus and the college's recycling center, powered by a wind turbine; to the newly refurbished Franklin Environmental Center at Hillcrest; to the full bike racks outside every building; to the dining-hall waste that is composted and then reinvested in the college's landscape and greenhouse."

Curriculum, Academic Life, and Unique Programs of Study

Students must meet a set of general distribution requirements in keeping with the traditional liberal arts education. Specific requirements include the successful completion of two intensive writing seminars by the end of the sophomore year. The academic calendar includes a January term wherein all students engage in one course of study involving an in-depth project under the tutelage of a professor in the particular department of interest. Middlebury has a longstanding reputation for the excellence of its English, writing, history, philosophy, and language departments. More students major in these disciplines than is typical of most liberal arts colleges today, especially in foreign languages. The study-abroad programs are utilized by two-thirds of the student body, a percentage well above that of most of the top liberal arts colleges in the country. On campus, Middlebury offers more than 850 courses and forty-four majors, maintaining a 9 to 1 student-faculty ratio and an average class size of 16. Middlebury's overseas offerings include common and unusual locations from Alexandria to Hangzhou, Irkutsk to Yaroslavl. The college also is affiliated with the Monterey Institute of International Studies in Monterey, California, which offers "programs in business, language learning and education, peace and security, sustainability and the environment, and trade and development."

For English and writing majors, the famed Bread Loaf summer program is readily available for advanced study and training. Language-immersion summer programs are also first-rate. As one administrator notes, Middlebury is "unique" because it is "a small liberal arts college that is also an international university— with summer programs in eight languages, schools abroad in five countries, and graduate programs in English literature in four locations." Another popular major is the interdisciplinary environmental studies program, which appeals to students who want to combine their love of the outdoors with concern for environmental problems. The six-story, $47 million science center makes this field and science studies in general stronger than ever. Middlebury has a host of intellectual opportunities for the serious student that will continue to expand over time. As one administrator notes, discuss-

ing change and its causes at Middlebury, "the boundaries of knowledge have been altered; boundaries separating the traditional disciplines have shifted or crumbled. More than half our students now study abroad, now take double, joint, or interdisciplinary majors. Our students and their families expect certain services: career advising, counseling, every club or sport under the sun. None of these changes has been the result of a single act or decision by president or faculty. And yet these forces are prompting us to change. Technology is another example, prompting us from the outside to adapt to it, and, where possible, to exploit it on the inside. We do our best to choose how to respond to these various demands and to define a vision that embraces these choices and makes clear our priorities."

Major Admissions Criteria

Middlebury has joined the ranks of a growing number of outstanding liberal arts colleges that make submission of the SAT optional or flexible in consideration for admission. Applicants are required to submit either the SAT, ACT, or three SAT subject tests in different areas. This is in keeping with the admissions committee's emphasis on the high school transcript, meaning the strength and demands of the four-year course load, grades, teacher recommendations, and the applicant's personal application. Middlebury developed a creative approach several years ago to the dilemma of too many talented applicants for too few available spaces. It admits approximately 100 first-year candidates who will begin their studies in the winter term. This expands the admitted pool and fills the campus vacancies created by the large number of present undergraduates who leave for study-abroad programs.

Top admissions criteria listed by one dean: "The quality of the work done in high school is number one. The standard measurements of ability are number two, but the quality of the person is a must." "Middlebury selects students for admission who show the potential and desire to succeed in a rigorous academic environment," notes the college. "For that reason, whether they come from small-town America, a big city, the suburbs, or from outside the U.S., Middlebury students find that they have much in common."

The Ideal Student

The student body has long been typecast as traditional-minded, white, upper-middle-class preppies who want a good education in a friendly, social, outdoors environment with good teaching available. There was no arguing the truth of this generalization in the past. However, efforts by the trustees and educational leaders of the college to broaden the base of the student body and the intellectual tone of the community is evident in the new buildings and academic programs, the expansion of the student body, and outreach on the part of the admissions team to attract more nontraditional men and women of talent. Some 45 percent of Middlebury students receive financial aid thanks to its healthy endowment and need-blind/full-need-met admissions, which are being used to diversify the community. The class of 2012, for example, contained about 19 percent U.S. students of color and 10 percent international students. Half the class were public high school graduates, 8 percent legacies, 6 percent Vermonters, 72 percent non–New Englanders, and 8 percent first-generation college students. The class represented 49 states and 36 countries. Middlebury is a very appealing place for the smart high school graduate who desires

immersion in a community of outgoing, physically active, moderate-to traditional-thinking students who take their education seriously but not obsessively. There is less engagement in community service or political activism here than in a number of the other campuses described in this book. As one dean notes, "Perhaps Middlebury's beautiful physical environment attracts students with an appreciation of nature, the outdoors, and adventure."

Student Perspectives on Their Experience

Students choose Middlebury for its academic programs, its location in the Vermont wilderness, its athletic programs, and its size. "The school excels, beyond all else I think, in offering a faculty and staff that are approachable," says a senior. "Professors, deans, custodians, security officers, advisors, and even trustees are all there to support and provide. They interact with the students and want to meet and engage us. This creates an atmosphere in which I found I was able to learn from people on many different levels, not just in the classroom." A junior says, "I think that in general Middlebury provides an amazing environment to learn in. Its small community feel heightens the college experience. Also, its internationally oriented academic program prepares students for participation in the current global workforce." Another junior adds, "The intimacy of the college is great—I've had dinners at professors' houses and I feel that teachers and deans are genuinely interested in helping us succeed. Classes are small and professors are encouraging. This school is just about perfect! We don't call it 'Club Midd' for nothing. . . ."

The students who do best at Middlebury are those who are involved in academics, but who look beyond the classroom as well. As a senior notes, "The same people who do well at other schools do well here: those who are focused on their work and what they want to get out of their education. . . . The students who prosper and thrive at Middlebury are not always those who have the highest GPA, but those who have a very grounded perspective on themselves and where they are, who know what their interests are, and who take advantage of the rich resources—infrastructural, academic, social, and human." It is the interactions between people and through these resources that help students succeed. "The students who come here are smart and always deeper and more rich with experience than at first meets the eye. Learning from different strengths and using the infrastructural facilities is the greatest catalyst to success. If you're looking for more solid examples: interactions at the dining halls; dramatic improvements at the athletic complex; a Center for the Arts designed to encourage collaboration between music, theater, and dance; a new science center that incorporates the geography and psychology departments as well as physics, chemistry, bio, etc. . . ." A junior adds, "The person who does best is the one who finds the happy medium between academics and social life. I think what helps most of all is participating heavily in as much as possible. Grades help, but I think in the end it's connections that open doors." A senior says those who do best are "students who are *active* in all senses of the word! Those who are willing to work hard academically, who are excited to think, learn, and work with others, who like outdoor activities, who are excited to try new things. Those with a lot of energy and enthusiasm for life." A sophomore points to "the person who always expresses his mind and is willing to try new things. Middlebury helps this person de-

velop in his own direction. Middlebury is like a guide that can point someone in a thousand directions; the student just has to experiment to find the right path. Ambition is the number one reason students succeed at Middlebury. It does not have to be ambition to do well academically, it simply has to be an ambition to follow what you want in life."

Again, faculty and classes get raves from students and form the core of their experience at the college. Says a junior, "My professors have all been wonderful, and I know that a small college like Middlebury can offer much more personal attention than a larger school. I took an Anthropology of Religion class with fourteen other students. Our professor shared stories of his trips around the world and led amazing hands-on activities that would have been impossible with a larger group. It honestly changed my life. Every class and every professor was outstanding, and it's great to be on a first-name basis with your teachers. When you're connected, it makes learning a lot more fun."

As Middlebury continues to change over time, particularly in its social and residential mission, it will confront its alumni and their views on what is right for "their college." The administration is well aware of the need for positive change and looking to the future. Students pick up on this as well. As one comments, "I would like to change the relationship between the college and some of the more disgruntled alums who see many of the recent changes as a threat to the college they once knew. This is a dilemma faced by every college when initiating change, but is truly a sign of how connected the alums feel to Middlebury." Another student, however, says, "Ironically enough the things I would change are the changes. I like the school the way it is and disagree with the direction it seems to be taking." Some feel the college is slow to change, slow to react to student concerns. And diversity at the college, while not extensive, is, according to many students, quite well developed in many respects and belies its public image to some extent. "Though Middlebury has a reputation of being one of the most homogeneous colleges, the diversity on campus is greater than is commonly perceived. The college is currently taking great steps to change that image and to increase cultural, ethnic, racial, socioeconomic, and other forms of diversity on campus. Predominantly the current conception of Middlebury College is a white kid with a Jeep Cherokee from a New England prep school."

The commons system, one new aspect of campus life, was controversial when implemented, but at least one student is a strong supporter of its potential: "The commons has been very important because I see my voice making a difference. People believe one voice cannot do much, but I found that is wrong—it can change things. I helped create a more relaxed and respectful security system. I helped plan a dining hall, lounges, study rooms, a green, and underground parking in and around one of the current structures. I helped create the renovation ideas for an improved quality of life and a freshman dorm. Most of this was done through the commons and it has been important to me this year."

"Middlebury is a college that provides opportunity," says a senior. "A student who wants to do something can do it here. The environment is idyllic. The people are friendly, the natural surroundings sublime. You can get out of an education here what you put into it." Another senior says, "The school provides ample opportunities for any person's interests. It also admits students with a wide range of interests and strengths and the friendships that are formed are amazing. Middlebury provides a close-knit community that is felt and understood by everyone. Everyone seems to share the feeling of happiness and contentment, regardless of how stressful the academic load— which is very stressful. Somehow, the school provides an environment of relaxation. It also provides places where students are encouraged to hang out together, places for all students to enjoy regardless of interests." One graduate, and winner of a Watson fellowship, says, "Everyone should know that at Middlebury students are enthusiastic and excited not only about their academics, but about their activities. Every student on campus is committed to some aspect of Middlebury. There are numerous chances to share, to help, to play, to learn, and to create."

What Happens after College

Says one administrator, "We keep close track of our graduates. We talk with them when we travel and when they return to campus. We ask about the ways in which Middlebury prepared them for the lives they now lead. And to a remarkable degree the responses are highly positive. Our students, in other words, are our students for life. To be sure, they have no way of knowing whether, had they decided to attend a different college, things might have turned out differently for them. They know only Middlebury. So comparative data can tell us only so much (and very little). But we do know what they think of Middlebury, and on the whole, their success and their attribution of that success offers us a good measure." A dean says that Middlebury graduates "seem to be prepared to do just about anything they want—and they do just about everything after leaving Middlebury. Increasingly, there are no geographic boundaries to what they decide to do, whether this involves work, study, or travel." A high percentage of alums will choose graduate school over time, and meanwhile will pursue over sixty different fields of endeavor within a year of graduation from the college. Law and education are two prominent areas of choice, as are English and foreign language and international arenas. Middlebury graduates do well in securing scholarships and awards.

Mount Holyoke College

Newhall Center
50 College Street
South Hadley, MA 01075-1488

(413) 538-2023

mtholyoke.edu

admission@mtholyoke.edu

[
Number of undergraduates: 2,185
Total number of students: 2,188
Ratio of male to female: 100% female
Tuition and fees for 2008–2009: $37,646
% of students receiving financial aid: 60
% of students graduating within six years: 83
2007 endowment: $615,376,000
Endowment per student: $281,250
]

Overall Features

Mount Holyoke, founded in 1837 as the first college dedicated to educating young women, has a long and honored history. It has stayed its course to deliver a liberal arts education of the first rank and foster in its students a strong social conscience. The classroom and the campus life are centered on the commitment to social service as a future career or through volunteer activities. This tradition helps explain the extraordinary generosity toward qualified students who cannot afford the price of the education Mount Holyoke offers. Three-quarters of the student body receives some form of financial assistance, truly an astounding ratio compared to that of most colleges in the nation. The direct consequence of this generosity is the diverse population of students of academic talent the college enrolls annually. The young women who elect to attend Mount Holyoke tend to be serious learners who cele-brate the idea of a woman's college where they can be leaders in and out of the classroom. They recognize that all of the resources of the institution are there for them to take advantage of, and this they do. True to the college's traditions, Holyoke women tend to be engaged in social causes and activism, be they political, gender, racial, or economic.

At the same time, a Holyoke student is more focused on her studies and her personal interests than on the traditional party life enjoyed at other colleges. There is plenty to do off-campus in this college-laden area. Some traditions do not fail: the men of Dartmouth, Yale, Amherst, and Williams still venture forth to the Holyoke campus for dates on the weekend. Since there is no history of sororities on campus, personal life centers on the dormitories that students help to run. Quiet hours are elected by students in the residence halls to ease the academic workload. Faculty are respected for their expertise and ready availability to students. All classes are small in size and active participation is expected of all students. An honor code is in place and sets the tone for student attitudes. Exams can be scheduled individually and are self-proctored.

Mount Holyoke has a longstanding reputation for excellence in the sciences among liberal arts institutions. One-quarter of the student body majors in the life sciences such as biology and chemistry as a foundation for a profession in medicine, teaching, or research. The strength of the English department, however, makes it the most popular major on campus. The campus is very attractive and will appeal to the aesthetically traditional student. The classrooms, dormitories, arts, and athletic facilities are outstanding. Another historical tradition is the presence of excellent music groups—especially the various singing groups—and the music and theater programs. There is a superb equestrian facility and a golf course. But if any academic course or activity is not available on campus, a student can take advantage of the Five Colleges group, which is composed of Mount Holyoke, Smith, Amherst, Hampshire, and the University of Massachusetts. Mount Holyoke is also a member of the Twelve College Exchange, which enables students to spend a full year of study at one of the cooperating colleges.

Notes a dean, "Mount Holyoke has always been an international campus, with a high proportion of international students and many students studying abroad. We now have a Center for the Environment, a Center for Global Initiatives, and a Center for Leadership and the Liberal Arts." President Joanne Creighton says, "Mount Holyoke is about to launch the new Nexus program, which integrates academic study with real-world experiences. A student

will be able to elect a Nexus program concentration as a supplement to her liberal arts major. Through this program, she will explore emerging areas of cross-disciplinary inquiry; apply her knowledge to problems in the world through experiential learning, including internships and off-campus research and community engagement; and complete her undergraduate education with a capstone project."

Argues one senior administrator, "Our identity as a woman's college—the oldest in the world—has been a hallmark of our history. Like most of the women's colleges, we have long had a socioeconomic diversity in our student body that other colleges are only now striving to match." Adds the director of communications, "Talking across difference enhances the education for all involved. The college has also long been committed to social justice. Our 2,200 students hail from forty-eight states and nearly seventy countries. One in three students is an international citizen or African American, Asian American, Latina, Native American, or multiracial. Our faculty speak over fifty languages. Half are women and one-third are persons of color. We have many campus programs aimed at helping all of the members of our community feel valued and welcome and aimed at ensuring that all students who matriculate at the college are successful academically. While the commitment to a diverse community on the part of the college is real and impressive, we have not yet reached nirvana on this count. We continue to work at it."

What the College Stands For

The president of the college says, "At the heart of the enterprise is our commitment to linking the liberal arts with purposeful engagement in the world. The liberal arts, 'the arts of thought, perception, and judgment: the arts that foster humanity and civility of spirit . . . it is these arts that Mount Holyoke College places at the center of its life,' to quote from the Principles of the College. Here students get an education that will transform their lives and prepare them for roles they may not have even imagined. Here they develop the confidence and competence, connection and community. Here agents of change are born because idealism is palpable and motivational."

An associate dean says, "Our distribution requirement (general education) asks students to take courses from seven different disciplines distributed across the sciences, social sciences, and humanities. We also expect foreign language and multicultural study. Our forty-nine majors are designed to move students deeply into the study of a particular discipline or into interdisciplinary work."

The president of the college points out that "Virginia Woolf famously argued that, for their intellectual and creative talents to flourish, women needed rooms of their own. A women's college, as a place for and of and by and about women, is an academic culture of women's own, a place infused with its own powerful traditions, norms, and values."

Here are the benefits of attending Mount Holyoke, according to one administrator: "Invigorates intellect and stimulates curiosity and creativity; hones communications and leadership skills—speaking, arguing, and writing; increases technological savvy and global awareness; increases a student's probability of getting into her first-choice graduate school; provides outstanding career preparation; heightens social consciousness and builds character; fosters strong sense of self-knowledge, confidence, and purpose; fosters lifelong friendships and connections around the world; enhances the ability

to make a difference; confers the prestige of being 'a Mount Holyoke woman.'" The president notes, "As the 'founding sister' of the historic Seven Sisters and as the longest-standing higher education institution for women in the world, Mount Holyoke spawned a distinct, influential, and powerful 'other' educational tradition, which, while drawing from the dominant male tradition, bears significant differences from it. This tradition has worked remarkably well in inspiring and enabling women to achieve their fullest potential and to make a positive difference in the world."

The mission statement is short and clear: "Mount Holyoke College reaffirms its commitment to educating a diverse community of women at the highest level of academic excellence and to fostering the alliance of liberal arts education with purposeful engagement in the world." To implement this mission, the college has put forth a set of principles that are deeply reflective of the values of a liberal arts education and notions of serving the public good. The opening and closing paragraphs outline the highest of college ideals: "All human experience is education in some sense, and a liberal arts education does not exclude the sorts of learning that derive directly from the process of living itself. A liberal education differs from other varieties of education, however, because it places at its center the content of humane learning and the spirit of systematic, disinterested inquiry. . . . Mount Holyoke is, specifically, a liberal arts college, concerned in a special way with the three ideas those three words bring together: freedom, learning, community of purpose. As a society of those who have come together for the sake of learning, the faculty and students profess and study those special arts of mind and spirit that they believe can free people—at least from ignorance, and perhaps from other poverties. The liberal arts are the arts of thought, perception, and judgment; the arts that foster humanity and civility of spirit; and it is these arts that Mount Holyoke College places at the center of its life."

Curriculum, Academic Life, and Unique Programs of Study

A reasonably flexible set of distribution requirements is in place to meet the college's aim of exposing students to the breadth and, in their major, the depth of the arts and sciences. Faculty are expected to do research and writing in their discipline to make them better teachers, and all faculty are required to teach. Mount Holyoke strongly encourages students to undertake independent study projects, research projects, or tutorials in a subject. Much of the success of the science majors in moving on to graduate programs comes from the opportunities to work closely with their professors on research projects. There is a wealth of opportunities for off-campus study programs sponsored by the college itself or in conjunction with the Five Colleges group and the Twelve College Exchange. A January term allows students to undertake an internship or independent study in any field of interest.

Major Admissions Criteria

The college says, "Mount Holyoke College seeks to enroll women who are outstanding in intellectual ability and personal qualities; differing in interests and talents; desirous of a liberal arts education; representative of many public and independent secondary schools; and with varied social, economic, and geographical backgrounds. The college looks for evidence of character, originality, and maturity, as well as sound academic training." The college is very individually focused in its admissions process, and candidates will be looked at personally and carefully. The college says, "At Mount Holyoke, we make every effort to get to know you as an applicant. The Office of Admission begins with your transcript, as this reveals your performance over time in a variety of subjects. Next, your essay is examined along with [a] graded paper and short answers. Students will write extensively at Mount Holyoke, so the caliber of writing is important. Through a student's writing, the admission office hopes to understand her more fully. Then, activities and involvement and letters of recommendation are reviewed. The interview report adds another dimension to the application and can be especially useful in evaluating a borderline candidate. There are many factors taken into account when deciding whether or not an applicant is admissible and each of them is examined carefully." One administrator says, "We look for smart, ambitious women who dare to challenge themselves, take risks, and want more from college than the average coed." In 2001 Mount Holyoke joined the ranks of liberal arts colleges that have made submission of standardized tests optional in the admissions process.

The Ideal Student

Mount Holyoke is dedicated to the task of training women for intellectual and social leadership in the larger community. Its environment is one of commitment to learning and active engagement in student government, the arts, and community service. There is life after studies, but the party-animal category of student will not feel at home here. The casual student may find the academic requirements more than she really wants. The demanding Holyoke program is less about being an all-women's college than it is about the traditions of receiving a top education and putting it to use in a meaningful way. Holyoke alumnae are well known for their abilities and their active roles in the community. The network for assisting one another is impressive. The most satisfied Holyoke student is the young woman who has a genuine passion for learning and for exploration, who wants to participate energetically in campus organizations, and has a social conscience that leads her to engage in some cause. Those who have their sights set on graduate school as the next step to a professional career could not do better than to study at Mount Holyoke. A student willing to work hard in a more limited and low-key social environment than that available at many coed colleges, but in a place where she will be supported in every way, and a location where she will have access to ample cultural opportunities, but not those of a big city, will enjoy the enveloping safety and challenge of the Mount Holyoke College experience.

Student Perspectives on Their Experience

Women choose Mount Holyoke for its academic programs and reputation, sense of community, and most often because it is a women's college. Says a senior, "I looked at Mount Holyoke first of all because I knew of its excellent academic reputation, but then I started learning about the benefits of attending women's colleges and became more intrigued. I finally made my decision after visiting the campus for the weekend. I was inspired by the vibrant community of women and felt very comfortable." A recent graduate says, "Initially I wasn't interested in going to a women's college because I assumed it would limit my academic experience. Once I

stepped on campus and met the dynamic women who live and learn here, I knew my assumptions were wrong and Mount Holyoke was the place for me."

The college is praised for its intellectual climate and ability to challenge and support students. A senior says, "Mount Holyoke is great at challenging every student to pursue their own interests, to think critically, and to make informed choices." A sophomore says that Holyoke "provides an endless amount of academic stimulation, support by students and faculty, and leadership roles and activities. Students are all women, so the role of women and their impact on society is highly stressed. The college places importance on the individual students." A recent graduate says, "Mount Holyoke produces intelligent, multitalented women with a strong sense of themselves and their place in the world." Another adds, "Mount Holyoke is a place where traditions are honored and progression is guaranteed."

Those who do best are "women who are intellectually curious and motivated, women who are open-minded, eager to try new things, women who are willing to share and learn from each other." Another says they are "strong, confident students who have done exceptionally well in high school, women who are overachievers, who are talented in multiple aspects." Others point to "anyone with an open mind, and the hunger to be challenged every day in every aspect of campus life"; "women who are willing to work and believe and continually challenge themselves and the world"; and "students who can work independently and balance their activities, sports, and social life with academics."

Student complaints, as might be expected, center around social-life options on campus and meeting men. Other areas of concern are residential availability and options. Nevertheless, one student argues, "I found that the party scene got boring rather quickly. Thank God I was at MHC where there are dozens of other options! Students can attend concerts and dance and theater performances, special lectures, film, poetry readings, and festivals as well as large and small parties." Students encourage prospective applicants to be balanced in their pursuits, and not to try to be "superwoman" by doing too many things at too high a degree of intensity.

The honor code is praised for setting the tone of campus life and creating flexible options for students in taking exams, and international and overall diversity at the college is seen as strong and positive. "The school does an excellent job of creating a diverse, supportive, comfortable environment," says one graduate. "They also look to tomorrow and prepare you for the future. Mount Holyoke can best be summed up with the saying we are 'uncommon women on common ground.' MHC has every country, every state, every religion, and every socioeconomic background within our gates. And we take a lot of pride in this fact. We really try to celebrate diversity, not shy from it." Programs of note include the innovative Speaking, Arguing, and Writing Program.

Students feel empowered and connected at Mount Holyoke, with and by each other, the faculty, and alumnae. "Teachers at Mount Holyoke," says a graduate, "are fully invested in their students. When they see talent in a student they seek to develop and challenge it." A sophomore says, "What makes Mount Holyoke special is definitely the women who are here. I have been so amazed by the brilliance, talent, ambitions, and dreams of my peers." Another adds, "Mount Holyoke gives students a sense of individuality. All of the teachers I've ever

had, even those in a seminar class, still recognize me, at least by face, and many even by name. It imparts a feeling of security that I haven't seen elsewhere. Mount Holyoke, for the first time in many women's lives, enables women to stand up as individuals and speak their mind, accomplish amazing feats, and excel in all areas of their lives."

Here are some other comments of note from students we surveyed: "Mount Holyoke celebrates diversity in an 'all-women's' setting. It provides women of nontraditional college age to fully engage in academic and social life with other students." "We have self-scheduled exams during finals period. It's based on the honor code and relieves some stress during the hardest time of the year. We have milk and cookies (every night in every dorm!) which brings students together and bonds dorms." "Ninety percent of the women at Mount Holyoke weren't looking for an all-women's college. We fell in love with the place and the people. Coming here has changed my life—if I had to do it over, I would most definitely choose an all-women's environment again." "Whenever there is an issue that a group of students feels is important, there is a campus dialogue, but the meetings are always optional so students don't feel they're being bogged down with every issue. The administration is also excellent and sponsors breakfasts for students to chat with the president." "I appreciate that trustees ask students about their concerns through student senate meetings and surveys and takes action to improve conditions, and I like our traditions: Mountain Day, Elfing, DisOrientation, Senior Pub Nights, etc." "I think what Mount Holyoke does best is allow their students to grow. Everyone leaves this school stronger, more eloquent, vivacious, and bold. And seeing strong seniors as a first year makes you want to do better."

"Mount Holyoke students are very focused and independent. At times the campus feels like it is a great ball of stress. The college should not baby students; however, it could try to foster a more collaborative environment between students, not just professors and students." "Mount Holyoke is a secret gem. Your fellow students are future global leaders and your professors are nationally and world-renowned authors and researchers. Your perspective on life will change from the people you meet. The focus on campus is on promoting you, as future women leaders. You will become a more confident, independent woman."

What Happens after College

Notes one administrator, "Among liberal arts colleges, Mount Holyoke ranks first in producing women who went on to receive doctorates in the life and physical sciences. Our students also do well in the world of work, and our younger alumnae are often particularly drawn to nonprofit human and social services professions." The director of communications says, "Typically 75 percent of our students go on to graduate or professional school within ten years. Because our students know how to think critically and communicate well, they are prepared to pursue any career they choose." President creighton concludes, "Our students go on to productive careers and lives of engagement in myriad ways. . . . Our alumnae let us know how much they value their Mount Holyoke educations: They remain loyal and involved throughout their lives, and some 81 percent donated to our most recent campaign."

Mount Holyoke sponsors an AlumNet program with thousands of alumnae career mentors for students. Participation in summer and January internships helps Holyoke students develop career interests and skills. Students win a number of major awards and fellowships, and have access to a good deal of active recruiting on campus and in Boston and New York. Key areas of interest and success for Holyoke graduates include such service-oriented areas as education, social work, public service, the law, and nonprofit organizations. The college notes that it "has produced more women Ph.D.'s in the sciences than any other institution in the United States. Our alumnae are found in the capacity of research positions, as engineers, industrial chemists, physicians, astronomers, and software designers. At the same time, our focus on the arts results in many alumnae going on to become performers, directors, or playwrights such as the Pulitzer Prize winner Wendy Wasserstein." Alumnae also pursue careers in business, management consulting, investment banking, and insurance. The college sees its large number of international students on campus and the high participation in study-abroad programs as contributing to graduates interested in and prepared for internationally oriented careers.

Northwestern University

1801 Hinman Avenue
PO Box 3060
Evanston, IL 60204-3060

(847) 491-7271

northwestern.edu

ug-admission@northwestern.edu

Number of undergraduates: 8,176
Total number of students: 16,425
Ratio of male to female: 47/53
Tuition and fees for 2008–2009: $37,125
% of students receiving need-based financial aid: 42
% of students graduating within six years: 93
2007 endowment: $6,503,292,000
Endowment per student: $395,938

Overall Features

Northwestern combines the features of a mid-size private university and a large public research university. Like the big state institutions in its region, Northwestern comprises a large number of specialized preprofessional schools at the undergraduate level as well as a number of outstanding professional graduate schools. At the same time, the undergraduate enrollment of 8,150 is well below that of the state schools, creating a friendly, more personalized campus environment. Northwestern has long enjoyed a reputation as one of the top academic institutions in the Midwest, and over time has taken its rightful place among the top national universities with deep resources of programs, facilities, and faculty.

There are six distinct colleges: the largest is the Weinberg College of Arts and Sciences with 4,256 students; the School of Communication; the McCormick School of Engineering and Applied Science; the Medill School of Journalism; the School of Education and Social Policy; and the Bienen School of Music. More than twice the size of a majority of strictly undergraduate liberal arts colleges, Northwestern is able to offer a deep and rich array of more than eighty academic concentrations and a number of interdisciplinary programs across the different schools. The university has more than doubled its application pool in recent years, positioning it as one of the most competitive and popular schools in the country. The largest portion of the student body still comes from the Midwest, followed by major representations from the mid-Atlantic, western, and southern states, in that order.

The appeal is the outstanding reputation of its overall academic programs, especially the schools of journalism, communication, and engineering. Its location along the north shore of Lake Michigan in an attractive small city, with Chicago literally down the road and easily accessible by train and car, is another attraction for serious students from all regions of the country. An East Coast sophomore's motivation for enrolling in Northwestern echoes many of her peers: "Where else could I find a university of this academic reputation with so many special programs of study that is located in such a great place and offers so much school spirit and fun?" Students describe the academics as very demanding and their peers as professional and graduate school oriented. They also refer to a fairly high level of stress over the workload and tough grading by the faculty.

The university operates on a quarter system, which means students have to keep up with their work since the terms are shorter and exams occur and papers are due before vacation breaks. A West Coast junior confirms what has been described as "the midwestern work ethic" that defines the Northwestern environment. Housing options are diverse and not necessarily guaranteed. About 4,150 undergraduates live in university residence halls; some 900 live in fraternity or sorority houses; and 2,600 live in off-campus housing in and around Evanston. A number of features contribute to "an atmosphere of polarization," in the words of one student. This is because students tend to identify more with their particular college rather than with the university as a whole; there are two separate campuses with distinctive personalities; a large percentage of students belong to fraternities and sororities; and the presence of a cohort of varsity athletes, many on scholarships, who compete as representatives of the only private university member of the Big Ten athletic conference, one of the most competitive in the nation.

There are eight men's and eleven women's varsity teams and numerous intramural programs on campus. Overall, students are happy with their experience. They praise the quality of the faculty and the resources of the campus. There are over 250 campus organizations to choose from, Division I intercollegiate athletic events to attend, and lots of social outlets and good culinary options in Evanston and Chicago. Many students take advantage of the urban centers nearby to engage in the diverse internships made available to them through the auspices of the university. Northwestern students like their experience because of the outstanding academic programs; the specialized schools and preprofessional training; the social mix; and the opportunities to explore on and off this large-size college with its university size resources and offerings.

In keeping with its longstanding policy of making the university accessible to outstanding students of all socioeconomic and cultural backgrounds, Northwestern recently implemented a financial-aid program that eliminates student loans and replaces them with grants for students with the greatest financial need. This will mean that need-based scholarships will be in the form of so-called gift aid and thus eliminate the burden of accumulated debt over the course of the undergraduate years for the neediest families. With its strong financial resources, Northwestern is able to provide more than $70 million in university-funded grants to 60 percent of its undergraduates.

What the College Stands For

Northwestern states the purpose of the university and its mission in no uncertain terms: "Northwestern is committed to excellent teaching, innovative research, and the personal and intellectual growth of its students in a diverse academic community." This commitment is translated into the many fields of study provided to its students by a highly qualified faculty that is expected to carry out original research and to educate young men and women. Diversity of academic programs, learning opportunities, and student body are the university's key goals. The mission of the university is to instill broad analytic skills by allowing students to experience the work of a scholar by undertaking research with a faculty member, and by experiencing a meaningful internship in the larger community. In the words of the dean of the College of Arts and Sciences, "as professions continue to change, our graduates find they are able to adapt to—and take advantage of—unexpected opportunities after leaving Northwestern."

Curriculum, Academic Life, and Unique Programs of Study

Northwestern's deans and faculty believe strongly in the importance of a sound foundation in the arts and sciences for all students no matter what specialized field of study they may pursue. While each of the six undergraduate colleges sets its own set of required courses to attain a diploma in their discipline, all students are expected to gain exposure to the humanities, social sciences, arts, and sciences. A prospective student should review the requirements for any of the schools he or she is considering, and can count on being expected to meet a number of elective course requirements. With fifty major concentrations available in the College of Arts and Sciences alone, students easily find many choices of subjects to fulfill the distributional requirements. Two quarter-long courses are required in natural sciences, mathematics/logic, social and behavioral sciences, history, philosophy/religion, and literature/fine arts. All first-year students participate in two freshman seminars. These emphasize critical

reading and writing skills, and analysis of specific topics in small groups. As another example of the university's philosophy of education, students in any of the concentrations in the School of Communication (communication studies, sciences and disorders, performance studies, radio and television, theater, dance) are required to attain a solid foundation in the liberal arts and thus must meet certain distribution goals. For those considering a career in teaching and educational leadership, the School of Education offers a secondary teaching program that certifies graduates to teach grades 6 through 12. Engineering concentrators, journalism, and music majors may combine various disciplines within the vast range of programs in the six colleges. Internships are tailored to the student's interests in all of the schools.

Study-abroad opportunities abound for undergraduates. Some 500 students spend a term or year studying in another country through Northwestern's own programs or recognized affiliate programs sponsored by other universities.

Major Admissions Criteria

In the most recent year 21,390 students applied, 5,872 were accepted (27 percent), and 1,981 enrolled. Applicants must submit either the SAT, with three subject tests recommended, or the ACT with writing. Middle 50 percent test scores for entering students were: SAT critical reading 670 to 750; SAT math 680 to 770; ACT 30 to 34.

The admissions office states that "the most reliable predictor of academic success at Northwestern is a strong academic performance at the secondary level in an enriched curriculum. The mean high school class rank of enrolled freshmen has been at the 94th percentile. The qualities we look for in each candidate are independent thinking, a sense of humor, self-confidence, energy, enthusiasm, and an interest in activities, people, and ideas." Accordingly, the admissions committee weighs into the evaluation of each candidate involvement in cocurricular activities, the content of personal essays and quality of expression in the application, and teacher and counselor recommendations.

In addition to a generous need-based financial aid program, ninety endowed athletic scholarships are awarded. There is an early-decision plan with a January 1 deadline, but students should note that Northwestern traditionally only admits or rejects students applying early decision, rather than deferring a significant number as many other Hidden Ivies do.

The Ideal Student

Descriptive words that present undergraduates use to describe the successful Northwestern student are *mature*, *independent*, *smart*, *motivated*, *focused*, *diverse*, *confident*, *competitive*, *balanced*, and *fun-loving*. This is not an environment for an individual who needs and expects a good deal of "hand-holding" and direction. In the opinion of one recent graduate, "Self-starters who will explore the extraordinary number of academic programs, special concentrations, internships, social activities, and the nearby cities of Evanston and especially Chicago will find everything and more that they could ever ask for."

Student Perspectives on Their Experience

A survey of undergraduates that we carried out indicated that the overriding attraction of Northwestern was its overall academic reputation and specialized schools. Its size, location, and social opportunities were icing on the academic cake for them. The nature of the experience academically and socially can be influenced by the undergraduate college in which one is enrolled. The curricular requirements, and the flavor of the students and faculty, vary from school to school. A senior in the College of Arts and Sciences has liked the fact that he could concentrate on his major in the social sciences and at the same time get heavily involved in the music and theater activities although he was not majoring in these departments. A female athlete likes the fact that "I can be a serious athlete in this very competitive conference and be respected at the same time as a serious student by my friends in both areas." Some students wish the workload was slightly less intense at times, leaving more time to take advantage of some of the many organizations and social events on and off campus. Overall, students' satisfaction with their lives in this multidimensional environment is very high.

What Happens after College

Thanks to the highly selective student body, the academic demands across all six of the undergraduate colleges, the preprofessional course work many students undertake, the opportunities for specialized studies and research, and internships, Northwestern students do well in gaining admission to top graduate school programs in the traditional professional fields of law, medicine, business, education, engineering, computer science, communications, and the arts. Approximately 30 percent of recent graduating students have gone directly into graduate studies. The most popular graduate programs are at Northwestern, particularly for medical, engineering, and law school students. Other popular destinations are Stanford, Michigan, Wisconsin, Columbia, Johns Hopkins, Yale, and Duke. Some 10 percent of graduates of the College of Arts and Sciences and the School of Communication engage in volunteer and community service activities. The great majority of graduates gain employment directly. The university maintains an active and fully resourced career and placement office, and several students commented to us on how helpful the staff proved to be in their career planning.

University of Notre Dame

220 Main Building
Notre Dame, IN 46556

(574) 631-7505

nd.edu

admissions@nd.edu

[
Number of undergraduates: 8,369
Total number of students: 11,132
Ratio of male to female: 52/48
Tuition and fees for 2008–2009: $36,847
% of students receiving need-based financial aid: 47
(class of 2012)
% of students graduating within six years: 95
2007 endowment:$5,976,973,000
Endowment per student: $536,918
]

Overall Features

What high school student is unfamiliar with Notre Dame's fame due to its great football teams and overall excellence in major intercollegiate athletics? Its athletic prowess and tradition has made this midsize university located in a small midwestern town a known presence across the country. At the same time, how many future young scholars know very much about the outstanding educational programs and community traditions that make Notre Dame a first-rate university? Thanks to its loyal following of alumni, it has emerged from its founding in 1842 by the Order of the Holy Cross as a tiny, struggling regional liberal arts college into one of the most well-endowed research and professional institutions in the nation. Today, Notre Dame is a full-blown university comprised of undergraduate colleges of liberal arts and letters, sciences, engineering, and business management. Its graduate schools of business and law have attained a national reputation

in recent years, as have some of its doctoral departments. While its undergraduate colleges are more career-oriented than the typical small liberal arts school, there is a strong tradition of ensuring that students develop the skills that come from exposure to a broad range of arts and sciences.

In several ways, Notre Dame is a throwback to the more traditional nature of colleges prior to the upheavals the sixties wrought on most campuses. The student body is conservative in its political, social, and religious views. The vast majority are career-oriented, white, middle-class young men and women. An insight into the nature of the student body is the fact that the largest enrollments are in the colleges of business and engineering. This is one of a dwindling group of colleges that holds to parietal rules in the dormitories and on campus. All dormitories are single-sex and students must leave the rooms of the opposite sex by midnight during weeknights and 2:00 a.m.

on weekends. Priests and nuns live in every dormitory and, while alcohol is allowed in the rooms, sexual activity is not, leading to severe disciplinary action if found out. There is a chapel in every dormitory for students to attend mass each morning. There are no fraternities or sororities on campus, so the dormitories are the centers for social life, particularly for students not involved in the hugely popular intercollegiate or intramural athletic programs. The isolated location of the university makes the campus a strong community and the focal point of the college experience.

Notre Dame is clearly a Catholic-centered university. Eighty-five percent of the students identify with the Catholic religion. A course in theology is required of all undergraduates. As it aspires to national prominence as an intellectual community, Notre Dame must contend with the natural state of tension that exists between the intellectual independence, areas of research, and teaching that define a great research institution and the strong Catholic traditions and philosophy that can restrain open-ended inquiry and research. There appears to be somewhat less control at present by the Order of the Holy Cross and an orthodox religious philosophy. There are fewer Catholic faculty today, which expands the points of view espoused in the classroom. Nevertheless, there is a vibrant and committed community of Catholic leaders and teachers who direct this growing university. True to its history of welcoming any student who is serious about his or her education, the university has a very generous financial-aid program in place. Almost one-half of those enrolled receive aid to make it possible for them to attend. A good number of academic scholarships are available as are athletic scholarships ships across all the sports teams. The spirit of the campus, the quality of

the faculty, the great physical resources for learning and extracurricular outlets, and the strong counseling and advising support system results in one of the highest retention and graduation rates among all the top universities. This prairie college has come a long way from its earlier days of anonymity until a poor Norwegian lad by the name of Knute Rockne—who attended on a full scholarship—took over the reins of the football program and put the college on the map.

What the College Stands For

Says Notre Dame's university spokesman, "At its core, the hallmarks of a Notre Dame education are the acquisition of knowledge and the formation of moral character. As a Catholic university, Notre Dame has a responsibility to improve the human condition by finding answers to societal problems. That cannot be accomplished by simply teaching what is already known. We must also engage in discovery in order to positively impact the world."

As one administrator recently noted in an address to the Notre Dame faculty, "The foremost goal of a university is to have great learning in the classrooms and on campus and to have brilliant publications emerging from the halls of our faculty offices. Whatever administrative initiatives we may undertake are minimal compared to this daily activity of learning and scholarship." Notre Dame's focus on combining undergraduate teaching and research scholarship showcases the university's identification of itself as a national research institution. The university's identity "has three major facets: Notre Dame is a residential liberal arts college, a dynamic research university, and a Catholic institution of international standing. Much of what is great about Notre Dame derives from the intersection of these three factors."

The dean goes on to spell out his view of the centrality of Notre Dame's religious values to the university's educational mission: "Notre Dame's spirituality in general and its Catholicism in particular enrich the liberal arts experience, with its ideal of educating the whole person. Prayer and liturgy belong to our students' college experience. Spiritual questions arise in all our disciplines. Perhaps only in a religious setting, where reflection on God, or, metaphysically stated, the absolute, is prevalent do we address life's most fundamental questions, which are increasingly bracketed at nonreligious liberal arts colleges. Religion also brings to the liberal arts ideal a strong existential component. At Notre Dame learning and morality, knowledge and virtue overlap. Students on our campus pursue theology not as the disinterested science of religious phenomena but as faith-seeking understanding. Our students study history and the classics in order to learn not simply *about* the past, but also *from* the past. They read literature and are exposed to the arts because of their moral value, in the broadest sense of the phrase, and because we believe that beauty is the sensuous presentation of truth, not idle and meaningless play. Students employ the quantitative tools of the social sciences not simply as a formal exercise with mathematical models but in order to develop sophisticated responses to pressing social issues. In a world in which scholarship has often become antiquarian, disenchanted, and even cynical, at Notre Dame our work is shaped by our values and our existential aspirations. The generous commitment of the Holy Cross religious order to our campus and the ways in which they serve as role models reinforce this integration of learning and character, of col-

lege and community, of faith and life." As another dean puts it, Notre Dame's central mission is "to provide a high-quality education for the Catholic leaders of tomorrow." Notes a university vice president, "Our Catholic character provides the framework of faith, a sense of the moral purpose of education, and a pervasive experience of community. Add to that the excitement of the most storied football program in the nation, along with other nationally prominent athletic teams, and you have a collegiate experience unlike any other. In addition, the Notre Dame alumni network is among—if not the—largest in the higher education, with graduates in all corners of the world ready to assist their fellow Domers."

Notre Dame sees itself as having a dual identity as well in its position as a "modern research university" with an interdisciplinary emphasis. The university encourages a focus on community, engaging in discussion, and developing "virtues." Notre Dame recognizes the positives of its strong athletic (particularly football) program, such as increased alumni involvement, support, and financial aid availability, as well as the negatives, including poor press around recruiting scandals, some students on campus more interested in sports than intellectual pursuits, and nonathletic students possibly feeling left out. These are areas that the university is working to improve. Other areas of focus include emphasizing academics at the university and improving Notre Dame's reputation and recognition in that area, targeting resources toward stronger faculty hires and student recruitment, increasing diversity among students and faculty, and developing academic strengths across the undergraduate and graduate curriculum in addition to the traditionally strong departments of theology, philosophy, and medieval studies, for example.

Curriculum, Academic Life, and Unique Programs of Study

Notre Dame's academic program is somewhat unique in its curricular structure and requirements. All first-year students, irrespective of their future field of concentration, are enrolled in the First Year of Studies program, which has its own dean and faculty. They must complete a ten-course program of study that includes an interdisciplinary seminar and a grammar/composition workshop. Second-year students enroll in one of the four specialized colleges of choice: arts and letters, business administration, engineering, or science, where they will have to fulfill specific course requirements. Unlike the strictly undergraduate colleges of note, Notre Dame does rely on graduate students in the arts and sciences to teach or lead discussion groups in undergraduate courses. Here they are similar to the larger university model. Overall, the curriculum is structured and content-oriented in contrast to the more flexible options of many other colleges. The university spokesman identifies several key program areas at Notre Dame: "As freshmen, all students are enrolled in the First Year of Studies, an innovative program that features a broad curriculum that enables them to build a well-rounded academic foundation without committing to a major. This program is one of the main reasons for Notre Dame's 97 percent retention rate between the freshman and sophomore years and its 95 percent graduation rate, a standard exceeded by only Harvard and Princeton. The sense of community at Notre Dame is palpable. Some 80 percent of students live on campus, forming social, spiritual, and scholarly bonds that begin in the classrooms and residence halls and continue for a lifetime. About 90 percent of our courses are taught by full-time faculty."

Notre Dame is exploring ways "to help bridge academic and residential life, including offering selected classes in the dormitories, increasing faculty-student contact in the dining halls, and giving students greater responsibility for cultivating intellectual life." The university has begun a program in peer tutoring and has emphasized improvement in "advising, including the informal mentoring of students." A dean notes that "the College Council passed a policy for departmental honors that encourages more ambitious and enriching intellectual experiences for undergraduates, including the writing of more senior theses. The Arts and Letters Student Advisory Council was revived and has proposed several ideas for enhancing student-faculty interaction and improving services to students."

Discussing Core, one of the university's central educational requirements, this dean reports, "Students may continue to take the traditional Core, which focuses on God, self, society, and nature, or select any one of the new versions of Core. The thrust is to invite faculty members to stretch beyond their disciplines, but with topics that are particularly inviting to them and which may resonate with their research and development interests. Much like the concentrations that have developed in the college in recent years, this option for Core creates clusters of faculty members working together on interdisciplinary topics that address value questions with students."

Major Admissions Criteria

Notre Dame has become increasingly selective, with about 25 percent of 14,503 applicants admitted for 2008. The university sought and found a bright applicant pool, with almost three-quarters in the top 5 percent in their class and a 1,350–1,470 middle 50 percent SAT. Of note is the fact that 80 percent of admitted students are involved in community service. As one dean says, top admissions criteria are "academic course load, grades in those courses, and test scores." The admissions office notes in its materials, "The most competitive students in our applicant pool have taken the most rigorous high school curriculum available to them, have excelled in it, and have risen to the top of their high school class. It is important to realize that student applications are assessed both in the context of his/her particular high school and in the context of an extremely competitive applicant pool. No minimum grade point average or class rank is required to apply or to be admitted to Notre Dame. Know that our admissions committee will use every means possible to understand your application and to build the best case possible for your admission. . . . All of our admitted students display passion for, commitment to, and leadership in their activities outside of school. Whether a world-class pianist or a well-rounded senior class leader, Notre Dame students get involved, stay involved, and facilitate the involvement of others. Find activities you love. Dedicate time to them. Take responsibility for them. Then, tell us about them. . . . [W]e find your essays to be the most enjoyable part of the application reading process. Why? Because we learn about important decisions you've made, adventures you've survived, lessons you've learned, family traditions you've experienced, challenges you've faced, embarrassing moments you've overcome. We do not offer an interview as part of the admissions process, so it is through your essays that we are able hear your voice, learn your sense of humor, empathize with your struggles. We get to know *you*, beyond lists of courses, numbers, and activities. Reflect. Have fun. Share yourself with us."

The Ideal Student

It should be self-evident that a prospective candidate needs to embrace the special traditions that identify Notre Dame. While it is not necessary to adhere to the Catholic faith, it is important to respect Christian values and teachings. It is also important to enjoy the sports-oriented, conservative student body whose long-term goals are to find success in a traditional career. The social or political activist and the free-spirited individual who prefers a more open curricular style of learning will not be happy, nor will the student who wants to experience an urban life for four years.

Students should expect and prefer a strong sense of community and college identity, as well as close personal support and supervision. As one dean notes, students succeed in part because of "an infrastructure that is so supportive of personal success. The halls are the 'backbone' of the campus, with eighty-four percent of the students living on campus. There is no Greek system. Full-time counselors/professionals are in the halls and in other units on campus (First Year of Studies Office, Campus Ministry Office, Student Counseling Center, etc.) . . . Notre Dame is a highly selective, residential academic community within a faith tradition."

Student Perspectives on Their Experience

Students often have trouble distinguishing and choosing between Notre Dame and two of its top rivals for attention, Georgetown and Boston College. One student who was admitted to all three described her choice this way: "The one thing that I saw as a negative about Notre Dame was location. Part of what was so appealing about the other schools was that they were set in the middle of great and exciting cities and there could never be a shortage of interesting things to see and do. I have come to realize, though, that the campus of Notre Dame is one of its greatest assets. It is exactly how I envisioned a college campus and one of the things that I love about Notre Dame is that it is all-encompassing. People often speak about the Notre Dame family, and there is really no better way to characterize life here. It is a community of people who live and learn together and it affords an opportunity for people to develop relationships unlike any others and to share unconditionally. I have friends that I not only study with, but eat, sleep, and pray with, too. Notre Dame puts things into perspective and brings great balance to what can be an extremely hectic and overwhelming time in one's life."

Another student describes her process of choosing Notre Dame for its spirituality after visiting a number of other prestigious schools: "In the process of visiting these colleges, I realized just how important my spirituality would be during my college years. I knew that no matter what school I would choose, I would make spiritual growth a priority. Soon after the East Coast swing, I visited Notre Dame with my mother, and from that first day of the visit, I knew that I would be most happy at Notre Dame."

Students at Notre Dame are service-oriented and want to get involved. Says one junior, "A criticism of Notre Dame could easily be that it is isolated and out of touch with the world outside the Dome. This problem, however, continually challenges Domers to reach beyond the isolation of our gilded existence. Having a Center for Social Concerns on campus and especially programs like Summer Service projects instills in the students the importance of put-

ting what they're learning in the context of the real world of people and places."

And there is clearly a connection between this service and community focus on the university's religious character: "Three things immediately come to mind as being among the most important things about my ND experience. Dorm life is unique and really shapes the character of the campus. The dorm plays a huge role in the day-to-day experience here and has afforded me wonderful opportunities to develop friendships. An especially important part of that is having mass in the dorms. The Catholicity of Notre Dame, although never imposed, is omnipresent. It also takes a multitude of forms. This unique brand of spirituality is also among the most important things at ND to me. It is wonderful to be in the shadow of the grandeur of the Basilica, to make late-night runs to the Grotto after an SYR dressed in a formal and Nikes, or to go to 10:30 p.m. mass in the dorm in pajamas and be a eucharistic minister in socks! A part of that sense of Catholicism also takes on a different spin in the presence of the Center for Social Concerns on campus. The sheer physical presence of the building serves as a constant reminder of the world beyond the Dome. The Summer Service project has challenged me to integrate the classroom with the outside world. It is all at once emotionally draining and thought-provoking, frustrating, and inspiring. Like much of the Notre Dame experience, it guarantees to be an adventure that will both challenge and enliven."

Another junior says, "PLS (the Program of Liberal Studies) is the greatest program ever conceived, also Emmaus and the NDE retreat are essential to the Notre Dame experience. NDE retreat was the best weekend of my life, and everyone who goes on it says the same. It was so special." A sophomore says, "Notre

Dame does a remarkable job of encouraging people to become involved with social work during their undergrad years. There are thousands of opportunities for volunteer work and thousands of students get involved. They do Summer Service projects, do service trips over fall or spring break, or do random projects during the school year on their own time. Several hundred students dedicate their lives to a year (and sometimes two or three years) of service immediately after graduation. This is a remarkable, admirable quality of Notre Dame students. Graduates aren't simply greedy for great-paying jobs. They are concerned with creating a beautiful, compassionate world, and they want to be active in making positive changes with their skills."

Students who do best at Notre Dame balance academic, social, and extracurricular life. "It can be overwhelming to be suddenly surrounded by students from around the country who were all the best and brightest in high school," says a junior. "Most people at Notre Dame adopt a work-hard/play-hard philosophy. They are motivated and zealous about everything they do and discover their niche within the community. Everyone at Notre Dame is motivated and I think we all help each other to be the best that we can be. When you are surrounded by people who share your drive and enthusiasm, it makes things easier and more enjoyable. There really isn't the competitive, cutthroat atmosphere that so often hovers over people who want to succeed, and instead Domers operate as a family and community.

Another student praises "people who are motivated, creative, and who aren't afraid to make a few mistakes (or lose a few hours of sleep!) in the process of learning and experiencing will be most successful. People who come with an open mind, a positive attitude,

and an excitement to get involved with all sorts of things are going to have a fabulous time at Notre Dame. The key is to *get involved*. Without getting involved with something that you love, you could spend your entire time binge drinking or playing video games in a dingy dorm room. There is so much going on! Look at posters! Chill with friends! Join a club! Volunteer at a homeless shelter! Go listen to guest speakers! See a play! The list is endless. For people who are willing to experience and challenge themselves, Notre Dame is going to shape them into an incredible individual with a beautiful attitude toward life, a great view of the world, and the courage to do something great with their own talents."

Students disagree over the diversity, or lack thereof, on campus. Says a junior, "Notre Dame is easily (and erroneously) criticized for a lack of diversity. There are a multitude of clubs and programs that explore and celebrate various cultures, ethnicities, and backgrounds. Also, each dorm has a multicultural commissioner who is responsible for organizing events within the dorm and increasing awareness and education about different issues." But another notes that "the school has failed in its acceptance of homosexuals, just research and you'll find out. Not enough financial aid to *really* make the school diverse. Most of the cultural differences are in the athletes."

Other critiques of the university center on financial aid, the role of athletics, and the pros and cons of dormitory parietal rules. "There is so much money available to the school, and yet so much of it goes to things other than financial aid for the students who desperately need it. I think it's fine that Notre Dame has an incredible athletic program, but sometimes it becomes too much of a priority. There are so many athletes who get full rides to Notre Dame, but there are zero scholarships given for academics. Not only do athletes get full rides, they get free tutors for all four years, free books, free everything when they go play in different places. I understand that athletes work very hard to maintain good academics, but there are certainly nonathletes who are working just as hard and getting absolutely no rewards for their work."

Incoming students should be aware of the strict dormitory rules at the university, and how they impact social life: "Gender relations at Notre Dame are very strained. I think all of this stems from the parietal system, which is basically a sexual curfew for all students. Due to the single-sex dorms combined with some very conservative, Catholic sexual guidelines (a.k.a. no sex before marriage) opposite sexes must be out of the other sex's dorms by midnight on Monday through Thursday, 2:00 a.m. on Friday and Saturday, 12:30 on Sunday. It's a great way to get parties settled down on the weekends, but it's ridiculous for all other purposes. If people are going to have sex, they will have sex regardless of the Catholic traditions of the school or any sort of parietal system. In essence, the parietal system seems to prevent the natural friendships that develop between men and women when they live in the same dorms. Guys can't just drop by my room without calling up. And even if I'm just casually visiting a friend or actually studying (which happens a lot in real life at ND) the guy has to leave my room. . . . Parietals make the opposite sex a little bit too mysterious to allow for a healthy amount of dating and casual friendships that develop outside classes, parties, and activities."

Much of the bonding and social interaction at Notre Dame occurs in the dormitories, facilitating a friendly and familiar atmosphere on campus: "The housing system is wonderful at Notre Dame. Students are randomly placed in a dorm in which they will stay for all their years on campus. Almost all students stay in the same dorm for all four years, although half the senior class moves off campus. This facilitates further community growth, while preventing any sort of exclusive attitudes among students in the way a frat system might. Every student automatically belongs to a dorm which has a specific personality and set of traditions, but there is no rushing or stressing involved. Everyone belongs and everyone can be involved."

Students leave Notre Dame feeling loyal members of a family. "Notre Dame is an institution that is very much alive—it lives through the people who love it and have made it their own. Everyone who comes here invests a piece of themselves into it. Each new class reaffirms that character—it is not a matter of maintaining some archaic rite or ancient lore, but of evolving and growing, grounded firmly in history and tradition. This is what makes Notre Dame great. Notre Dame becomes a way of life and charges its graduates to carry within them the spirit of the university and an undying love for the school and family." A sophomore says, "There is something magical about Notre Dame. You can sense it in the students, in the tourists, and in the air. A humble pride (or in the case of football, a blatantly proud pride) is a part of the Notre Dame experience. I felt it and was attracted to it. . . . I chose Notre Dame because I wanted to spend four years of my life growing, experiencing, and learning in a place that creates compassionate, successful, educated people. I wanted to go to a school I could fall in love with both inside and out."

The "family" feeling at Notre Dame is prominent: "There is a true sense of community among the campus as a whole. And from my experiences this year, I saw the family respond to so many needs with endless amounts of love. Just as any family drops everything to help someone in a desperate situation, I was astounded to see the way students responded to several tragedies during the year—one boy with sudden brain surgery, a girl with cancer, deaths among close relatives, etc. Absolute strangers made banners, wrote cards, came to prayer services—people actually cared about fellow students." A sophomore raves that "professors are amazing, honest, and excited to help you grow. The religious figures are also very supportive and crucial in making Notre Dame a thriving environment. The most important asset to student success, however, is the student body itself. Students create the unparalleled charisma and passion that is present at Notre Dame. Please do not come to Notre Dame just because you think it will look great to some company in the future. Come because you are drawn to the passion of the student body, to the bold pride in the importance of spirituality and compassion. Choose your school because it will be perfect for you, not because it will look perfect to other people. Don't choose your school because you want to impress relatives at your graduation party or students in your high school. It won't make you happy."

Here are some other interesting comments from students: "Some words and phrases that come to mind when I think of Notre Dame: *family, faith, mission, community, challenging, enriching, holistic, personal formation, educating the whole person (socially, spiritually, emotionally, intellectually, psychologically), expanding views of the world, opportunities for growth, service, outreach, involvement . . .*" "Even though Notre Dame's student body hasn't traditionally been the most diverse, it seems like each new incoming class is more diverse

than the one before. Students here tend to be open-minded and embrace differences and diversity." "Notre Dame professors are excellent. They are always willing and always want to meet their students." "As a Catholic university, Notre Dame can be a bit conservative and not as welcoming of alternate points of view. As an atheist, I sometimes feel a bit isolated and left out. I've been atheist since high school, so I knew what I was going to have to face when I came here, and so far, most students have been very understanding and supportive, but I still find some who aren't as understanding, especially when they assume everyone here is Catholic, or at least a theist. It doesn't help that the university doesn't even recognize the existence of atheists on campus. Other religions have student groups to promote their faiths, and while I don't see why one would need a group for atheists, I often get the feeling that the school itself denies our very existence." "Those who do best at ND are the students who can see beyond the stereotypical/popular reputation the school has and learn how to access and exploit the various resources the school has to offer. People with such independent motivation are the students who typically find and utilize the best abroad programs, the most enriching service activities, and develop the greatest sense of the academic mission of the university. Those who aren't trying to find these things come across them naturally because of the vocal nature of many of the university's strongest programs, but people who are earnestly looking for academic and personal challenges find them more readily than those who aren't."

What Happens after College

Service-oriented positions and graduate education are the main choices for Notre Dame graduates. Some 10 percent leave Notre Dame and engage in some form of community service, such as the Peace Corps or the Jesuit Volunteer Corps. Some join the military as commissioned officers. A large percentage immediately enter law, medical, dental, and other graduate schools. Notre Dame graduates also regularly earn prestigious scholarships and fellowships, such as the Rhodes, Marshall, Truman, Stevenson, and Mellon awards.

The director of the Career Center notes, "The Career Center at Notre Dame was ranked number two in the nation by the *Princeton Review* in 2007–08 and received the Presidential Team Irish Award for outstanding service to students on October 4, 2008. General Electric ranked Notre Dame as the number-one recruited school in the Midwest. In 2008, the Career Center conducted 187 workshops, coordinated eight career fairs and advised 4,900 individual students." According to the Director, "Our students find jobs in all fifty U.S. states and over twenty foreign countries through-

out the world. Our undergraduate Mendoza College of Business was ranked number three in the nation by *BusinessWeek* magazine with median salaries of $53,500. Placement rates for Notre Dame graduates pursuing careers was 97 percent within three months of graduation. The breakdown of career outcomes for 2,000 graduating seniors on an annual basis looks like this: 55 percent pursue and obtain full-time jobs in a diverse array of industries, business, education, and government; 31 percent are accepted to graduate and professional schools; 4 percent enter ROTC programs; and 10 percent pursue and obtain one-to two-year positions in volunteer service, community service, social change, etc."

Says an enthusiastic graduating senior of his experience, "I'll be heading to medical school next fall. I feel extremely well prepared for whatever lies ahead. Regardless of where you're heading after graduation, Notre Dame has outstanding advisors who are committed to helping students achieve their professional goals."

Oberlin College

Carnegie Building
101 North Professor Street
Oberlin, OH 44074-1075

(800) 622-OBIE

(440) 775-8411

oberlin.edu

college.admissions@oberlin.edu

Number of undergraduates: 2,762
Total number of students: 2,774
Ratio of male to female: 44/56
Tuition and fees for 2008–2009: $36,282
% of students receiving need-based financial aid: 52
% of students graduating within six years: 83
2007 endowment: $816,135,000
Endowment per student: $294,208

Overall Features

The Protestant missionaries who established Oberlin in 1833 for the express purpose of training teachers and future missionaries who would go out into the world and do good works would be pleased with the present-day tone of the college. Oberlin was the first male college to admit women and among the first to recruit African Americans to the campus. Its long history of social action and intellectual progressive thinking has continued to define the academic and social tenor of this distinguished institution. A considerable portion of its resources are allocated to financial aid, in keeping with its commitment to educating students of all backgrounds and persuasions. Over half of the undergraduates receive aid and a full fifth of the student body are minorities. Somehow Oberlin combines a liberal, open-thinking environment with a highly academic program while maintaining a spirit of goodwill and co-

operation among the students and faculty. The campus is always buzzing with passionate debate and action over issues of concern. This is clearly a campus with a heightened concern for the right to hold unpopular opinions and foster multiculturalism in all its modern forms. It informs the social life, the classroom content, and teaching. Special-interest groups along gender, racial, and political lines abound. There are a good number of affinity houses for students who wish to identify themselves with a special category ranging from an intellectual interest to race and ethnicity to sexual preference.

The glue that holds this collection of diverse individuals together is the ethos of the community. There is respect for both the assemblage of all the different people and their beliefs and for the exceptional education available. Oberlin attracts outstanding faculty who delight in teaching and mentoring talented, independent, nonconformist learners, who are in the majority. Great weight is given to the quality of teaching and the personal direction of students as they explore their particular interests. There is a good deal of work assigned by most professors, which students are expected to keep up with while they go about their many outside activities. Oberlin is unique among the top liberal arts colleges for the presence of an outstanding music conservatory on campus. This enhances the cultural life of the college community, which is significant since Oberlin is located in a small town in Ohio.

What the College Stands For

"Think one person can change the world?" asks Oberlin. "So do we." Oberlin stands proud about its egalitarian ethos and historical legacy of progressiveness in American education. The college's current mantra is "Fearless." The college's statement of goals and objectives, adopted in 1977, outlines its goals for its students: "To equip them with skills of creative thought, technique, and critical analysis which will enable them to use knowledge effectively; to acquaint them with the growing scope and substance of human thought; to provide for their intensive training in the discipline of a chosen area of knowledge; to ready them for advanced study and work beyond the college years; to foster their understanding of the creative process and to develop their appreciation of creative, original work; to expand their social awareness, social responsibility, and capacity for moral judgment so as to prepare them for intelligent and useful response to the present and future demands of society; to facilitate their social and emotional development; to encourage their physical and mental well-being; to cultivate in them the aspiration for continued intellectual growth throughout their lives."

Oberlin seeks diversity in its student body and its intellectual approach to learning and sees mutual benefit in the interaction between its College of Arts and Sciences and its Conservatory of Music. As one administrator put it, "Different people would probably have different answers" to the question of what Oberlin's educational goals are. But he suggested, "To help people learn how to learn; to create talented writers, scientists, historians, etc.; to develop students' passion for learning; to educate people to make a difference, make their mark; to challenge students to the fullest."

Curriculum, Academic Life, and Unique Programs of Study

The curriculum reflects the progressive philosophy of the college in a number of ways. Students are required to complete three courses in the four traditional disciplines, but have great flexibility in choosing from a diverse and generous palette. A course that focuses on multicultural topics and issues is required of all students. A January term also is required of all students, which allows them to complete an independent project under the guidance of a faculty member or to undertake a meaningful internship off-campus. No one would question the presence in most classes of a liberal (some would say political correctness defines the attitude) perspective by which most courses are taught. While many students are keenly interested in the topflight social sciences, humanities, and classics offerings where they can study and debate the hot issues of the day, many others major in one of the several outstanding science departments. Oberlin has a well-deserved reputation for its biology, chemistry, physics, biopsychology, and neuroscience programs. Similar to a number of its sister liberal arts colleges,

Oberlin prepares a disproportionately large number of students for graduate studies in medicine and science teaching and research. Studying abroad is a popular option for many students, which the college encourages through its own programs or through cooperative arrangements with other major institutions. In its College of Arts and Sciences alone, Oberlin offers 47 academic majors, and 42 minors and concentrations. Students are also able to earn a dual degree in the arts and sciences and the music conservatory. The music conservatory retains its own faculty and intense curriculum that prepares its highly selective student body for performance or teaching careers. A beautiful art museum on campus offers students the chance to work and study in the fine arts. Students in the college of liberal arts can readily elect courses of interest in the conservatory. The college offers 22 Division III athletic teams, and some 140 clubs and organizations. Other interesting programs and initiatives include: a Creativity & Leadership Project, launched in 2006 to help students bring innovations to market; participation in Farm to Fork and Eat Local Challenge programs, to encourage sustainable agriculture and using local and organically grown foods; the Bonner Center for Service and Learning to promote service programs in which more than 1,000 students participate each year; and the Lewis Center for Environmental Studies, which is a model for green buildings nationwide.

A key factor that alleviates the demanding academic pace is a policy that erases all grades of C or lower from the official transcript and allows students to take as many of their courses on a pass/no pass option as they wish. This is a popular policy since a majority of Oberlin seniors continue their education in a wide variety of graduate programs.

A strong community-service program, an honor code, the student research opportunities and winter term, and the double degree program in liberal arts and the music conservatory "have been important," says an administrator, "in terms of both academic challenge and stimulation and also Oberlin's goal of helping to create leaders and good citizens." The college fosters diversity and divergent views on campus, providing a great deal of support for individual students. "Oberlin prides itself on its diversity. We were the first coed school and the first college to admit students regardless of race. Our history has helped us to maintain diversity, but it is also a major priority for us in admissions and administration. Oberlin continues to have a strong multicultural resources center and a huge variety of student organizations that promote all kinds of diversity."

Major Admissions Criteria

The admissions office at Oberlin looks for "a strong academic background, and also a willingness to be challenged, a love of learning, and an open mind. The people who do best are students who challenge themselves, are willing to learn new things, and work cooperatively rather than competitively." The three most important admissions criteria, according to one member of the admissions office, are "high school grades, choice of classes, and overall intellectual nature and 'fit' for Oberlin." The admissions office notes the following information about its process: "Oberlin admissions counselors engage in a holistic review process when reading and evaluating each application. This means many factors are considered, with no one component of the application singled out as most important. We seek an incoming class that represents a variety of talents, viewpoints, and achievements. We prefer a high school record demonstrating proficiency equivalent to four years of English, four years of mathematics, three years of the same foreign language, three years of laboratory science, and three years of social studies. These are recommended but not required. For your teacher evaluations, we strongly recommend they be written by teachers you recently had for academic courses—the people most familiar with your writing, analytical, and quantitative abilities. Although we require you to provide standardized test scores, no applicant is ever granted or denied admission to Oberlin solely on the basis of test scores. Recommendations from private instructors, coaches, clergy, mentors, extracurricular activity advisors, or employers that show evidence of your writing, analytical, and quantitative abilities." Since 2000 the college has become increasingly competitive for admission, with only 36 percent of 5,778 applicants admitted to the College of Arts and Sciences in 2008. Average SAT scores were 700 in critical reading, 674 in math, and 689 in writing. Just 25 percent of 1,227 applicants to the Conservatory of Music were admitted in 2008. Twenty-four of these students were admitted to the double degree program. Only 9 percent of students at Oberlin are from Ohio, and 31 percent from the mid-Atlantic region. Twenty-three percent are from the Midwest. Eleven percent are from New England, and 19 percent from the West or Southwest, while 9 percent are from the South. Seven percent are internationals.

The Ideal Student

This is an easy call for the high school senior in search of the right college match. Oberlin is not the most appropriate place for the conservative personality or the very traditional learner, nor is it likely to appeal to the committed athlete who wants a competitive intercollegiate sports program. Those who are attracted to the concept of a highly diverse community of individuals with as many points of view as there are students will find Oberlin an exciting opportunity. Respect for the serious intellectual education to be obtained from the excellent faculty will result in a most fulfilling four years that will leave a lifetime imprint on one's actions and view of the world. As one administrator noted, successful students at Oberlin have a "love of learning, passion for ideas, a willingness to be challenged, and an open mind (both socially and academically)." Environmentally, students must be comfortable with life in a small, pretty, self-contained town, even though the college is only thirty minutes from the growing metropolis of Cleveland.

Student Perspectives on Their Experience

Students choose Oberlin because of its academic reputation, its "legacy of social engagement," contact with faculty, music program, and location. As one junior notes, "Oberlin offers opportunities to work with faculty closely in an academically challenging environment. Also, the presence of the Oberlin College Conservatory promotes frequent, high-level musical performances, as well as opportunities for non-Conservatory students to take music lessons from Conservatory students and faculty." Professors earn much praise for their accomplishments and availability. Says one student, "Many of Oberlin's faculty members are amazingly masterful in their fields. Learning from these professors is a wonderfully enriching experience." A sophomore says, "My decision to attend Oberlin was based on my visit there. As soon as I stepped out of our car, I knew that

Oberlin College was the place for me. It just felt right. I felt incredibly comfortable and had no doubt I would do well. There was an energy to the campus that was infectious. I was excited to think I could be a part of it. My college search ended that day and I applied early decision."

Those who do best succeed because they are smart, involved, hardworking, and tolerant. "Students who do well at Oberlin are reasonably smart students who are willing to work hard enough to be academically successful. The Oberlin environment prevents no capable students from succeeding. Close contact with faculty, numerous opportunities for undergraduate research, and the absence of a 'cutthroat' competitive environment help students succeed." A sophomore says, "The people who do best at Oberlin are those who take a personal interest in the things that are going on. Whether those be on campus such as class, student organizations, or committees; or off campus such as local, state, national, or international events, the students who want to get involved are those who do best." A junior says, "Those who use the four or more years at Oberlin to explore themselves and get to know people, leaving the campus regularly to interact with the broader community, do best at Oberlin College." A graduate of identifies "the kind of person who cares about learning and helping other people. Someone who takes action. Someone who knows how to study intensely and work hard for reasons beyond 'making the grade.' Someone who makes things happen rather than lets them happen." This student counsels her peers: "Don't feel bad about yourself because everyone around you seems as talented or more talented than you. Instead, find inspiration through these incredible people. Work hard and find reasons to care about what you're doing." A sophomore says that Oberlin "creates a safe environment where students are able to explore not only themselves but also any potential interests they may have without fear of social reprisal. Individuality is cherished and a person is valued for who he or she is. The school also does an outstanding job at piecing together a group of exceptionally diverse people in its student body."

Diversity is seen as significant and positive at Oberlin. One junior says that "all types of diversity exist. Oberlin is extremely accommodating. For example, there are special dorms, such as Third World House, Afrikan Heritage House, and a collection of language program houses that promote diverse cultures." Nevertheless, some students would like to see even more recruitment of students of color, less self-segregation by students (a common complaint across college and high school campuses), and the development of more curricular diversity and flexibility. But the college "creates an atmosphere where students take initiative and do incredible things, such as winter term projects," and where there are experimental college courses taught by students. One student praises "Oberlin's tolerance and respect toward marginalized people, especially gay and lesbian students. Oberlin has always been ahead of the times by being a place that welcomes and supports people who are discriminated against and persecuted by the majority of American society." A sophomore says, "Oberlin College is the epitome of diversity. If you could make a concentrate out of Earth, as you would an orange, Oberlin would be it. Every kind of diversity, whether cultural, ethnic, racial, socioeconomic, intellectual, religious, geographic, or experiential, exists in abundance on campus. Because the people (not only students but professors and staff) are so different, everyone is very open-minded and sensitive to the needs of others. What makes Oberlin so special is the people. The students, the professors, the administrators, and the staff are some of the most amazing people I have ever met. We are all vastly different but are brought together in this one place where we are encouraged to grow both as individuals and as a community. I have just started my second year here but I already feel I will never truly be able to repay Oberlin for everything it has helped me become. If you're not afraid to be challenged or shocked, and want to get involved, come join us because Oberlin College is where you belong."

A junior reflects the college's accomplishment of its positive message about changing the world: "The amazing, talented, inquisitive, and socially concerned students who go here make the college special. Students tend to help one another rather than view each other as competition. The Oberlin Student Cooperative Association (with a big emphasis!!) offers the unique opportunity for students to control and operate programs that have dramatic, positive effects on the community. The lessons you learn at Oberlin encompass more than academic courses. You can find hope here, knowing that with dedication and hard work you can make positive change."

What Happens after College

Oberlin students enter myriad careers and fields after graduation. As one administrator puts it, "While about 65 to 70 percent of our students eventually go to graduate school, the jobs and fields they pursue run the gamut. They are prepared to do just about anything. The most popular career choices include education and communications." Indeed, higher education (17.5 percent) is the number-one occupation of Oberlin graduates (Oberlin has one of the highest numbers of Ph.D.'s among the small colleges), while over 12 percent each enter health professions or business and commerce, 9 percent go into public service, 8 percent into music performance, and 7 percent into K-12 education. Oberlin maintains an active alumni network, and over 1,600 internship opportunities through its Business Initiatives Program, factors which help students secure jobs after graduation.

According to the Office of Institutional Research, "Though more Oberlin graduates earn a Ph.D. than graduates of any other liberal arts institution and roughly two-thirds of our graduates attend graduate school, we have seen a decline over the years in the percentage of our graduates who immediately attend graduate school full-time (21 percent in 1994 vs. almost 16 percent in 2008). Other colleges are experiencing declines as well but more of their graduates are heading to full-time graduate study than at Oberlin. The universities in this group have held fairly steady." A survey of seniors in 2008 showed that 54 percent were headed to full-time employment, while 6 percent were volunteering, and many others were undecided or doing a variety of other things. Less than 1 percent were in the military. In terms of long-term plans, a full-third of graduates intended to pursue a doctorate, a quarter intended another master's degree, and about 7 percent were focused on law school, 5 percent on an M.B.A., and 3 percent on medical school.

Pomona College

Summer Hall
550 North College Avenue
Claremont, CA 91711-6312

(909) 621-8134

pomona.edu

admissions@pomona.edu

[
Number of undergraduates: 1,521
Total number of students: 1,521
Ratio of male to female: 50/50
Tuition and fees for 2008–2009: $35,625
% of students receiving need-based financial aid: 53
% of students graduating within six years: 95
2007 endowment: $1,760,902,000
Endowment per student: $1,157,726
]

Overall Features

One of the oldest independent colleges in the West, Pomona was modeled after the traditional New England schools like Amherst and Williams. Over the course of its 100 years as the leading liberal arts college west of the Rocky Mountains, it has taken on its own special identity as a reflection of its mission and its location. In addition to stressing the importance of small classes and a commitment from the faculty to teach students the critical skills that distinguish a liberal arts education, Pomona prides itself on the highly selective and unusually large number of racial and ethnic minorities in its student population. Few private institutions can match the 40 percent minority census of Pomona. The largest number of minorities represent the many different Asian heritages. Pomona has a significant endowment that allows the admissions committee to practice need-blind admissions in their decision-making process, a growing rarity among high-priced private colleges today. The result is a student composition of highly talented young men and women of a wide range of economic and racial backgrounds. The student body is a diverse, liberally minded, intelligent cohort of 1,500 undergraduates. Pomona's small population ensures a close community of students and faculty and small classes.

At the same time, the potential for social and intellectual claustrophobia in such a setting is mitigated by the unique consortium that comprises the five Claremont Colleges. As other colleges were created in the Claremont group, they modeled themselves loosely on the Oxford University college system. Each of the five schools has a particular academic focus and pedagogical style. Claremont McKenna emphasizes the social sciences, Scripps the humanities and arts, Harvey Mudd the physical sciences and engineering, and Pitzer the behavioral and social sciences.

Students have access to all courses and activities sponsored by the other colleges. Thus a small college can expand for the enterprising student to an intellectual and social community of over 5,000. A new recreation facility and campus center have become the center of action for students and tend to keep them on campus.

Students and administrators consider the college a vibrant and serious place to develop intellectually and personally. There are only a handful of fraternities on campus and no sororities. They do not have a major impact on the social tone of the campus. The large endowment has created an extraordinary set of majors to choose from, an excellent faculty, and many extracurricular organizations to join. Athletics are given a good deal of support as well: Pomona fields twenty varsity teams that compete in Division III. Several, such as beach volleyball and inner-tube water polo, reflect the culture and locale of sunny Southern California. Pomona administrators, teachers, and students believe they have the best of all worlds: a small, caring academic and social community with excellent resources for learning and a student body that is representative of the demographics in the western United States; at the same time Pomona has the advantages of a mid-size university because of its membership in the Claremont group. All of this exists in a locale where the sun always shines. Although Los Angeles is only forty miles away, some students state that they rarely venture into that sprawling urban center because they have what they want on campus or do not own a car, which is the only way one can get around.

What the College Stands For

Pomona maintains among its faculty and staff an intense commitment to undergraduate education in the liberal arts tradition, focusing primarily on quality teaching and interaction between students and faculty. The college sees itself as part of a very workable college consortium that greatly expands academic, social, and extracurricular opportunities for its students. Its location near Los Angeles and its proximity to the Pacific Rim are held up as major strengths and defining aspects of the college. Pomona is committed to diversity on campus and backs that aim up with a lot of aid, financed by its large endowment. About 10 percent of students come from a family with income less than $40,000, and a substantial number are first-generation college students. Pomona has eliminated loans from its aid packages. A senior administrator expresses his vision of the college in this way: "I would say that Pomona has a number of powerful elements of uniqueness and distinctiveness. Among America's premier liberal arts colleges, we are the only one in the western part of the country geographically proximate to the Pacific world and Latin America and located on the edge of arguably the most dynamic city in the country, Los Angeles. Southern California is one of the richest teaching laboratories in the world, and we use our placement here to great advantage. Pomona is also distinguished from other leading colleges by its role in founding the Claremont Colleges, five colleges and a graduate school within walking distance of each other. This enables our students and faculty to combine the benefits of a small college with those of a larger setting, giving us a critical mass of faculty in selected fields, adding to the diversity of students and cultures, and providing a university-level library far superior to that at any individual liberal arts college I know. Students find that their neighbors in the other Claremont Colleges also enhance social life."

Another senior dean says, "Our focus is solely

on undergraduates and our primary educational goal is to provide an undergraduate education that is distinctive and better than any other. Within that context, we seek to develop students' critical capacities and to stimulate them to achieve academically at the highest level. In all of our educational endeavors we stress adaptability and flexibility, i.e., knowing how to go about finding solutions rather than solely developing proficiency in given solution. We seek to prepare our students for leadership and for service and to participate in a number of roles in the larger world. We seek to develop an eager embrace of pluralism, both domestically and globally. Importantly, we focus on the development of a global consciousness and an ability to understand and to participate in a variety of cultures and environments."

Says one administrator, "Pomona's educational goals are to promote in our students very high levels of academic achievement, to cultivate in them the capacity to live responsibly and creatively in a diverse society, to prepare them for advanced study, and to equip them for positions of leadership. We pursue these goals not only through formal instruction but also through the character of residential life on the campus."

Says the long-serving dean of admissions, "With an average size of fourteen, most classes here are taught as seminars, in which the professor serves not as the source of all knowledge, but as a participant in a common search for understanding. In the lively discussions that are the heart of these classes, students are free to draw their own conclusions and express and defend their own ideas."

Curriculum, Academic Life, and Unique Programs of Study

Pomona sets high academic standards through rigorous course work at the same time that it gives students a good deal of freedom to fulfill the classic disciplinary areas of study. There is a three-semester foreign language requirement that can be waived by demonstration of competency. All first-year students take a critical-inquiry class or an interdisciplinary seminar. All seniors are required to complete a thesis or project related to their major. In addition to the large array of courses and concentrations available, students can create their own major field of study under the tutelage of a professor. About one-fourth of students at Pomona major in an interdisciplinary program. The underlying goal is to ensure completion of the ten essential skills of learning and thinking the faculty believe are necessary for further study and leadership roles in the larger world. As the dean of admissions notes in a letter he shared with us which he sends to his advisees each year, "In the end, we want students to write effectively, communicate well orally, and learn how to acquire new skills and knowledge on their own. Students should learn to think analytically and globally, formulate creative and original ideas and solutions to problems, evaluate and choose between courses of action, learn to lead and supervise tasks and people, relate well and comfortably to people of different races and nations, and function effectively as a member of a team. They must know how to use quantitative skills, and computers and how to place current problems in perspective, identify moral and ethical issues. Students should learn to understand themselves, their abilities, and interests; know how to function independently without supervision, to gain in-depth knowledge of a field; plan and execute complex projects, read or speak a foreign language; appreciate art, literature, music, and drama; and learn how to acquire broad knowledge in the arts and sciences, and develop awareness of social problems. They must develop self-esteem and self-confidence, know how to establish a course of action to accomplish goals, synthesize and integrate ideas and information, and really understand the role of science and technology in society. And they should touch upon all of this in four years." To accomplish these goals, Pomona has a set of general educational requirements that students fulfill by taking one course in each of five areas: Creative Expression; Social Institutions and Human Behavior; History, Values, Ethics and Cultural Studies; Physical and Biological Sciences; and Mathematical Reasoning. Strong departments are English, economics, history, and life sciences. Pomona has a well-deserved reputation for preparing students for medical school successfully due to the excellent teaching, research opportunities, and small classes. A science major is not likely to find himself learning chemistry or biology in a lecture class of 200 to 300 students. Says one dean, "Pedagogically, we are more and more emphasizing discovery-based or research-based learning as well as collaborative learning." Members of a typical senior class will move on to many of the nation's best graduate schools of medicine, business, law, and the social sciences. The majority of students have a purposeful attitude regarding their studies; at the same time they regard themselves as somewhat laid-back in style and in concern for grades and competition.

One administrator raves about the changes and developments on the campus in recent years: "Our facilities are better than ever: a wonderful new campus center; the renewal of science facilities with the construction of the Andrew Science Building for Physics, Mathematics, and Computer Science, followed by the complete renovation of Seaver North chemistry building; the Hahn Building and the award-winning renovation of the historic Carnegie Building put all our social-science teaching in a completely modern, well-equipped space; the construction of Seaver Theatre, which has been described as the best undergraduate teaching theater in the western United States; and the complete renewal of athletic facilities. The dean of admissions says, "We literally are across the street from the other colleges. Two thousand two hundred courses are offered among the Claremont Colleges each year, with 650 at Pomona. The library collections of the Claremont Colleges include more than 2 million bound volumes. There are nearly 300 student organizations among the Claremont Colleges. And more locally at Pomona College is the remarkable geographic mix, the even distribution of academic interests reflecting the even distribution of academic strengths in the offerings of the college, the depth of the physical and financial resources to support this enterprise and the rather unusually large number of 'happy students' at an institution with such a level of academic rigor and selectivity."

Students at Pomona benefit from numerous academic and residential-life programs that connect them to the campus, to faculty, and to internships. These include college facilitation of faculty living on or near campus, training for faculty in advising, matching of appropriate faculty advisors with students, a student sponsor program matching sophomores with first-year students, numerous clubs and resource centers, and the Summer Undergraduate Research program, which is funded by the college.

Major Admissions Criteria

Pomona's admissions process has become increasingly selective with recent statistics indicating that 86 percent of the freshman class were in the top 10 percent of their high school class and had median SAT scores of 740 in critical reading, 750 in math, and 730 in writing. Pomona seeks diversity and does not consider financial need in its admissions decisions. According to the dean of admissions, "Frankly speaking, a student who is a great tester and presents perfect grades in a perfectly challenging curriculum may be a great match or may not. If s/he is a brutal competitor in the classroom, selfishly trying to move ahead without regard to classmates but focusing only on 'success' perhaps measured only by a grade, we may not get too excited. We are looking for students who are interested in a residential college, interested in developing intellectually though collaboration with very bright peers and faculty. We do seek those who are inclined toward generosity and real engagement with their classmates and who are looking to appreciate those with perspectives which may be quite different from their own." Top admissions criteria are "academic ability (measured broadly), intellectual curiosity, and a propensity to engage this kind of academic community (represented by essays, references, activities, and life!)."

The Ideal Student

The challenges one will encounter at Pomona belie the beautiful setting and pace of Southern California. Pomona selects intelligent academic performers who want a first-rate education and who will take a good deal of initiative in planning their studies and overall experience. They also look for the more mature candidate who will embrace the great mix of student types and the open-minded attitude that prevails. This is not an environment for the social phobic who will feel comfortable only in a conservative residential community of similar types of students. The individual who

looks at all the opportunities available to him will find Pomona an ideal place to spend four years. There is a plethora of social, political, and athletic activities in which to find one's niche while gaining a first-rate education and a jump-start to graduate school.

The college is challenging but supportive, as one administrator describes: "Students succeed at Pomona because they have it in them to do so: We choose them with this in mind. But other things help, as well. Pomona's faculty (for all their scholarly accomplishments) are first and foremost teachers. Students benefit from taking advantage of this and making the most of the ease with which faculty can be engaged. Students also benefit from our financial-aid policies, which are need-based, and our commitment to meeting their full documented need. This takes a source of anxiety off the table. Student culture helps sustain our students, because while very achievement-oriented it is *not* aggressively competitive: People help and support each other, and there is a lot of student collaboration with faculty and with each other. Finally, I am impressed by the system of safety nets we have in place for our students, overlapping patterns of advisors, mentors, sponsors, clinics, and the like, some of which are functional, some of which are identity-based, and some of which are residential. A student who is in trouble has a very high probability of getting picked up before things get out of hand."

Student Perspectives on Their Experience

Students choose Pomona because of its small size, location, connections to the Claremont Colleges, faculty interaction, diversity, and academic reputation. Says one senior, "I chose Pomona in part because it is a small college. Going to Pomona I feel as though I belong to a family or community. The administration, faculty, and staff care about me. I don't feel this is true about every college. I enjoy having small classes and access to professors. I also appreciate the fact that all classes are taught by professors and not graduate students. Because there are no graduate students here there is no need for competition between them and undergraduates for resources, professors' time and attention, and research opportunities. Although Pomona is small it still has a lot to offer by way of majors, clubs, and other opportunities. We are close enough to Los Angeles that students can take advantage of all a large, diverse city has to offer. I also wanted the opportunity to compete in athletics at a competitive level in college without sacrificing academics. Also, being in Southern California is great for training."

Another senior says, "I chose Pomona for several reasons. First, I was looking for a small, academically rigorous liberal arts school. Pomona offered this and more. Because it is part of the Claremont Colleges consortium, it offers all the benefits of a small school with the resources of a larger university. The Claremont Colleges, for example, pool their resources for the main library, which contains about two million volumes. Additionally, I can take advantage of the guest speakers and social functions that occur on other campuses while still taking small classes with a lot of student/professor interaction. The second main reason I chose Pomona over other highly selective liberal arts colleges was because of its noncompetitive academic environment. Pomona students do not compete against each other but rather work collaboratively and collectively both in and out of the classroom. Third, extraordinary diversity among the students and faculty drew me here. Not only does Pomona have the highest per-

centage of minority students of the colleges I considered, it is committed to diversity of all kinds."

Pomona challenges and encourages its students without overwhelming them. "Pomona is best at allowing and facilitating the growth of individuals in many different ways, academically, athletically, and ideologically," says a senior. "This was also important to me when I was picking a college. I wanted a place where I could feel comfortable." A freshman from the East says, "I feel like everybody loves to learn simply for its intrinsic value, and enjoys a number of other activities as well. Pomona is just a happy place." A junior captures the environment this way: "I chose Pomona because it combines what's wonderful about learning with what's wonderful about beach volleyball. Pomona focuses on students as individuals, and concerns itself with the health of our souls as well as on our minds."

Those who do best at Pomona are "students who are willing to work hard, ask for help, and are open to new ideas. After students have graduated from Pomona they are not only expected to have retained an enormous amount of information about their major but also to understand the way it is connected to all other disciplines. Nothing exists in a bubble." A senior says, "The students who seem to do best at Pomona are those who were well-rounded in high school and want to continue exploring a variety of interests in college. Balance, therefore, is a key factor for those who are happiest at Pomona. Open-mindedness is another characteristic of highly successful Pomona students. No matter where a student is coming from, he or she will encounter a variety of different people with views and experiences that are unfamiliar. A willingness to learn from others' unique experiences will also lead to much success at Pomona. Finally, motivated students are bound to excel here. Pomona is very tough academically, and a lot of motivation is needed to get the most from your Pomona experience. Since these are some of the qualities that the admission staff is looking for, it is no wonder that Pomona students and alumni are so successful."

In addition to their peers and professors, students find support through a number of campus resources and programs. "When I arrived at Pomona," says a senior, "I had already been assigned to a sponsor group. This is a group of fifteen or twenty freshmen and two sophomores. It becomes your base. They match people up according to interests, so many of my closest friends were in my sponsor group. It's like a family unit. The sponsors are your mentors and can help you with basically anything. I was also assigned a mentor from the Office of Black Student Affairs. She was basically another person that I could go to, and she also introduced me to a lot of people. Both of these programs helped to ease my transition. I was also assigned an academic advisor who helped me pick out classes and eventually choose a major."

Here are some other comments worth considering: "This is not a preprofessional school. This is one of the reasons I like it best—students are not in cutthroat competition for banking jobs, and professors encourage students to take advantage of their time at Pomona to stretch their academic imaginations and pursue any interest. But if you're looking to be prelaw or to start planning a career path early on, you might not find too many like-minded students." "Pomona is very *eclectic* and *vibrant* but can also be a bit unreal or *surreal*, and it's always sunny and *warm*, and all in all (a) very *loving* environment." A junior says, "Students

with an intellectual passion and a passionate intellect do best here. Everyone I know has some academic interest they truly believe is cool and are anxious to share with the world—just ask the comp sci majors—and I know many students whose interests are varied and multifaceted. A dance and bio double major; an econ and English double major; and an environmental-analysis major with an emphasis in gender and women's studies. And even the more standard of majors are undertaken with a real spirit. I once heard someone describe how he loved his chosen history major because he got to sit around and tell stories all day. When I talk to my friends at other colleges, they are often surprised that Pomona students might have a subscription to the *Economist* or the *New Yorker* or that it's often difficult to find a copy of the *New York Times* if you get to breakfast too late. Students care about what they're learning in a way that interests them outside of the classroom, and speakers, debates, and lectures are usually well attended." An international student says, "The professors are the greatest treasure of Pomona College. It is extremely easy to get in touch with any professor and engage in a discourse on any given topic. The school does a great job at promoting an atmosphere where such interaction is welcomed." Another international student agrees: "I really enjoy the annual mysterious dinner. Students sign up for the dinner, and a professor will bring four to five students out for a dinner. The awkward part of the dinner is that students have no idea which professor they would be eating out with. The best part, however, is that normally professors and students become great friends after the dinner." He warns, "Coming to Pomona means that you may come with a dream of becoming a surgeon, and graduate with a bachelor's degree in social justice. We offer many possibilities here, and your discoveries will shape your future career."

Finally, a word of advice about name recognition, which could be applied to the Hidden Ivies in general: "Prospective students considering Pomona often struggle with what seems to be Pomona's lack of a reputation. Often I hear prospective students say, 'No one at my high school has ever heard of Pomona.' True as this may be, Pomona's academic excellence speaks for itself. The overwhelming amount of personal and career success that Pomona alumni achieve is proof that the college has an outstanding reputation. Keep in mind that graduate schools and top employers *do* know about Pomona!"

What Happens after College

Pomona's graduation rates are among the highest nationally, and the college's attrition rates among the lowest. Graduates go on to succeed in many different professional and graduate programs, including medical schools, law schools, and M.B.A. and Ph.D. programs. While a higher percentage eventually pursues graduate degrees, many go directly to graduate programs. The admissions dean notes, "In 2008 for the second year in a row, Pomona College topped all liberal arts colleges in the total number of Fulbrights awarded to its graduating class, with the class of 2008 receiving sixteen of the prestigious awards." Pomona has a well-developed Career Development Office to "help students discover and develop satisfying careers through career counseling, internships, a national on-campus recruiting program, cutting-edge technology, volunteer experiences, and extensive information about graduate schools and career-related topics." About one-quarter of Pomona's graduates head immediately to graduate or professional degrees, and about three-fourths will do so within five years of graduation. Pomona ranks high in the number of students who will pursue doctorates. Pomona College graduates are found in these fields: business and professions (24 percent), law and government (15 percent), medicine and health (15 percent), education (13 percent), arts and media (10 percent), science and technology (10 percent), and other professions (13 percent).

Reed College

3203 SE Woodstock Boulevard
Portland, OR 97202-8199

(533) 777-7511

(800) 547-4750

reed.edu

admission@reed.edu

[
Number of undergraduates: 1,408
Total number of students: 1,437
Ratio of male to female: 41/59
Tuition and fees for 2008–2009: $38,190
% of students receiving need-based financial aid: 49
% of students graduating within six years: 77
2007 endowment: $455,705,000
Endowment per student: $317,122
]

Overall Features

Whenever reference is made to the progressive model of higher education, Reed is always one of the top colleges mentioned. Reed was founded in 1908 to offer a more flexible, individualized approach to a rigorous liberal arts education. This is an important point to make because many people assume that the academic structure and student culture that personify this liberal institution of today grew out of the student revolts and experimentation of the late 1960s. In fact, Reed has remained true to its original mission to provide intelligent, intellectually passionate young men and women with a first-rate education in an atmosphere of free inquiry and reflection. Any changes in programs or requirements over the decades have had to meet the test of this philosophy. Too casual an interpretation of the quintessential Reed student and the culture of the community can lead to some wrong conclusions.

While students take pleasure in presenting themselves as free spirits, nonconformists, and intellectuals in search of the eternal truths, they are very serious students who relish meeting the rigorous demands of the curriculum and the high standards set by their professors. Classes are quite small and the opportunities for independent research and writing are considerable. A premium is placed on the quality of teaching and advising since intellectual dialogue and study is the key point of the Reed experience. Teachers are evaluated by their professional peers and students on a regular basis in determining tenure. One chooses to teach at Reed because of a passion for one's discipline and the desire to transfer this to students rather than to engage in advanced research as one's primary activity. Typical Reed students gain such a mastery of the subjects they choose to study by virtue of the intellectual skills required and the opportunities to work hand in glove with their teachers that the track record for entrance to the top graduate schools is extraordinary. For example, Reed ranks second among all liberal arts colleges in the production of Ph.D.'s and third in the percentage of graduates in science and engineering who earn Ph.D.'s. Only Cal Tech has a higher ratio of graduates who earn doctorates in all fields. Thirty-one Reed alumni have won prestigious Rhodes scholarships for advanced studies at Oxford University. A significant number of candidates qualify for medical school entrance. These are impressive results, indeed, for a college of only 1,400 undergraduates. There are no fraternities or sororities on campus, and the athletic programs are noncompetitive intramural programs. A great many students will study abroad for one or two semesters irrespective of their field of concentration. Reed operates on an honor code system that gives students the freedom to monitor their own behavior academically and socially. A student judicial review board deals with infractions and makes recommendations to the administration. Only 60 percent of students live on campus due to housing limitations and the availability of alternatives in Portland, a small, diverse, and interesting West Coast city. All first-year students are guaranteed housing, however. The fact that one-half of all students receive financial assistance reflects the diverse mix of individuals who make up the Reed community.

An administrator says, "Reed seems to be gradually becoming more open to hearing diverse intellectual perspectives that are represented through collaborative programming with students, academic departments, and the Multicultural Affairs office. In addition to offering insights to offices and departments on campus, Multicultural Affairs also coordinates a peer mentor program that serves as an anchor for students from diverse families, as well as main-

tains a small multicultural resource center."

Recent additions to Reed's 116-acre campus in the interesting city of Portland include a new technology center, as well as renovations and additions to various science buildings, the library, the athletic center, and the studio arts building. Students have access not only to an eclectic and livable northwestern city, but also extensive outdoors opportunities, from mountain biking and windsurfing to skiing and snowboarding at nearby Mount Hood and Reed's own ski cabin.

What the College Stands For

"There is a dissonance," a senior dean told us, "between the curriculum at Reed, which is formal, and the social environment, which is not." This is the central identity of Reed College, an institution with a demanding set of core courses and a thesis requirement. These aspects of the college create an "intellectual community" where all students are completing the same readings at the same time as all other students. Students are encouraged to work together and be vocal in classes taught in a "conference method." The college seeks and nurtures intellectual students in a very liberal and progressive social atmosphere. Reed sees itself as a "unique, intellectual place, with a long-standing and serious commitment to the liberal arts," with more kinship academically to such institutions as Columbia University, the University of Chicago, and St. John's than to Bard, Bennington, or Hampshire. As its catalogue says, "The goal of Reed College is to provide an education in the liberal arts and sciences with emphasis on the highest intellectual and scholarly standards. The Reed education pays particular attention to a balance between a broad study in the various areas of human knowledge and a close, in-depth study in a recognized academic discipline."

"From the beginning," say college officials, "Reed has sought to provide its students with the most serious and rigorous undergraduate education in the liberal arts and sciences. The college was explicitly founded in reaction to the then prevailing model of East Coast, Ivy League education. At Reed, there would be no varsity athletics, no fraternities, no exclusive social clubs: The college would be nonsectarian, coeducational, and egalitarian in spirit. The focus would be strongly academic and intellectual. The idea was to create an intense intellectual community, devoted to the life of the mind, in which students would be expected to attain the highest degree of undergraduate scholarship." That philosophy has remained in the forefront of Reed's identity and culture today. And in the future? "Reed will continue to remain uncompromisingly committed to the highest undergraduate intellectual standards in American higher education. To that end, the college is striving to do even better what it has always done exceedingly well: to support an environment where faculty and students engage in an academic program of rigorous collaborative learning. To enrich the learning environment, for example, the college has reduced the student-faculty ratio to ten to one by adding tenure-track faculty positions without significantly increasing the size of the student body. A senior dean says, "The program is demanding and rigorous, but also celebrates and represents the joys and intrinsic rewards of serious intellectual engagement. Reed College and Reed students are devoted to the life of the mind. Our goal is to develop in each of our students a powerful, disciplined, critical intellect,

and to nurture in each of our students a lifelong passion for learning."

Curriculum, Academic Life, and Unique Programs of Study

Reed fosters student independence of intellectual exploration and opinion, but it does so within a curriculum that ensures exposure to the great ideas and defining events in the history of the civilized world. Students will share in the experience of Humanities 110, a two-semester course based on the great books concept. This entails a heavy reading load of primary sources, which range from the Greek and Latin classics to St. Augustine's *Confessions*, taught by a team of twenty faculty. A majority of students continue on to other humanities topics that cover the Renaissance, modern European history and culture, and Asian civilization. Beyond this foundation, students are free to choose from a large selection of arts and science courses. There are stages in the four-year education that measure what a student has learned in his field of study. Juniors must pass a qualifying exam before moving on to their culminating work, a senior thesis of considerable depth of thought and content. If the results of the junior qualifying exam are not satisfactory, a student can be required to undertake further study or change his major if the faculty think it advisable. Seniors must also sit for a comprehensive two-hour oral review by their department's faculty that is in essence a defense of their senior thesis. It is no wonder that Reed undergraduates are prepared to tackle the rigors of graduate school studies, which require the same skills and process of learning.

"Over the years," says an administrator, "the college has adhered to these principles with extraordinary fidelity. The fundamentals of the academic program have remained unchanged: a highly structured curriculum, including a required first-year course on classical Greek and imperial Roman civilization; substantial distribution requirements ensuring genuine breadth of study for every student; a required junior qualifying examination, in which each student certifies his or her mastery of the major field of study; a year-long senior thesis, required of every student and culminating in a two-hour oral defense before a committee of the faculty." Philosophy is a popular and highly regarded department. Reed offers an excellent science curriculum for the serious student who is contemplating medical studies, engineering, or research. As an expression of its determination to emphasize self-learning and self-discovery rather than grade accumulation and competition, Reed faculty issue grades in their courses but do not release them to students unless their performance is lower than a C.

While Reed's core curriculum has remained largely unchanged over the years, the college's approach to the core material has evolved. According to one administrator, "For over fifty years our first-year students have been required to study, among many other works, Herotodus's history of the Persian wars. But the ways in which we teach this book have changed dramatically, and now include newfound attention to the issues of material culture, slavery, gender, multiculturalism, and the like. The faculty is professionally active and engaged, and when they teach materials they are naturally disposed to rely on the latest, most advanced scholarly theories and techniques. . . . This faculty controls the academic program. In doing so, it has consistently and enthusiastically reaffirmed the traditional pedagogical values of Reed College while at the same time bringing the academic program to the cutting edge of contemporary scholarly and intellectual life." Reed offers twenty-two departmental and twelve interdisciplinary majors, as well as dual-degree programs in such areas as computer science (with the University of Washington), engineering (with Cal Tech, Columbia, and Rensselaer), and forestry–environmental sciences (with Duke), and numerous study-abroad and internship opportunities.

Another important aspect of Reed's academic life is the college's approach to grading, through which, in the context of a campuswide honor code, it tries to discourage inter-student competition and encourage meaningful evaluation: "Virtually all work at Reed is graded in the standard way, and, as a statistical fact, there has been essentially no grade inflation at Reed for at least fifty years. But we have an unusual approach to talking about grades. Grades are never affixed to papers or exams, which are returned with lengthy comments only. Student report cards do not provide specific grades, but only indicate if work is satisfactory or not. Grades are never posted on faculty members' doors. There is no dean's list and no honor roll; students do not graduate cum laude, and there is no such thing as a valedictorian. The idea is simple. We believe that grades are important and that work needs to be rigorously evaluated in the traditional manner. But students should study not primarily for the grade but for the satisfaction of the work itself. This view is best communicated by a set of practices that keep grades more or less private and that do not make a public fetish of them. The results are extraordinary. Grade-mongering is rare at Reed, and students really do commit themselves to intellectual activity for its own sake." Overall, the college maintains "a philosophy of openness and egalitarianism with respect to community life. The campus runs according to an honor principle that governs everyone. There are no closed groups on campus; anyone who wants to join a group or organization can. Students are treated as adults and play a full and equal role in governing their own lives."

Major Admissions Criteria

The college notes, "We realize that there is no perfect combination of test scores and GPA that would predict success at Reed, but certainly a proven track record of academic accomplishment serves students best in the admission process. In reviewing applications, we look for students who are passionate about learning and who have articulated that passion through course selection and achievement. Strong verbal and quantitative skills and demonstrated writing ability are also important considerations. We look for a student body that is diverse in its range of backgrounds, interests, and talents while uniformly committed to a demanding and rigorous undergraduate preparation." The three most important admissions criteria are identified as: "Quality of academic preparation (as shown by courses taken, grades achieved, and standardized test scores); writing ability (application essays and graded paper from a high school class); and evidence of passion for learning, a hunger for intellectual nourishment." A dean says, "Reed looks for students who evidence strong academic ability *and* passion for learning. Reed students are independent thinkers prone to challenge conventional thinking and often eager to take discussions of ideas beyond the classroom and lab. Reed is for students willing to take intellectual risks. It is a community of open minds."

Reed's applicant pool is small, self-selecting, and talented. Exhibiting an average GPA in high school of 3.9, 3,485 students applied for the 2008 freshman class. Of these, 1,132 were accepted and 330 matriculated. As the dean of admissions notes, "Since 2001, applications to Reed have doubled, resulting in an admit rate that has declined from 71 percent to 32.5 percent." Sixty-five percent of the class entering in 2008 were in the top tenth, and 86 percent in the top fifth of their high school class. Fifty-nine percent graduated from public schools, and twelve percent were first-generation college students. The middle 50 percent SAT ranges were 660 to 760 in critical reading, 630 to 710 in math, and 650 to 740 in writing, for a combined average SAT of 2,068. The average ACT composite score was 30. A quarter of the class came from the Northeast, 20 percent from California, 16 percent from the Northwest, 13 percent from the Mountain West and Southwest, 10 percent from the South, 9 percent from the Midwest, and 7 percent internationals representing 14 countries. More than a quarter of all Reedies are students of color. Although Reed is generous with financial aid, it is not entirely need-blind, and becoming so is a goal of the college.

The Ideal Student

Be careful in judging Reed from any cursory description of its approach and style of education or from a campus visit. What can be perceived as a laid-back or countercultural gathering of men and women will not get to the heart of this small community. Reed is an intellectual turn-on for the bright and independent thinker who relishes the exercise of reading the great writers and thinkers or the opportunity to conduct scientific research with dedicated professors. A strong work ethic also defines the successful Reed student. There are a number of academic requirements and hurdles that must be mastered in order to graduate. This is not the right community for a conservative person, socially or intellectually. The faculty pushes its students to think for themselves and to challenge existing concepts and traditions. There is little interest in social conformity, partying, or traditional athletics. The potential candidate for Reed is a lover of books and ideas who is comfortable in his or her washed-out jeans.

College officials say that successful students at Reed "love to ask questions; have strong personal organizational skills; are self-motivated; have a study style well adapted to a rigorous academic pace; enjoy a conference-style learning environment; have a willingness to consult or ask for help when needed; are willing to engage in dialogue with faculty and others; are sustained by an intense curiosity about learning for learning's sake; are able to say no, set limits, stay on track; and have a sense of humor and balance in what can be a stressful culture." Another notes Reed's combination of social and academic environments: "What is most unique about Reed in today's made-to-order world is its unfaltering commitment to tradition. A quick glance at Reed's traditional curriculum proves that the college maintains the classical academic rigor on which it was founded, but one look at the independent-minded student body confirms that Reed is the perfect school for the quirky intellectual. Reed attracts students with a high degree of self-discipline and a genuine enthusiasm for academic work and intellectual challenge, but beyond that there really is no typical Reedie— the community proves to be a place where geeks, glamour queens, and geniuses coexist in academic bliss."

According to a senior Dean, "Reed is not an alternative college, unless it is alternative to still require a common freshman experience (the Humanities 110 course) and make a senior thesis a graduation requirement. Reed is distinctive in offering a traditional liberal arts education to a student body of alternative thinkers." Another Dean says, "Students who are not intellectually oriented, who do not cherish the life of the mind, and who are not especially interested in figuring things out are unlikely to be successful at Reed." And from another administrator, "Reed is a very self-identifying school that does not appeal to all students. The students that fit and enjoy the Reed experience are going to do well. Those who do not want to work in such an academically challenging environment will leave and choose another school. We also have a very traditional curriculum, so if you, say, want to study Hispanic Studies, we don't have that, and a student interested in this will probably choose a larger institution with more variety of majors."

Student Perspectives on Their Experience

Students choose Reed because of its culture, intellectual stance, academic program, location, and faculty. As one graduate says, "It's really the profs that make it so wonderful. They are absolutely accessible, to the point of having students to their houses for dinner *and* accepting invitations from students for dinner at the students' houses. They are brilliant, open, and ready to talk about academics, school, life, or the weather. I think it helps that the profs are here to teach, and that there is no publish-or-perish mentality. They make it great." Adds a senior about her visit to Reed as a high school student, "The campus was beautiful, the staff and students were remarkably friendly and open, and I sat in on an English conference led by Professor Pancho Savery. He walked in, sat down, read a passage from Ralph Ellison's *In-*

visible Man, put the book down, looked around the room, and waited. Eventually, a student made a comment and students proceeded to lead an amazing discussion for an hour and a half. It was pretty exciting."

What kind of students do well at Reed? "Generally speaking, those who do best are self-motivated and have a profound interest in learning," says a senior. "Grades are actually somewhat of a taboo subject among the student body—we just don't talk about them. The pressure to succeed, if there is any, becomes internalized because there is no external competition (with other students). The only benchmark of performance that one has is one's previous performance, and so we are challenged to excel." Another student identifies "people who are open-minded and who want to take responsibility for their own education. Those annoying kids who always answered the questions and/or questioned the teacher." Students succeed because of "flexibility and a willingness to put their nose to the grindstone and *work*. There's no way around it. The only way to succeed in academia is to have a modicum of intelligence and a large will to work; Reed is a true ivory tower of academia."

Reed has a reputation for having students who are lone intellectuals, and some students, including this junior, clearly reflect that opinion of the resulting campus life: "For one thing, Reed's social scene is, on the whole, pretty weak. Reed students come here because they're individualists and nonconformists, and thus big social events tend to be pretty dysfunctional. Nobody here has fun in quite the same way as everyone else. Also there's sort of a mystique here of the angst-ridded, homework-laden, confused young adult. People pretend to be much more bitchy than they actually are. This translates into a campus full of students who for the most part don't look anyone else in the eye and walk with their heads down, even when they pass by people they know from their classes. On the whole, Reed students aren't friendly, or at least they cultivate that appearance. They tend to have a few good friends and remain pretty neutral to everyone else. It's good that the campus isn't full of cliques and so forth, but it would be nice if people smiled more often around here."

A senior says, "Life at Reed can be very isolating at times. We've never quite gotten a handle on how to balance our extremely heavy academic load with other healthy pursuits. The college tries to help students in this respect, but the social climate and the average mentality of Reedies makes this nearly impossible." Students take solace not only in their individual academic pursuits, however, but also in the many available social options available in Portland. Other areas of the college that students would like to see improved include more diverse curricular offerings taught by more faculty and greater student diversity. "That said, however, Reed does have a lot of diversity. Not in skin color or socioeconomic diversity, to be sure, but in a diversity of political opinions, hair colors, tastes in music, sexual orientations, and geographic origins."

A particularly valued annual Reed event is Paideia. "This has to be the coolest student-run event *anywhere. Paideia* is ancient Greek for 'learning' or 'transmission of culture.' The event itself is held during the ten days before the beginning of spring semester, so it's usually the last week in January. During Paideia the student body springs for classes to be held on campus. Usually there are between two and three hundred, and the budget can run to $10,000. Most are one-shots, although some are three—or four—session deals. The classes are about anything and everything. There are lots of cooking classes, there are also classes on Plato. And jazz. And furniture-making. And Tuvan throat singing. Newspaper-hat folding (true class—taught by an older student's six-year-old son). There have also been acrobatics and clowning classes. Essentially, anything goes. They are taught by students, staff, faculty, alums, and random Portlanders. It must be noted that our college's 'most distinguished' alum is Dr. Demento, and he puts on video versions of his famous radio show for the students every year. Those usually fill the hall to capacity, no matter what the capacity actually is. Paideia is just a whole lot of rip-roaring wholesome fun. It's a nice way to learn all kinds of cool stuff and to ease back into the swing of the school year while hanging out with friends."

The honor code is recognized as an important foundation for campus life: "The gist of the honor principle is that Reed students (and faculty, administration, etc.) are expected to be honorable without being forced to be honorable. If you live in a nonsmoking dorm, you don't smoke in there. You can take your final exam back to your dorm room to do it and you are expected not to peek at your books for help, and as far as I know, people are honest about this. They behave this way because they are supposed to, not because there is a formal system of rules and punishments that makes them do it. When someone on campus does something dangerous, stupid, mean, or juvenile, it's the students that usually mete out the punishment, not the administration. As a result, most students here have a lot of respect for the honor principle and get pissed when someone stupid violates it. This way, people learn to be responsible, rather than chafing and rebelling against school policy."

A student further explains the impact of the grading policy on academic life at Reed: "It removes the competitive pressure. It's hard to compare comments. The benchmark of excellence then becomes yourself. The only way I can judge my performance is by looking at my previous performance, any personal goals I set for myself, and more important how I feel about the work I've turned in. Because grades are actually recorded, we have a transcript to show for our efforts that is readily understandable by outside agencies, such as grad schools. The policy has a significant effect on the community; instead of talking about our grades, we talk about what we're learning in class. Across disciplines, across major divisions (i.e., science/humanities), our student body discusses what it studies. It creates an unparalleled intellectual atmosphere on campus, one that I relish."

This environment is also facilitated by the conference style of teaching. Says a recent graduate, "People have to be self-motivated and willing to defend their beliefs at Reed. Classes are all conference style, with the professor taking a backseat to student discussion, so you can't just sit and take notes and you *must* do the reading, otherwise everyone would just sit around the table and say nothing. Also, there are no 'sacred beliefs,' meaning nothing is taken for granted. One must support all premises as well as arguments."

Reed challenges its students academically, and in terms of their preconceptions. Says a senior, "The thing that Reed does best, and has done consistently throughout its history, is challenge its students. Or rather, it allows its students to challenge themselves and each other. From day one you're interacting with ideas that are completely different from those that you have known. These ideas come from fellow students, yes, but also from the past 3,000 years of the intellectual history of mankind. Through being confronted with difference, the minds of Reedies acquire a tremendous critical skill (finely honed through thousands of hours of in-class and out-of-class practice). This skill is applied to the important texts of our heritage but also to current social issues. By the time Reedies graduate, our minds have gone through a tremendous transformation—and we've developed a set of views and opinions that are our own and have been fully thought out. Somewhere along the way Reed also manages to instill a deep love of learning (if there wasn't one to begin with), which leads almost two-thirds of our alums to grad school."

Another senior says, "Reed will teach you how to think. You learn critical skills—how to write down what you think in anywhere from five to one hundred pages, how to pick out the important parts of what you're reading, and how to discuss questions creatively and intelligently. Reedies can pretty much discuss any subject because we have at least a cocktail-party knowledge of everything. Well, not *everything*." And a final piece of advice for prospective students: "Reed is different from every other college and university that's out there. That difference is neither good nor bad. But if Reed is the kind of place a student is looking for, it's paradise. And if you're not into how we do things, it's a very painful experience. I give tours of the campus and one of the things I tell students who seem unsure about coming to Reed is to choose wisely—you won't change Reed, Reed will change you, in more ways than you can imagine. But that's not an experience that everybody wants, needs, or is looking for."

Here are some other comments of note: "I came to Reed because I wanted to be part of a community that valued intelligence and academia as well as weirdness in social contexts. In high school, I was the kid with my hand in the air when no one else wanted to talk in class, but

at Reed people are dying to have intellectual discussions in and out of class. I love it." "Reed students absolutely must be able to manage their own time well. No one is telling you to do your reading, but there are hundreds of pages assigned every week." A sophomore says, "I love the fact that I call my professors by their first name and that they remember my name, too. I'm a people-oriented person and my performance in anything I do is influenced by my relationship with other people involved, and I think the intimate class settings and the professors being our friends—as well as instructors—allows me to do well." Asked for some words to describe Reed, a junior responds, "To be obnoxious: *Nietzsche, Homer, Foucault, Durkheim, Marx.* With sincerity: *serious, quirky, academically traditional, creative,* and *analytical*." "The honor principle is the most important policy or practice at Reed. Over the course of the year, the Honor Council regularly hosts events that hundreds of students come to in order to discuss the honor principle's importance and relevance. They are some of the most well-attended events on campus." "Reed is a very serious place, but I don't think we take ourselves too seriously. We are passionate about what we are learning, but we are also a little silly. There is no better place to go if you want to become (as our dean of faculty famously said) 'a lean, mean analytical machine.' But there is also no better place to go if you like making ten-foot high bicycles out of spare parts, getting student-body funds to make your own ice cream, or discovering new and interesting people, subjects, and Portland landmarks every day of the week."

What Happens after College

The majority of Reed graduates (some 80 percent) will eventually earn advanced degrees.

Reed alumni report involvement in these fields: business and industry (36 percent); education (35 percent); government service (8 percent); miscellaneous (7 percent); health care (5 percent); law (4 percent); arts and communication (2 percent); and community service (1 percent). From Pulitzer Prizes to MacArthur "genius" grants, Reed graduates have distinguished themselves in diverse fields, including business and academia. "It is often said," an administrator points out, "that the liberal arts prepares students for anything. Indeed, a Reed education can provide limitless opportunities. A liberal arts curriculum, with its inherent focus on critical analysis and oral and written communication, provides a breadth of general skills not found in a specific career-oriented program. Students who study the liberal arts and sciences at Reed acquire an array of experiences, perspectives, and interpretive methods with which they can tackle any problem, within any career." The college provides connections to many internship, summer study, and undergraduate research opportunities that facilitate students' career search and development.

Rice University

6100 Main Street
PO Box 1892
Houston, TX 77251-1892

(800) 527-OWLS

(713) 348-7423

rice.edu

admission@rice.edu

[
Number of undergraduates: 3,001
Total number of students: 5,193
Ratio of male to female: 53/47
Tuition and fees for 2008–2009: $30,486
% of students receiving need-based financial aid: 39
(class of 2012)
% of students graduating within six years: 91
2007 endowment: $4,669,544,000
Endowment per student: $899,199
]

Overall Features

Rice is one of the newer members of the group of America's elite universities. William Rice, a transplanted New Englander who made his fortune in Houston, donated the funds in 1891 to endow a college modeled after a combination of Princeton and Oxford with an emphasis on the liberal arts curriculum in a residential living environment. The first class entered in 1912. Thanks to William Rice's generosity the university created a huge endowment over time, which explains its relatively small student population for such a topflight teaching and research university. It ranks second only to Princeton in its endowment resources for each enrolled student. The tuition is the lowest in the nation of all the top-rated independent universities. Rice has deliberately maintained less than a two-to-one ratio of undergraduates to graduate students to keep the teaching focus on the former. The small overall size and low student-to-faculty ratio have

created an exceptional institution, one that can retain a world-class research faculty that is also dedicated to teaching and mentoring undergraduates through small classes and research and internship opportunities. With a large faculty of 611 full-time teachers, Rice is able to structure its curriculum into six distinct undergraduate colleges: architecture, engineering, humanities, management, music, natural sciences, and social sciences. The professional schools have long enjoyed a top reputation because of the small size of the programs and the excellent faculty and facilities. The humanities program has expanded in strength and size to match the opportunities of the other programs.

The student body is one of the most selective and talented cohort of men and women to be found in any of the major universities. Of the 2007 first-year class, 21 percent were National Merit scholars. For the class entering in 2007, 8,968 students applied, 2,251 were admitted, and 742 enrolled. SAT middle 50 percent ranges were 640 to 750 in critical reading, 670 to 780 in math, 640 to 730 in writing. The ACT composite was 29 to 34. They would enjoy an average class size of just fourteen students, and a student-to-faculty ratio of five to one. Rice meets 100 percent of families' financial need, and does so without any loan component of the aid package for families earning below $80,000. For those earning more who still have need, Rice caps total loan debt at $10,000. Also, Rice offers 30 percent of students merit-scholarship assistance. Thanks to its endowed funds, Rice can offer admission to any student who qualifies irrespective of need. The generous aid program combines with the low tuition to make Rice one of the most attractive institutions for outstanding students of all backgrounds and means. Over one-third of the student body is of color, and all fifty states and some 32 foreign countries are represented among the 16 percent of students who are internationals. Almost one-half of the students are still from Texas, however. The admissions committee is working to widen the geographic representation to bring still further diversity to the campus.

The atmosphere is a serious one. Students enter Rice to avail themselves of the outstanding programs of study and renowned faculty, 98 percent of whom hold a Ph.D. in their respective disciplines. A majority of undergraduates major in engineering, natural sciences, social sciences (tied for 2nd), and the humanities, in that order. The music and architecture schools purposely remain very small and highly selective. An honor code guides the academic life of the school. Students take exams free of proctors and are expected to present only their own work. Violators must come before a student judicial board. The community treats the honor system with great respect, according to students and administrators.

The residential system is unique among American universities, with the exception of Harvard and Yale, since it was modeled on Oxford University. The university believes this is the heart of the Rice experience for its undergraduates. There are eleven residential colleges or clusters, each of which houses 225 men and women. Two of the residential colleges are due to open in fall 2009. About 75 percent of undergraduates live on campus in the residential colleges, and others keep their affiliation to the colleges even after moving off campus. Each cluster has its own dining halls and social commons room. A faculty member or master is in residence and helps set the tone and activities of the cluster, together with the students who govern most of the life of their college. The particular college a student experiences will play a major role in defining his or her growth and development socially and intellectually. Each cluster supports intramural athletics and social events. Students are assigned to their cluster for all four years as a means of forming strong ties with their peers.

Rice participates in intercollegiate athletics at a highly competitive level in Division I-A. Many students engage in intramural sports or in the large number of clubs and community service organizations that are supported by the university. There are no fraternities or sororities on campus because of the house system. Rice is a handsome campus four miles from the center of Houston, which is now the fourth-largest city in the country and growing at an accelerating rate. This is fueled by the technology, aerospace, and business industries located in the area. Rice's faculty and student body benefit from the opportunities for internships, research, and funding these groups provide for career and academic projects.

According to an admissions dean, one change to watch for at Rice is modest growth in the next few years: "The board of trustees has unanimously approved an increase in undergraduate enrollment by approximately 30 percent over the next decade. Rice's undergraduate enrollment will be increased to approximately 3,800 students within the next decade. For the past couple of years, and moving forward, each entering class has been slightly larger than the one before it. The class entering in fall 2008 has 790 students; we project an entering class of 850 for fall 2009."

What the College Stands For

Rice's educational leaders believe they have not wavered from the original vision and mission to combine the finest quality teaching of highly motivated and intellectually curious young men and women with significant areas of research, all to the ultimate purpose of serving the public good. The concept that scholars who are engaged in research and experimentation in their discipline are the most exciting teachers drives the size, curriculum, and modes of learning at Rice. Bringing the methodology of scholarship and discovery into the undergraduate classroom will enhance the skills and interests of their students. The public will then be served by the new knowledge that will result from the facilities and support with which the university provides its faculty, and the education students receive will ready them for positions of leadership and innovation in both the public and private sectors. To emphasize the point, two of the faculty in chemistry who have been honored with Nobel Prizes have taught first-year chemistry. Their work was recognized

for its contribution to society. How can a budding scientist not be enthralled to interact with teachers of such prominence? As the university states, "In a good university, the traditional functions of teaching, research, and public service can scarcely be separated—effective teaching demands continual research, and the development of the intellectual life of young people is a major public service." The guiding spirit is the value of arming students with a combination of a broadly based education with advanced study in a chosen field of specialization. Engineers and scientists should become versed in the great literature and thought of Western civilization, which will guide them as technical leaders in society. Those who will engage in commerce or law or education should be familiar with the fundamental principles of science and mathematics so that they can engage in intelligent dialogue with the scientific community. Certainly Rice's generous financial-aid program ensures that any student with a serious commitment to learning will have the opportunity to receive this first-level training. The university sees itself as an agile community of teachers, researchers, and learners that can adapt to new frontiers of knowledge because of its financial and human resources, its small size, and its commitment to serve the nation.

As an admissions dean puts it, "As a leading research university with a distinctive commitment to undergraduate education, Rice University aspires to pathbreaking research, unsurpassed teaching, and contribution to the betterment of our world. It seeks to fulfill this mission by cultivating a diverse community of learning and discovery that produces leaders across the spectrum of human endeavor. Our campus is among the more diverse, highly selective colleges. We actively recruit a diverse applicant pool that allows us to build a freshman class which is diverse in terms of academic interests, geographic background, ethnicity, talents, and political and personal views. Also, our residential college system assigns all students at random to their home for four years, so there is no one residential college that is all athletes, artists, international students, etc. Each residential college is a microcosm of all of the Rice student body, and each college celebrates the diverse backgrounds and beliefs of its residents."

Curriculum, Academic Life, and Unique Programs of Study

We call attention again to the particular strengths of the academic program: six distinct colleges of study within a small undergraduate college and a smaller graduate school. Historically, engineering and the natural sciences have enrolled the largest number of students. Yet the colleges of the social sciences and humanities have grown in enrollment, while those of architecture and music are quite limited in numbers. Whatever field of concentration a student opts for, he or she will be required to take foundation courses in the broad spectrum of arts and sciences before moving on to the field of concentration. As the school notes, "The academic philosophy at Rice is to offer students beginning their college studies both a grounding in the broad fields of general knowledge and the chance to concentrate on very specific academic and research interests. By completing the required distribution courses, all students gain an understanding of the literature, arts, and philosophy essential to any civilization; a broad historical introduction to thought about human society; and a basic familiarity with the scientific principles underlying physics, chem-

istry, and mathematics. Building on this firm foundation, students then concentrate on studies in their major areas of interest."

Rice prides itself on being on the cutting edge of research and teaching in chemistry, physics, and engineering. Some of the specialized fields of study offered are biochemistry/ cell biology, bioengineering, neuroscience, and space physics. Opportunities abound to develop combined majors of one's choosing; it is not unusual to major in a science and minor in a foreign language or religion and philosophy, for instance. The faculty encourages students to engage with them in independent learning projects or scientific research. The joining of intelligent and goal-oriented students with a world-class faculty teaching small classes creates a serious and demanding learning environment at Rice. Most teachers grade on a curve system because of the talent pool in their classes. This has the unpleasant effect of setting up a competitive edge. Also unique for a small undergraduate institution is the presence of several centers for specialized studies, such as the Center on Race, Religion and Urban Life; the Shell Center for Sustainability; the Institute of Biosciences and Bioengineering; and the Rice Space Institute. A number of joint degree programs are also available. Rice is also expanding its own international studies programs for its undergraduates in France and other countries following its vision of preparing leaders in all fields for the global economy and intercultural world in which they will work.

Major Admissions Criteria

An admissions dean states that two sets of questions are asked that will determine which candidates will be selected from such a large and qualified pool:

- When you are exposed to intellectual opportunities, how do you respond? Have you taken the most challenging courses offered in your school or reached out to enhance your academic interests in other ways?

- How have you performed in a demanding learning environment? Do your grade performance and teacher comments indicate a true desire to learn, to make demands on your intellectual abilities and extend your knowledge to the maximum?

The committee must then choose those students who are most likely to succeed in the particular college for which they have applied. Another dean states, "We seek students who demonstrate high levels of academic talent, creativity, motivation, and leadership. Students who do well at Rice are self-starters who are passionate about the life of the mind and the world around them; they take advantage of the many opportunities in classrooms, labs, and student life by knocking on doors, speaking with professors, and being active in their residential colleges. The most important factor we look at is the academic profile: How well will this student do academically at Rice? How fully will she use Rice's academic resources? Will he be working at the level expected by our faculty? We evaluate the academic profile by examining the student's school and curricular level, grades earned, standardized tests, teacher and counselor recommendations, and the student's written application. Secondary to the academic profile, but still very important, is the evaluation of the student's extracurricular and nonacademic profile: leadership; breadth, depth, and achievement in life outside the classroom; character and unique life experiences." Musicians, artists, and architects must demonstrate through portfolios and performance certain talents and strengths, in addition to intellectual ability, while future engineers need to show mathematical and science aptitude and course foundation of a high caliber. Standardized test scores are not the primary factor in the decision process, but they are considered as a partial reflection of academic preparation and ability. In part they are a necessary screening device. The ultimate question asked of each applicant is how he or she has acted on his or her interests and abilities—this will determine who will best use the many advantages Rice provides. Because of the university's commitment to educate young men and women of many disadvantaged backgrounds, the admissions committee gives weight to applicants of color and ethnicity. The committee tries to ascertain the backgrounds of its candidates through the important personal statements required on the application and through personal recommendations from school and outside supporters. Rice is determined to keep its doors open to deserving students who can take advantage of its learning resources.

The Ideal Student

Do not consider Rice if you are not serious about studying and learning. These activities come first for all students on campus. Social life is squeezed into a demanding and busy daily life. Due to the university's outstanding science and engineering departments and the professional goals of a majority of undergraduates, the overriding tone of the community is conservative to moderate, one that does not take much time to consider the politically correct issues and activist behavior that define many of the other top colleges and universities.

It helps greatly for students to be focused on a field of interest, to have a desire to work hard, and to enjoy closely interactive teaching. The sciences and engineering programs and social sciences are the dominant ones on campus, while a small cluster of students will be highly focused on their professional studies in art and architecture or music. There is not a great deal of casual course selection or a laid-back attitude toward studies. Extracurricular activities are plentiful and can be fitted into one's schedule comfortably. *Traditional* is the operative word to describe the attitudes and lifestyle of the majority of Rice students. There is a southwestern flavor to the campus because of the location and the dominant percentage of Texans in the student body. Jeans and boots will be more acceptable than Birkenstocks in this setting.

Student Perspectives on Their Experience

Students choose Rice for its academic programs, faculty, college system, and size. "The quality of the architecture school was my primary factor," says a sophomore, "however, the social scene created by the college system tipped the scales in favor of Rice over other schools." A senior lists "no Greek system, the college system, and academic reputation" as her primary reasons for selecting Rice. Rice is praised for its provision of individual attention to students and its focus on undergraduates. "The college system creates a grouping of relatively small, tight-knit communities that provide many friendships and possibilities for involvement but also foster a sense of belonging and pride."

Those who do best at Rice are "the people who see that Rice offers more than academics and enrich themselves in the many diverse opportunities for involvement available," according to a sophomore. Student success is assisted by "the support of peers, the reasonable guidance of professors, and the ability to remain calm and rational when faced with deadlines and stress." Another student identifies "peer help and individual initiative" as important.

Diversity at Rice is promoted by student organizations, recruitment and retention of a talented and diverse student body, and the residential college system. "Many different clubs exist which act to strengthen bonds within certain groups," says one student, "as well as sponsor activities to celebrate aspects of the group with the entire university." The university benefits from "the communities created by the college system and the opportunities, relationships, and unity therein."

Here are some additional comments, from students featured on Rice's Web site, that offer a flavor of the university: "I went to the symphony two nights ago. I'm going to the ballet in a week. There is a ton of culture and all kinds of stuff here that would probably go against a Texas stereotype," says a physics and astronomy and math major. A history and policy studies major says, "What makes Rice great is the quality of its undergraduate teaching and the closeness, the relationships that can be built from small classroom environments, and relationships with good teachers that really care about the students." Says a Pennsylvania native, "Maybe it's just a southern thing or a Texas thing, but that whole hospitality nonsense, it actually is true. I didn't really understand it or believe it before I came here, but the people are really nice. The students are actually open to meeting all different kinds of people. It's a really diverse school. I didn't really expect that in the South." An international student notes, "The fact that I got to attend lectures in a very close environment with Helmut

Kohl, Amos Oz, and Hosni Mubarak is something that I had never dreamed of."

What Happens after College

The university boasts that over 70 percent of its graduating seniors in any given year are admitted to their top choice of graduate school. This is entirely believable given the selectivity of the students in the first place and the rigorous training they will receive over their four years of study. The largest number of graduates continue their studies in engineering, followed by the natural sciences, business administration, and architecture. Rice graduates are awarded a large number of National Science Foundation fellowships in science and engineering. Premedical concentrators also have great success in admittance to medical schools. Similar to the students in the majority of other Hidden Ivies, Rice students succeed because of their close relationships with their teachers and the chance to do research projects or independent work under their aegis. The bottom line is that if one does well in one's undergraduate studies, the opportunity for an advanced degree at a top graduate school will follow naturally. The university is also aware that many graduates want to work upon graduating. A comprehensive career advising and placement office is in place; there is a proactive approach to encouraging major companies to recruit on campus or over the Internet. Says an admissions dean, "Our students nearly all have graduate school admission offers, job offers, or other firm plans by graduation. They win Marshall, Fulbright, and Rhodes scholarships, and they are in much demand with employers. One hundred and thirty-six employers conducted more than 1,040 interviews on campus last year."

University of Richmond

28 Westhampton Way
University of Richmond, VA 23173

(804) 289-8640

(800) 700-1662

richmond.edu

admissions@richmond.edu

Number of undergraduates: 2,719
Total number of students: 2,886
Ratio of male to female: 47/53
Tuition and fees for 2008–2009: $38,850
% of students receiving need-based financial aid: 40
% of students graduating within six years: 87
2007 endowment: $1,654,988,000
Endowment per student: $573,453

Overall Features

In the last two decades, the University of Richmond has transformed itself from a good regional college into a top-tier, national, highly selective liberal arts university. Due in large part to a series of extraordinary gifts from generous donors, Richmond has a large endowment that permits it to meet 100 percent of demonstrated financial need of its students, recruit and retain outstanding faculty, provide a beautiful campus, and pay close attention to the development of its students. Its reputation in the sciences is growing; the study-abroad opportunities are many; the undergraduate business school, which enjoys a national reputation and ranking, is the most popular undergraduate major; and the jewel in the crown may well be the Jepson School of Leadership Studies. The nation's first college program of its kind and now in its fifteenth year, the Jepson School provides a unique and challenging curriculum intended to prepare Richmond undergraduates for lives of ethical leadership across all spectrums of society.

With the advantages of its financial resources and the vision and energy of the university's leaders, Richmond is fast establishing a national presence, as evidenced by the increasingly selective student body and the increase in geographic and demographic diversity represented in recent classes. Founded in 1830 as a seminary for men, Richmond developed into a liberal arts college and created the campus of Westhampton College for women in 1914. In 1990 the academic missions of the two colleges were united to form the School of Arts & Sciences. While it reaches out assertively to broaden the base of talented students and faculty, Richmond maintains the coordinate campus structure in the belief that this enables both genders to develop leadership skills, active participation in social organizations, and a close community of friends with peers and faculty. The honor system has been a fundamental part of the university since 1830. Academic integrity and trust are essential elements of the campus community, and students appreciate the honor system's role in their Richmond experience.

Richmond may be a university in name, but in every sense it is an intimate community of students, faculty, and administrators interacting in a highly supportive environment. Declares a chief administrator, "The faculty and staff are dedicated to the personal approach on this campus." There is a coordinate system that puts the responsibility of academic and social development in the hands of residential deans who report directly to the deans for academic and student affairs. The faculty are committed to a personal approach to teaching and advising their students.

The coordinate residential system is "a very deliberate gender approach to academic and social development in a coeducational institution." The teaching faculty-to-student ratio is 9 to 1; the average class size is seventeen. There are over 100 majors and minors available to undergraduates. Richmond definitely qualifies as one of the most beautiful campuses in the nation with its 350 acres of traditional architecture, lakes, and rolling hills and lawns. Over 90 percent of undergraduates choose to live on campus all four years, thanks to the quality of the residences and overall facilities. The state capital, a city of 1 million residents, is close by and provides opportunities for off-campus socializing, community service, and internship involvement. Washington, D.C., ninety miles from campus, attracts students with appropriate academic and internship interests.

Aware of its traditional population of white, middle, and upper-middle-class students, mostly from the eastern seaboard, Richmond is fully committed to expanding its representation of

minority and nontraditional students, at present 14 percent of the student body, through active recruiting and expanded financial aid-awards. A variety of programs are offered to integrate these students into the university community. These include preorientation sessions, peer-to-peer student-monitoring connections, and extended overnight programs that provide opportunity for open discussion. There are also programs to support students who are identified as having high financial need. The Mentor Alliance Program connects financially needy first-year students with faculty and staff who support them in making a smooth transition into the university community. Part of the strategic plan for greater diversification of the student body is a concerted effort to increase the international representation to 10 percent of the student body.

Thanks to its healthy financial resources—Richmond's endowment places it in the top 2 percent of the nation's colleges—Richmond is able to be need-blind in its admissions process and meet 100 percent of student need-based aid. It is able to attract outstanding individuals through the combination of the need-based funding and a generous merit award program.

Athletics play an important role on this traditional and spirited campus. Richmond competes in the Division I Atlantic 10 Conference in nineteen major men's and women's varsity sports. The director of athletics states that the primary goal of the athletic department is to be a national leader in student-athlete graduation rates. Richmond has consistently ranked in the top 5 percent of graduation rates for Division I student-athletes: "In athletics, we attempt to recruit men and women who are skilled athletes, but who are also excited and willing to be challenged academically." The Greek system is very active on campus; 30 percent of men join fraternities and 45 percent of women join sororities. The Greek houses are a locus for weekend partying. There are 300 clubs and organizations on campus to provide plenty of options to meet individual interests. But most of the weekend activities tend to focus around the Greek parties.

What the College Stands For

The stated mission of the university "is to sustain a collaborative learning and research community that supports the personal development of its members and the creation of new knowledge. A Richmond education prepares students to live lives of purpose, thoughtful inquiry, and responsible leadership in a global and pluralistic society." The goal of the total program offerings is to provide students with opportunities both inside and outside the classroom to develop ethical judgment, leadership skills, and the tools to live a life of meaning.

"We want to develop well-rounded citizens. As a result, there is a focus upon collaboration between academics and student affairs," states a senior administrator. An overall strategic curricular plan for learning has been established with a goal of engaging students in all aspects of their intellectual and personal growth. Richmond's Bonner Center for Civic Engagement links service with learning and the academic curriculum. The university's senior academic leader presents this picture of the university: "The faculty are dedicated teacher-scholars who care deeply about student success. The combination of great resources and committed faculty and staff means that students receive the guidance, challenge, and opportunities they need to ensure they develop the best of their potential. Small classes, thoughtful advising, and numerous opportunities for student leadership in organizations from student government (under the

coordinate system that ensures both women and men are leaders) to Greek organizations means that students are well prepared for the world beyond college."

Curriculum, Academic Life, and Unique Programs of Study

The university offers 108 majors, minors, and concentrations. Students in all undergraduate programs must complete general education requirements. These include a writing course, proficiency in a foreign language, and one course each in historical studies, literary studies, social analysis, visual and performing arts, symbolic reasoning, and natural science.

Seventy percent of undergraduates participate in one of Richmond's 75 study-abroad programs. The Institute of International Education recently ranked Richmond fifth among baccalaureate colleges for the number of students studying abroad. There is an international-studies major and special funding for student research abroad.

Richmond has created a number of living/learning communities like the Sophomore Scholars-in-Residence program and the Live Learn Lakeview program where professors teach classes in the residence halls and facilitate related out-of-classroom activities. Students are encouraged to undertake research projects under the guidance of a faculty mentor. During the summer more than 150 students participate in a research project. As Richmond's faculty and resources in the sciences have expanded, more than fifty scientific research laboratories are accessible to students keen on gaining research experience year round. Nearly two-thirds of students get involved in community service projects each year. The Center for Civic Engagement coordinates the volunteer, service learning, and research activities. The Bonner Scholars Program provides scholarship support to nearly a hundred students who are committed to community service.

Major Admissions Criteria

In the most recent year, Richmond received 7,970 applications, accepted 2,522, and enrolled a first-year class of 755 men and women. Three hundred and seventy-five students applied on the early-decision plan and 213 were accepted. The average high school GPA was 3.5.

Either the SAT or ACT is required. The middle 50 percent range of SAT scores was 1,220 to 1,350 (critical reading and math combined). The ACT midrange was 26 to 30.

Sixty-eight percent of entering classes receive some form of financial aid.

Each year fifty merit-based, full-tuition scholarships are awarded. One hundred seventy athletic scholarships are also granted.

Admissions states that it is looking for students who want to make an impact, who are globally and open-minded, and who love to learn and question. Admissions officers seek students who want to gain the feel of a small college yet experience the academic, extracurricular, global, and experiential opportunities of a larger institution. The main criteria used to achieve this goal are: the rigor of the high school curriculum; a recalculated GPA in core courses; admissions test scores; personal interests and activities; teacher and counselor recommendations; personal essay writing; community service; and diverse backgrounds.

The Ideal Student

Those who do best at Richmond "are motivated individuals who know how to manage their very busy schedules, are not afraid to meet new people, and are eager to get involved in the many campus activities and organizations," according to a senior. Another appreciates the increase in the number of international students on campus and the different cultural perspectives they bring to the classroom and dorms. The tone and tenor of Richmond is a traditional one in terms of campus rituals, the composition of the majority of the students, and their professional aspirations. The ideal student wants to join in campus life and engage in a very interactive learning style. This is not the most appropriate environment for the proverbial loner or the more extreme social or intellectual activist. The small-size, intimate campus calls for an engagement in and out of the classroom, and will thus appeal to the student who wants this kind of collegiate experience.

Student Perspectives on Their Experience

A senior who has been heavily involved in leadership roles on the Westhampton campus observes that "Richmond gives students opportunities to mingle with faculty; get involved in over 250 campus activities; and to gain fellowship, grant, and scholarship funding for any and all interest areas." Students universally describe the college as "beautiful, inspiring, friendly, academically challenging, supportive, and fun." There is also a positive awareness by traditional undergraduates of the college's efforts to attract more minority students from other racial and lower socioeconomic backgrounds. States one student with pride, "There is a *ton* of money to help facilitate campus life for nontraditional students."

Drinking on the weekends is a campus issue, and continued efforts to provide more social options and counseling are major efforts of the administration. One student describes the situation as one in which "some of the students plan their weekends around drinking excessively." Richmond students and recent graduates are very enthusiastic about their experience, focusing particularly on the quality of their professors and their interest in them, the excellent campus facilities and resources to accomplish whatever interests the individual, and the campus spirit.

What Happens after College

Twenty-seven percent of graduates enter graduate school directly; 11 percent go on to selected learning experiences such as study abroad, fellowships, and internships; and 61 percent are employed in a wide range of business and professional organizations. The university's accomplished alumni include a NASA space shuttle astronaut; a Pulitzer Prize–winning author; Grammy Award winners; successful entrepreneurs and CEOs; professors, physicians and scientists; and a number of state and federal judges. Five graduates have been selected as Rhodes scholars and others have received Marshall, Goldwater, Truman, and Fulbright scholarships. The CPA exam pass rate for Richmond students frequently has ranked in the top ten nationally. Ninety percent of Richmond graduates are accepted into medical schools on their first application.

University of Rochester

300 Wilson Boulevard
Box 270251
Rochester, NY 14627-0251

(585) 275-3221

(800) 822-2256

rochester.edu

admit@admissions.rochester.edu

[

Number of undergraduates: 5,178
Total number of students: 9,108
Ratio of male to female: 50/50
Tuition and fees for 2008–2009: $37,250
% of students receiving need-based financial aid: 68
% of students graduating within six years: 81
2007 endowment: $1,726,318,000
Endowment per student: $189,538

]

Overall Features

The University of Rochester may well qualify as one of the great hidden major universities of the East. Its location in western New York State near Lake Ontario, six hours by car from New York City, has put it at a geographic disadvantage in the minds of many talented high school students and their parents. Yet there is much to commend this university academically, culturally, socially, and athletically. Located along a bend in the Genesee River, the university is composed of a handsome set of neoclassic and modern buildings on a self-contained campus two miles from the center of a modern city of one million residents, a city that has been rated as one of the best places in the U.S. to live.

Rochester is a relatively small, residential research university with a deep curriculum that includes sixty academic programs. A slight majority of undergraduates major in the humanities and social sciences rather than in the sciences and engineering. Perhaps to a greater degree than the more academically traditional, structured colleges, students at Rochester have to rely on close relationships with their professors, beginning with their pre-major advisor. This is because there are no preset lists of courses from which to choose, and thus students must navigate the entire course catalogue from their first semester on campus. Students are encouraged by their teachers to undertake independent study and research projects in their areas of interest. The professional graduate schools of business administration, music, engineering, medicine and dentistry, education, and nursing provide exceptional resources of faculty, libraries, and laboratories to carry on advanced projects and research. Rochester's metropolitan location offers many opportunities for internships across all kinds of professions and organizations.

Undergraduates actively manage over 250

student organizations that represent an extraordinary spectrum of interests, talents, and cultural and racial backgrounds. The great majority of students live on campus by choice over their four years in residence.

The Eastman School of Music is a performance-based academy and is one of the largest undergraduate programs within the university. Thanks to the early huge generosity of George Eastman, the founder of the Eastman Kodak Company, the School of Music and the science and medical programs are considered among the most outstanding university programs of their kind. Eastman is home to twenty student ensembles and boasts an outstanding teaching faculty and facilities. Singing groups, open to all students who love to sing, are a longstanding campus tradition. The Eastman Theatre is the venue for a multiplicity of music performances, from classical to jazz to choral, by its several university orchestras and ensemble groups and chorales. Rochester fields varsity teams in ten sports for men and ten for women. It competes in NCAA Division III and is a member of the University Athletic Association, a group of eight top-ranked academic universities. Among the many first-rate campus facilities is the $14.6 million Goergen Athletic Center, which includes an 11,000-square-foot fitness center.

The Interfaith Chapel, which has a 117-foot tower and 6,500 square feet of colored glass walls, is a statement of the university's commitment to welcoming individuals of all beliefs to its community. Rochester attracts a large share of students from international and minority backgrounds. Incoming students in the most recent first-year class were native speakers in fifty-one different languages. About 75 percent of the students are public school graduates and one-half are New York State residents. A majority of the

remaining students are from the New England and mid-Atlantic states.

What the College Stands For

Rochester cares about innovation and continuous review of its academic programs and personal development of its students. The major change in the undergraduate curriculum is part of Rochester's Renaissance Plan, which was launched in 1996. The goal was to attract, through the clusters approach to learning, talented students with a bent for intellectual independence and curiosity. Under the leadership of its energetic president and an experienced senior administrative team, Rochester has undertaken a plan for growth in enrollment, programs, faculty, and facilities over the next ten years.

The undergraduate enrollment will increase to 5,900 undergraduates and the strong financial resources of the university will be used to enroll greater numbers of talented students who, regardless of financial need, are an ever-better fit for the unique curricular approach. The goal is to increase the university's reputation as a place where students who carry more of an independent and personal vision for their education can find a perfect home. Representative diversity of interests and talents, and of cultural, racial, ethnic, and economic backgrounds, is important to Rochester. Admissions is need-blind and full need is met for all who qualify. Undergraduates represent all fifty states and sixty foreign countries.

Curriculum, Academic Life, and Unique Programs of Study

Rochester students experience one of the most unique undergraduate curricular programs among the highly selective universities. They will learn about three separate and distinct subject areas in depth through a combination of at least one major in a subject area within the three major divisions of humanities and arts, social sciences, and natural sciences and engineering, and at least two "clusters" of three related courses in other subject areas. The clusters may be expanded into minors of five or six courses or into second and third majors, which typically amounts to eleven or more courses. With this plan in place since 2000, there are no required specific subject areas and no general-education, distribution, or core courses. The focus on depth in each of the three subject areas is typically somewhat greater than that found at otherwise comparable research universities. Breadth is ensured by requiring that students completing a major categorized in a broad area—for example, the social sciences—will complete their clusters in the other two areas of humanities/arts and in math/sciences. The structure of this program leads not only to breadth and depth in an undergraduate education, but also a close connection with faculty in chosen fields on a one-to-one basis. The university believes that under this plan undergraduates assume personal responsibility not only to achieve success in the courses that they have chosen freely, but also to work out independently the combination of courses that reflect their academic interests and future goals.

The share of students pursuing full double majors in disparate subject areas has accelerated from 18 percent prior to the introduction of the new curriculum to 45 percent today.

The Take Five Scholars Program encourages students to take courses outside of their formal majors tuition-free for a semester or full year upon graduation. The Early Medical Scholars program offers guaranteed admission into Rochester's medical school, at the end of the four undergraduate years, to first-year students with outstanding academic profiles. New major concentrations in economics and business strategies, and in international relations, have recently been created under the aegis of the popular economics and political science departments, respectively.

Engineering is a major undergraduate school within the college of arts and sciences, offering concentrations in five specialties: biomedicine, chemistry, electricity and computers, mechanics, and its unique Institute of Optics. One of the newest and most impressive academic facilities is the Goergen Hall for Biomedical Engineering and Optics.

Major Admissions Criteria

In the most recent year there were 11,676 applicants; 4,900 were offered admissions, and 1,062 enrolled.

For admitted students the SAT middle 50 percent range on the critical reading was 620 to 730, and the math range was 670 to 760.

The ACT composite range was 30 to 32. Ninety-four percent of students graduated in the top quarter of their high school class.

The most important criteria in the selection of students are: demonstrated excellence in one or more areas rooted in strong personal motivation, as reflected in the high school transcript and test scores but also in extracurricular activities and writing samples; a mature sense of personal responsibility as gleaned in the teacher and counselor recommendations and personal interviews (more than 70 percent of incoming students have had a personal interview on campus or with a local alumni representative); and an appreciation for a residential community of peers with both similar and differing interests and perspectives. The dean of admissions addresses the special student qualities he believes are particularly relevant to the Rochester programs and environment: "Beyond wanting to feel sure students can manage their time and studies, our unusual secondary consideration is trying to identify candidates who have an independent streak. I would even say we celebrate students who have been more likely to chafe at restrictions and challenge authority in their earlier educational environments. The 'perfect 4.0 GPA' student is not always our primary pick, especially if he or she were achievement oriented largely to match their parents' expectations rather than intrinsically motivated to learn. The students who thrive best here generally have obvious 'edges' rather than being perfectly well-rounded."

The Ideal Student

Serious, independent, self-starting, balanced, normal. These are some of the leading characteristics Rochester students identify with. The self-selecting nature of the curriculum calls for thoughtful consideration of academic interests and enough self-confidence to make decisions. The workload is strenuous and calls for good organizational and time-management skills. Rochester is anything but a party-school atmosphere, or one where social life centers heavily around fraternities and sororities and athletic weekends. The ideal student is interested in his or her work and enjoys time with a cluster of friends on and off campus, as well as participating in some of the many campus organizations.

Student Perspectives on Their Experience

Senior satisfaction surveys in recent years showed a sharp increase in the share of students "very satisfied" or "satisfied" with their educational programs, along with similar increases in positive assessments of other aspects of their university experience such as student life, career preparedness, and development of a personal philosophy. The administration is very proud of the fact that the share of students who default on federal student loans after graduation is below 1 percent and has remained there each year from 2000 to 2007. Unlike many of its peer institutions, the Greek system plays a small role in the social life of undergraduates; only 20 percent of men and 15 percent of women participate.

Students find that there are many campus activities to choose from, as well as off-campus downtime in Rochester with its many cultural events, restaurants, and student haunts. Besides, according to several students, the workload is demanding and "does not leave that much free time for simply playing." The great majority of undergraduates describe themselves as liberal or moderate in their social and political outlook. Only about 10 percent consider themselves conservative. Observed one senior, "You can think any way you want, and act according to your own beliefs and find acceptance from almost everybody."

Said another, "What passes for liberal on this campus would be most likely be considered pretty moderate on a lot of other universities." Students perceive their peers as serious about their studies, committed to doing well and preparing for future careers. Most intend to continue their studies at the graduate level either right after college or some kind of work or internship experience. A recent graduate who is now in law school believes "that everyone at Rochester wants to do well, but there is not an intense competitive environment among the students." There is a good deal of appreciation and respect for the faculty in terms of their expertise and availability. One varsity athlete believes that "I made the right choice in coming to Rochester. I get to participate in a strong Division III program and know that I will graduate with a top degree in economics that will lead me to a successful career after graduate school."

What Happens after College

Students graduating under Rochester's innovative curricular approach have doubled their level of postgraduate study. A typical graduating class profile has changed from one-third entering graduate school to two-thirds. About 15 percent of graduates enter medical school or other health-related fields; a similar number enter law school and professional master's degree programs such as education, business, and public administration. An unusually high portion, 20 percent, enter Ph.D. programs. The remaining third enter various categories of employment, including a large number who join Teach for America and the Peace Corps. Popular destinations for first careers and internships in for-profit and nonprofit employment include New York City; Boston; Washington, D.C.; Rochester; San Francisco; and Los Angeles. Graduates are very successful in garnering major fellowships including Fulbright, Goldwater, and Marshall scholarships.

Smith College

7 College Lane
Northampton, MA 01063

(413) 585-2500

smith.edu

admission@smith.edu

Number of undergraduates: 2,569
Total number of students: 3,038
Ratio of male to female: 100% female
Tuition and fees for 2008–2009: $36,058
% of students receiving need-based financial aid: 62
% of students graduating within six years: 88
2007 endowment: $1,360,966,000
Endowment per student: $447,980

Overall Features

Smith takes great pride in its long and distinguished history as the largest of the women's colleges in America. Sophia Smith founded the college in Northampton, Massachusetts, in 1875 to provide intelligent, motivated young women with a liberal arts education the equal of that to be found at neighboring Amherst, Williams, Dartmouth, and Yale. The college has maintained its high academic standards and expectations for students but with greater flexibility for choosing a field of study, undertaking research projects and internships, and experiencing academic and social life on a number of other select campuses or abroad. Like its peer institutions of note, Smith has allowed far greater social freedom as well. Students live in one of thirty-six residential houses of a wide variety of styles and sizes. They accommodate anywhere from 10 to 100 women who are responsible for governing themselves and planning activities. All four classes, from freshman to senior, are mixed in the majority of the houses to provide more interaction among the students.

Like many of the other top liberal arts colleges, Smith governs its academic behavior by an honor code that appears to influence the students beyond the classroom. Smith women are self-governing and serious about learning and creating their own lifestyles. They pick up early in their college careers the Smith ethos of becoming independent thinkers in and out of the classroom as a preparation for a life of successful careerism and community service. The college leaders celebrate actively those alumnae who have achieved success in the professions and business world and the arts. Smith women are liberal and outspoken in their views, hardworking, and assertive in accomplishing the goals they set for themselves. Women's issues are an active subject of dialogue on campus and in the classroom. Courses on feminist thinking, women's literature, and history are offered, and numerous speakers are invited to campus to discuss these topics. A great many Smith students are very talented in or committed to the arts; many are competitive athletes. The faculty is excellent, with 96 percent holding an earned doctorate in their discipline. All faculty members are expected to teach and to keep up on research in their field, and the college notes with pride that the faculty win many prestigious research fellowships. Smith can afford a large faculty, some 285 full-time professors who teach in forty-one academic departments. The faculty-to-student ratio is nine to one, with many classes averaging five students. Students have 1,000 courses to choose from in fifty different areas of study.

The Smith campus of 125 acres is very attractive, and its facilities for academic programs, athletics, and the arts exceptional. The heart of the campus is the library, which contains 1.4 million items, making it the largest undergraduate liberal arts library in the country. The performing arts center is huge and technically very modern, allowing for every form of visual and performing arts study and practice. The science center is also up-to-date and addresses the needs of a strong science department and the many women who concentrate their studies in the sciences, including engineering. There is even a separate science library of distinction. An art museum and collection of some note is used by humanities and arts majors particularly. The college sponsors fourteen varsity sports for intercollegiate competition, including a major crew program— Smith has its own lake and boathouse on campus. There are indoor and outdoor riding rings for equestrian competition and a large indoor athletic complex. Smith is determined to be in the forefront of learning technology by offering a large number of computer facilities, training, and access for all students on a fully wired campus.

The president of Smith, Carol Christ, is the fourth woman president in a row for Smith, which has chosen female presidents since 1975. To attract young women of all socioeconomic and racial backgrounds, the college provides financial aid for 62 percent of those who demonstrate need. The minority population presently is 30 percent of the community. According to an admissions dean, "We enroll students from all fifty U.S. states and more than sixty other countries. Each year, in a class of 640 new students, some 550 different high schools are represented. Almost 30 percent of our domestic students are Asian, African American, Latina, Native American or multiracial, and 7 percent of the overall student body is international. In addition, 7 percent of our students are older students returning to school after an interruption in their education. In 2007–08, 23 percent of the student body received federal Pell grants." The average grant is very generous and well above average for the majority of independent liberal arts colleges. Smith's invested assets of over $1 billion put it in the rarefied company of the richest coeducational colleges in the nation. The traditional notion that women will not be generous givers to their alma mater is simply not true.

What the College Stands For

Smith seeks to educate its students in the foundations of the liberal arts and to prepare them for active careers that challenge the boundaries in "traditionally male" fields. According to the president of the college, "Our classes are small, so students get much personal attention from their professors, and they develop their ability to present and defend their ideas in small groups. We place particular emphasis on independent work. Our curriculum has an international emphasis; we seek to prepare women for global leadership." As one administrator puts it, the educational goals for Smith students are "to develop their leadership and other intellectual and civic virtues; to produce highly able and articulate graduates who are equipped in the best fashion of the liberal arts to pursue their vocations and other life goals; and to prepare women for career and other objectives, regardless of whether these have traditionally been regarded as 'male.'" Another senior official notes that "programs we have instituted that emphasize writing, speaking, and quantitative skill-building across the curriculum, will build on our traditional strengths by ensuring that all Smith graduates write cogently and persuasively; speak publicly with confidence and coherence; and answer complex questions through the careful analysis of quantitative data. Because of our new universal internship program, which is designed to bridge theoretical knowledge to the practical application of problem-solving in a variety of work environments, Smith students will graduate with considerable work experience and the demonstrated ability to interact effectively in a professional environment, the personal resourcefulness to provide solutions to problems in organizations, and the ability to connect theory and practice. Finally, Smith has long had a commitment to undergraduate research, and with the founding of the Kahn Liberal Arts Institute, which provides student fellows with the opportunity to conduct multidisciplinary research on a series of fascinating questions, Smith graduates will have considerable experience in using the tools of scholarship to answer vexing questions, with the additional benefit of having analyzed problems from the multiple vantage points that interdisciplinary research allows."

Smith sees student success as being helped by the college's ethos of educating young women to achieve their best in any field of endeavor, the community environment of the college, the faculty, and the school's strong alumnae network. "The very culture of Smith breeds success," says an administrator. "The stories of alumnae who have done extraordinary things are told and retold, and those alumnae act as mentors to current students (the 'old girl' network rivals anything the 'old boys' can come up with at other institutions). Smith celebrates academic success (in comparison to other places where athletic and social attributes often receive top billing). The housing system, small and more homelike than traditional dorms, provides a support system for students to excel. Smith has an extraordinary teaching faculty, who know students individually and are consistently demanding. And there are class deans and residential-life staff who keep individuals from falling through the cracks."

As is obvious from these statements, Smith, like its single-sex peers, sees in its all-female environment a very particular mission toward educating and instilling a high degree of self-confidence in young women. Students are helped, says one official, by "the belief from the first day of the first year that anything is possible. That is inculcated from the beginning."

Curriculum, Academic Life, and Unique Programs of Study

With its broad, sweeping curricular offerings, Smith gives women considerable freedom to choose what they wish to study. There are few distribution requirements and a multitude of courses to fulfill them. The only universal requirement is an intensive writing course. All other requirements are determined by the departments for major and minor concentrators. Students bear the major onus for deciding which of the many subject areas they will commit to. The faculty encourages building a major and minor curriculum or selecting one of the many interdepartmental majors. All students are encouraged to engage in research or independent study projects. Smith is probably the first college of note to have created the concept of junior year–abroad study for its students. Today a majority of women will take advantage of the college's own programs in Florence, Geneva, Hamburg, and Paris or those sponsored by other major colleges. A senior administrator notes, "The Smith curriculum is constructed to ensure that our students become informed global citizens; that they value tolerance and appreciate diversity and that they can apply moral reasoning to ethical problems and steward the earth's resources." A senior dean says, "In the context of an open curriculum, Smith encourages students to study across all fields of the liberal arts. Eligibility for Latin honors requires students to complete courses in seven areas of study."

In 2000 Smith established a pioneering engineering program. Notes a senior dean, "In the 1960s about 15 percent of Smith students majored in the sciences; the vast majority of Smithies in that era majored in the humanities and, to a somewhat lesser degree, in the social sciences. Today, more than 30 percent of the student body majors in the natural sciences, engineering or mathematics and we expect that percentage to increase markedly with the opening of Ford Hall, our new state-of-the-art teaching and research facility for engineering,

computer science, chemistry, and molecular biology." A senior administrator notes the establishment of other unique offerings: "A key strategic direction identified in the Smith Design for Learning is the creation of three interdisciplinary centers to facilitate student and faculty engagement with critical societal issues: the Center for International Studies and Cross-Cultural Communication; the Center for Community Collaboration; and the Center for the Environment, Ecological Design and Sustainability. As they develop and evolve, the centers will manifest Smith's longstanding commitment to active learning, hands-on research, and direct engagement with society's challenges."

A senior administrator says, "The Smith College Museum of Art is one of the best in small-college art museums in the country. Moreover, the college has historic strength in the study of art history and studio art. Similarly, Smith has a renowned botanic garden, and a landscape-studies program. Students have access to the resources and classes of five colleges, not one."

Concentrators in all fields of study are encouraged to broaden their intellectual and social experience either by studying abroad or by participating in an exchange year on the Twelve College program, the Pomona College exchange, or the Historically Black Colleges program. One-half of the student body will gain new experiences and learning through the elaborate internship opportunities Smith sponsors directly, and all students are guaranteed an internship if they so desire. These include the Semester-in-Washington Program and the Smithsonian Institute internship. A program run by the Career Development Office called Praxis provides opportunities and funding for a wide variety of summer internships for the purpose of combining theory and classroom learning with actual exposure in the field. Other interesting intellectual pursuits are available throughout the college, including a scholarly journal founded at Smith in conjunction with Wesleyan University, *Meridians: Feminism, Race, Transnationalism*, focusing on women of color throughout the world.

The academic schedule is divided into two semesters of thirteen weeks each and an interterm of three weeks. Students can immerse themselves in an independent project or internship that usually relates to their academic concentration. Still another exceptional opportunity for Smith women is membership in the Five Colleges group centered in the Amherst area. The five colleges sponsor student and faculty exchanges, joint faculty appointments, and courses in order to enlarge the curriculum options for students. In a typical year, 800 Smith women enroll in courses at the other colleges. Not to be taken lightly, social life is also enhanced by the many exchange programs Smith sponsors. Majors in electrical, computer, and environmental engineering are available to Smith undergraduates.

Major Admissions Criteria

The incoming class at Smith has gotten more competitive and more diverse in recent years. For the class of 2012, 3,771 women applied, 1,800 were admitted (48 percent), and 647 enrolled (36 percent). Middle 50 percent SAT ranges were 600 to 710 in critical reading, 570 to 680 in math, and 590 to 700 in writing, and the average ACT composite was 28. Sixty-four percent ranked in the top decile of their class and two-thirds graduated from public schools. New England provided 30 percent of the students, with 26 percent coming from the middle states, 14 percent from the West, 10 percent from the Midwest, 9 percent from South, and just 2 percent from the Southwest.

An admissions dean says, "Smith seeks talented young women from around the globe. Successful applicants will have taken advantage of the most demanding courses and programs available to them and will have performed well academically. We also look closely at applicants' written work as well as the evaluations of teachers and counselors in order to get a sense of the curiosity and intellectual engagement necessary for success at Smith. Finally, we look for involvement in extracurricular activities, focusing more on the depth and quality of a student's engagement than the particular activity she chooses."

A senior admissions officer states, "An admission decision is primarily an academic decision and so we're looking for students who are smart and have a strong academic record. More than that, a successful applicant exhibits intellectual promise and is a curious, eager learner who likes to be challenged. Smith is a residential college so we are also looking for students to become engaged members of a diverse community and who resonate to the idea of 'learning to make a difference.'"

According to this officer, "the most important criterion by far is a strong secondary school transcript, reflecting a demanding academic program in the context of the opportunities available to the applicant. We also look for evidence of academic promise along more qualitative measures, such as motivation, determination, curiosity, resilience, maturity, creativity, open-mindedness, social awareness and the ability to overcome adversity. Because Smith's curriculum is writing intensive, we look for students who can communicate clearly and write with fluidity and maturity. Finally, it is important that she demonstrate involvement outside of the classroom, whether athletic, cultural, religious, service, or employment (which could include paid work or assisting the family). Note: standardized tests are optional in the Smith admission process."

The Ideal Student

Choosing Smith College should not be based primarily on the fact that it is a women's college. This is secondary to the opportunity Smith provides to experience one of the great educations available in a residential liberal arts college. The faculty, the curriculum, the resources, and the many diverse opportunities for intellectual and personal growth should appeal to a self-motivated, independent young woman who enjoys the company of other intelligent, curious, ambitious, and independent thinking women. According to a senior administrator, "The students who do best at Smith are intelligent, motivated and intellectually engaged." The Smith student of today is not a traditional white female with good brains and good manners who is dependent upon the elite men's colleges of the past for her social life. Today's Smithie is liberal-minded, assertive in espousing her point of view, likely to be on scholarship or of a nontraditional background, and welcoming of her peers, who may have a different social or cultural orientation from her own. Most of all, she is dedicated to acquiring strong intellectual training and preparedness for the competitive world after college. She is also likely to engage in contemporary issues of concern such as social injustice, women's rights and privileges, and equal treatment whatever one's sexual habits and persuasion. As one administrator puts it, Smith looks for "very bright students who want to develop their strengths as women; activist types; and science fans." Academic conservatives or social wallflowers who do not enjoy an intellectual discussion or heated debate of sensitive social issues will not feel at home on the Smith campus. Students who do best at Smith are those who "will feel comfortable in a community of individuals; students who will speak their mind but listen with care and respect to opposite views and beliefs; and students who are compassionate, studious, positive." Another administrator says, "Strong-minded, competitive, courageous women do best here. We want bright, fully dimensional students who want to put their passions to work."

Student Perspectives on Their Experience

The college administration surveys the student body regarding its satisfaction with life at Smith, and nine out of ten seniors express satisfaction with their overall experience. Eighty-five percent affirmed that the academic training had increased their ability to write effectively, think analytically and logically, understand themselves, synthesize ideas, and gain knowledge of their particular academic field of study. These are exactly the skills that liberal arts colleges set as their goal. Smith has improved its graduation and retention rates in recent years, though some students do transfer from Smith for a variety of reasons. Several transfer students with whom we talked expressed great respect for the academic programs and the faculty they experienced for one or two years at Smith, but they felt they were ready for a larger, diverse, coed community in a setting with more action than Northampton could provide.

Smith has changed, however, as one graduate and current administrator notes, "Smith seems to be a more diverse and vibrant place than when I was here in the eighties. There is a sense of pride more loudly proclaimed on campus. More risks are being taken and with great effect—no stasis here." Current students choose Smith because of its academic programs, reputation, Five Colleges consortium, alumnae network, personality, and financial resources. As one student notes, "I applied to Truman State in Missouri and to Smith College. Smith was my dream school. A dream because of its reputation and cost. I was accepted to both and because of Smith's need-based financial aid, both colleges were almost the exact same in cost."

And while many students did not initially think they would attend a women's college, visits to Smith and discussions with Smith representatives and alumnae showed them the reality of life on campus today. "I heard about Smith at a college fair. The representative had a good handshake, she looked me in the eye and gave me a lot of impressive information in a short amount of time. I was very impressed with Smith's mailings, which were frequent and informative. I also liked how the students that were pictured were not all skinny, blond, and white. It appeared to be an open environment, where all walks of life were accepted. . . . I also fell in love with the campus. I am a big one for wide-open spaces, and Paradise Pond and the athletic fields are beautiful! The softball field is the best I have ever played on—the grounds crew is amazing. The housing system also caught my eye. Each house, for the most part, has its own kitchen. This means you can roll down to breakfast in your pj's and nobody cares. . . . The networking that Smith was known for also influenced my decision. I don't know what I want to do still, but I do know that somewhere in this world there is a Smithie who can help me get my foot in the door when I do figure it out."

A junior says, "I chose Smith because I was looking for a women's college, and early in the college search process Smith emerged as one of the leaders. At the time, I planned to be a women's studies major, and the obvious place to do that was one where women were put first and foremost. There was such energy there among the students on that second visit. Ironically, the thing that really sold me on Smith was the swim team. I thought of myself as a student first and an athlete second, but I knew that all the schools I was considering were academically strong. Smith distinguished itself as the socially superior school as well. As I attended classes and sat in the dining room and wandered around campus, I met people that I knew would be my friends if I gave them the opportunity. Though it is conceivable I could be as happy somewhere else, it's simply not possible that I could be one bit happier."

Students speak very positively about Smith's professors, classes, and their peers, and about the significant amount of work they must do. However, they feel welcome, challenged, and supported on campus. "I think Smith does an excellent job of teaching us how to communicate. It's great to have an idea or opinion, but if you can't express it verbally or through written words then it is wasted. Smith uses the presentation of papers to increase public speaking skills, and most professors demand a lot out of any paper. And man do we write a lot. I think you could get just as good an education at most colleges, if you applied yourself. However, at Smith you have to apply yourself even if you are wicked smart. Class participation is a must to receive an A in most courses because class size is so small. Smith also teaches us how to think and to continue learning throughout our lives."

Another student notes, "Smith professors and staff provide a sturdy and extensive support network for students. They are watchful of more than just our academic success; they are concerned also with our emotional, social, and spiritual well-being. Without being intrusive, they nonetheless are involved in our lives, by our decision. Few schools can boast such close relationships and important bonds between the students and the faculty and staff." The Five Colleges program works for students, even though it is not completely convenient to travel the fifteen to twenty minutes from one campus to the next. Says one student, "If Smith were not part of the Five Colleges, it would be a very different place. In my opinion, one of Smith's greatest strengths is its ability to fully exploit all the resources offered by the other area schools without becoming dependent upon them. I thought Five Colleges cooperation wasn't really as important as the admissions office made it out to be. Now, I am so happy that I was wrong."

Students who do best are "those who get involved, meet people, develop relationships with professors and administrators, apply themselves academically . . . those who are able to maintain a balance between interests, duties, wants, and playtime. Balance is key, but talking with your professors also helps. If they see you taking an interest, things just seem to run smoother. The Jacobson Center is available for peer editing and tutoring. There is a campus chapel for the spiritual side, and nature surrounds Smith. I think that if you look hard enough in yourself and decide what is truly important to you, you will find others with the same interests and form friendships and support groups for sharing ideas, plans, hopes and fears. But you do have to find a release somehow. You can't just study all the time." And from a junior: "The students who do best at Smith are those who put themselves to work seriously at getting the most from academic op-

portunities without sacrificing the social component of being a college student. At Smith, with some of the best professors in the country and facilities to match, it's all too easy to become concerned only with book learning. But Smith has a lot more to offer than that . . . namely, some of the most fascinating young women in the country."

Here are some other illustrative comments of note: "One of the things Smith does best is offer leadership positions to students." "There are always things to do on campus—movies, parties, concerts, theatrical performances."

"As a tour guide I am often asked what is it like to attend a women's college. I really enjoy it. It is a whole different way of looking at other people. I feel that I am among equals and my time here will allow me to avoid the ingrained cultural superiority of men. Because I have spent so much time looking at other people as equals, based on gender, I feel that I will cleanse myself of the patriarchical predisposition of our culture."

"A student will only succeed here if they take the initiative to seek out the help they need, whether that be of a professor, a learning center, a counselor, a good friend—all resources that are plentiful and well advertised on campus."

"Many young women come here concerned about how their romantic/social lives will fare whilst at school because there are no boys here. In my opinion this is easily fixed if one so chooses, because of the Five Colleges consortium. Sure, it takes a little more effort, but many women get through it and have successful/positive experiences. Other people come here concerned about the sexual orientation of their classmates. This is a stereotype about Smith that has been perpetuated for a number of years, at the detriment both to the students that go here as well as the credentials of the institution. In my opinion, the *perceived* sexual orientation of yourself or your peers is a poor excuse to attend/not attend one of the best academic institutions in the country."

"All of this diversity may be shocking for some when they first come to Smith, but I feel like it's all very manageable. With an open mind, it is possible to be accepting without being involved on any deep level with one's or someone else's diversity, as much as someone can dive into it head first, and really take advantage of all of the learning that can be gained from such a diverse community."

"The Smith campus is beautiful. You can take out canoes and kayaks for free from the boathouse and paddle up the Mill River and get away from it all. You can also borrow camping equipment from the college and go camping for free. You can go on a walk on the trail next to the river and take a break from work. You can also just walk downtown and window-shop in the cute stores in downtown Northampton. You can go rock climbing in our gym for free."

"At Smith I find myself constantly learning, not only in the classroom but also in my daily encounters. My life at Smith has taught me so much about life, love, and happiness outside of the classroom. Smith is a place of endless possibility and encouragement to follow your dreams in every direction. I have been following my dreams ever since I arrived and have never been happier."

"The professors at Smith yearn to help their students succeed. They honestly care about us and how we are doing. They become our mentors and our friends; they provide us with the guidance we need and support us. The students here at Smith also help each other succeed. We are all on the same page—writing papers together, studying for exams together, reading together, quizzing each other—and we support each other. As friends and people brought together at Smith, we become a community and that makes a huge difference in our success."

What Happens after College

Confirmation of the ambitious nature of Smith women and the outstanding education they receive is the track record of acceptance to top graduate schools and the large number of highly competitive fellowships they win. Over the past decade these include many Fulbright, National Science Foundation, Truman, Mellon, and even Rhodes scholarships, plus a host of other prestigious awards for study and travel at home and abroad. Smith has cherished and nourished its old-girl network that assists present and recent graduates to obtain jobs and internships in their chosen profession. Many Smith alums have told us of their positive experience in this regard. As one administrator puts it, Smith graduates are "prepared to shift with nimble grace to meet all challenges."

An administrator in the Career Development Office says, "I oversee Praxis, a program that provides stipends to students for unpaid internship experiences. The majority of students take advantage of this program, which enables me the chance to see, hear, and know that they are taking the knowledge gained in the vibrant classrooms at Smith and bringing it to the world outside of the college."

A senior dean notes, "According to a 2005 survey, 84 percent of Smith alumnae at ten years out reported their primary activity in their first year following graduation was employment for pay (66 percent) or being a student in a degree program (18 percent). Within five years of graduation, 66 percent of Smith alumnae pursue advanced degrees. By their tenth reunions alumnae have generally attained the following degrees: master's (40 percent), law (9 percent), doctoral (8 percent), medicine (6 percent), M.B.A. (5 percent). Our graduates speak consistently about the excellent writing, analytic, and critical skills they gained at Smith."

Says an administrator in the alumni office, "In many instances, Smith alumnae have made a significant difference on the national and international level. Notable alumnae include: Gloria Steinem (author and noted feminist); Margaret Edson (Pulitzer Prize–winning playwright); Hari Brissimi (first United Nations high commissioner on refugees); Jane Lakes Har-man (member of Congress); Tammy Baldwin (member of Congress); Niki Tsongas (member of Congress); and April Foley (U.S. ambassador to Hungary)."

The career development professional continues, "Smith students have a notable inner compass that guides them—a set of core values that influences their view of the world. They gain a global perspective at Smith that affords them a more complex and not-so-neat-and-tidy vantage point when looking at issues, questions, and problems. Smith graduates are strong, intelligent women who are not afraid to beg the important questions that bring to light how the world can be made a better place."

University of Southern California

Office of Admission
File 51158
Los Angeles, CA 90089-1158

(213) 740-1556

usc.edu

admitusc@usc.edu

[
Number of undergraduates: 16,091
Total number of students: 30,410
Ratio of male to female: 48/52
Tuition and fees for 2008–2009: $37,890
% of students receiving need-based financial aid: 36
(Class of 2012)
% of students graduating within six years: 85
2007 endowment: $3,715,272,000
Endowment per student: $122,172
]

Overall Features

In the heart of South Central Los Angeles sits a sunny, active campus; an oasis of gardens, fountains, and ivy-covered and modern buildings; premier athletic, arts, and science facilities; and a smart, diverse student body from all over the world. A mirage? Hardly. This is today's incarnation of the University of Southern California, whose Trojans dominate on the football field and in other athletic forums, and whose graduates form a "Trojan family" alumni network active in many professions.

USC is a university that fits many different kinds of students with diverse interests. A senior administrator describes the environment: "USC is a private research institution that maintains small class sizes and a ten-to-one student-to-faculty ratio. Students receive individualized attention in classes taught by distinguished tenure-track faculty, even in general education courses. They have access to a vari-

ety of academic and social opportunities, and can choose from more than 150 undergraduate majors and 130 minors, through our liberal arts college or through one of our sixteen professional schools. State-of-the-art facilities often rival those found in professional settings, and provide practical, hands-on training for a variety of fields. At USC, students may develop their own interdisciplinary course of study, and/or participate in a variety of honors programs, including the Renaissance Scholars and Thematic Option."

USC is not the same school of a generation ago. A college official says, "Over the last twenty years, USC has moved from being a very well-known regional institution to a truly world-renowned global research university. In this time period, the academic profile of our student body has increased dramatically, our retention and graduation rates have seen enormous improvements, our student body has become much more diverse (by every measure: ethnically, geographically, academically, socioeconomically), our endowment has quadrupled, our academic offerings have expanded, and our research activity has skyrocketed. In short, USC has matured."

What has promoted these developments? "Our changes over the last two decades have been the result of a very concerted effort on the part of the university to expand and improve all aspects of its academic and extracurricular offerings. An aggressive and comprehensive ten-year strategic plan was successfully completed in 2004 and we've since adopted another equally ambitious plan that will take the university well into the next decade." Looking forward, "to meet the challenges of the twenty-first century, USC will continue to leverage its strengths as a culturally diverse research institution with a Renaissance approach to education, in the heart of a dynamic world center. USC continues to unite basic and applied research to meet social needs and demands, through efforts such as the Institute for Creative Technologies and the USC/Norris Comprehensive Cancer Center. USC will grow as a learner-centered university, adaptive and responsive to the needs of its students, while expanding its global presence. USC will continue to reach out to international communities to attract the best students around the world and to deepen our understanding of Southern California's role in the global arena."

One of the largest of the Hidden Ivies, USC walks a challenging line between its focus on undergraduate student life and its development as a graduate research institution. The university is almost evenly balanced between about 17,000 undergraduate and 17,000 graduate students. Established in 1880—early days in the birth of Los Angeles as a major metropolitan area—USC's main campus and the focus of most undergraduate residential life and education is the University Park campus. Home to the College of Letters, Arts, & Sciences, as well as the Marshall School of Business, Annenberg School for Communication, and most of the other of USC's seventeen professional schools, the campus teems with trees, lawns, attractive buildings, and attractive students. Other LA campuses house USC's major health-science schools and programs: medicine, pharmacy, occupational sciences and therapy, and physical therapy, among others. USC maintains other California-based programs in Marina del Ray, Sacramento, Catalina Island, and Orange County.

The neighborhood surrounding the University Park campus is a residential urban community that supports a great deal of off-campus student housing, as well as stores and restaurants serving both students and residents. Crime, particularly muggings and theft of property, is an issue for students and the university, especially since such a large proportion of students (about half) lives off campus in a variety of apartments and houses. At the same time, USC promotes significant engagement and community service in the area, with many students volunteering, tutoring children, and serving in internships.

School spirit is high at USC and, in fact, much higher than one would expect given the size of the place and the social fragmentation that occurs in many city schools. How is the university able to maintain this balance? Several elements combine, including significant support for and participation in a major Division I athletic program that is led by a national-champion football team; a strong Greek system located centrally on campus ("the Row") that houses a fair number of students (about 20 percent of men and women belong to one of thirty-two fraternities or twenty-four sororities); some 600 student organizations active across many areas; a thriving on-campus arts scene with theater and music performances and art and architecture exhibitions; and an increasingly ambitious and talented student body.

Even with all these campus attractions, students make regular use of the greater LA area, from Venice Beach, Santa Monica, and Hollywood, to the nearby San Gabriel Mountains. Restaurants, clubs, bars, art exhibits, and major-league sporting events abound in the city; skiing, snowboarding, hiking, surfing, beach volleyball, and sunbathing are easily available. No wonder applications are running at a record high and USC has become the major private university option in Southern California.

What the College Stands For

The mission statement of the university reads, in part, "The central mission of the University of Southern California is the development of human beings and society as a whole through the cultivation and enrichment of the human mind and spirit. The principal means by which our mission is accomplished are teaching, research, artistic creation, professional practice, and selected forms of public service. . . . The integration of liberal and professional learning is one of USC's special strengths. We strive constantly for excellence in teaching knowledge and skills to our students, while at the same time helping them to acquire wisdom and insight, love of truth and beauty, moral discernment, understanding of self, and respect and appreciation for others. Research of the highest quality by our faculty and students is fundamental to our mission. USC is one of a very small number of premier academic institutions in which research and teaching are inextricably intertwined, and on which the nation depends for a steady stream of new knowledge, art, and technology. Our faculty are not simply teachers of the works of others, but active contributors to what is taught, thought, and practiced throughout the world."

A senior admissions dean says, "Students will need to have an understanding of how global events impact individuals at the local or regional level. They'll need critical-thinking and creative problem-solving skills that allow them to seek a better understanding of contemporary issues and how best to address those issues. They'll need to be able to work with diverse populations of people, often with widely divergent points of view. We want to prepare our students to be leaders, innovators and role models in all aspects of their lives."

In trying to capture what makes USC unique, he continues, "We are a values-driven research institution located in the heart of America's most diverse city. We are an institution that provides an undergraduate education modeled on the Renaissance ideal: enlightenment, the active search for knowledge, education that is both broad and deep, and the search for truth and beauty, among others. We are an institution with a conscience; a private institution working for the public good that believes in meaningful and respectful relationships with our neighbors as well as the importance of service learning. We have what is perhaps the most diverse student body (by any measure—racially/ethnically, socioeconomically, geographically, academically, religiously) of any independent institution in America. Although USC shares many characteristics of other fine American higher-education institutions, none combine them in the way we do."

Curriculum, Academic Life, and Unique Programs of Study

Rare among most universities, USC allows its undergraduates to study and earn degrees in all of its seventeen professional schools in addition to its main liberal arts college (the College of Letters, Arts & Sciences). Though the latter is the most popular home for undergraduate majors—about a fourth of all students—the large majority of USC students will pursue at least one preprofessional path, from business or communication to architecture, journalism, engineering, art, film, education, public policy, theater, or social work. What do we mean by "at least one"? USC encourages its students to mix and match majors and minors across any and all of its schools, creating great opportunities for students to expand their learning and combine disciplines in novel ways.

As the university puts it, "The university offers some 120 different minors—the broadest array of any United States university—and encourages students to develop 'breadth with depth' by studying in widely separated fields." USC truly does offer a dizzying array of undergraduate majors and minors, and students agree that they have much flexibility in pursuing and combining them. At the same time, the university does require a broad set of courses for graduation. In addition to a writing (one lower-division and one upper-division course) and a diversity (one course with a multicultural designation) requirement, and requirements specific to majors and minors, undergraduates must fulfill a set of general education (GE) re-

quirements. These entail one course each in: Western Cultures and Traditions; Global Cultures and Traditions; Scientific Inquiry; Science and Its Significance; Arts and Letters; and Social Issues. GE course requirements are filled in the College of Letters, Arts & Sciences.

Students may take a fair number of credits on a pass/fail basis. And they may take graduate level course work during their senior year. Additional program options of note include the Resident Honors Program, to which students apply during the junior year of high school and which accepts fifty of the most competitive students in USC's applicant pool. If successful, these early-admission students will spend their senior year of high school taking courses at USC. USC's Learning Community Program is designed for those who are undecided on their major. Groups of fifteen to twenty students take two courses together during freshman fall that are oriented toward one of about fifteen themes from which students may choose. Students also meet together during the year and participate in other programmatic activities. They apply to a Learning Community during the admissions process or once they get to campus during orientation.

USC's Office of Overseas Studies sponsors many semester or year programs in such locations as Argentina, Australia, Brazil, China, Egypt, the United Kingdom, Ghana, Israel, and Russia. Specialized study-abroad programs are available in such fields as business, communication, engineering, and architecture through USC's professional schools. On campus, a wide range of interdisciplinary programs allow undergraduates to make use of the variety of USC academic fields. These include such fields as several different ethnic American studies programs, animation and digital arts, business administration combined with cinema-television, or international relations; and minors in children and families in urban america (through the School of Social Work), engineering management, law and public policy, peace and conflict studies, 3-D or 2-D art for games (through the School of Art), and videogame design and management (through the School of Engineering).

USC is diverse in many ways, and the university attempts to promote this diversity and dialogue among groups and individuals in a variety of ways. As an admissions dean puts it, "The Office of International Services helps new and continuing students navigate immigration and academic requirements, coordinates orientation for new students, and helps them adjust to life in Los Angeles. The campus also offers a variety of clubs and organizations for students from diverse backgrounds, including El Centro Chicano, the Center for Black Cultural and Student Affairs, and the LGBT (Lesbian, Gay, Bi, Transgendered) Resource Center. Students living in university housing have the opportunity to join special-interest communities that foster awareness of diverse populations and promote a spirit of camaraderie and mutual understanding."

Major Admissions Criteria

According to a senior admissions dean, "Each applicant is considered individually on the basis of performance, academic passion, personal qualities, and potential for success in USC's environment. While candidates are assessed according to their grades, test scores, and the rigor of their high school course work, no strict formula is applied, nor are there absolute cutoffs for these. We are interested in the interplay of these elements, as well as in students' personal accomplishments. Extracurricular activities, essays and counselor/teacher recommendations, along with a student's commitment, leadership, service, or unusual contributions (to a field or to the community), play crucial roles in our decision. We want students who can bring their passion, their commitment and dedication, as well as their own unique 'voice' and a willingness to explore and embrace new ideas. We look at three general areas in our selection process. The first is academic record: the rigor of the high school curriculum, grades earned, and grade trends for years 9 to 12. We also consider the results of standardized test scores (such as the SAT). Next are the student's extracurricular involvement and accomplishments: long-term commitment and initiative, as well as demonstrated leadership and service. Finally, we consider personal qualities: demonstrated written and verbal communication skills, academic curiosity, entrepreneurial outlook, motivation, ambition, and idealism. While solid grades and test scores are important, successful applicants are well-rounded leaders and role models, motivated to contribute to their community and/or to a chosen field of study."

Admission to the class of 2012 was very tough. Some 35,900 students applied, 22 percent were admitted, and one-third enrolled to fill 2,766 first-year spots. The class includes 23 percent legacies, and 11 percent first-generation college students. Fifty-four percent are female. Twenty-four percent earned merit scholarships, including 128 full-tuition Trustee scholars, 332 half-tuition Presidential scholars, and 117 quarter-tuition Dean's scholars.

The average GPA of the class was 3.7. The middle 50 percent range on the SAT was 620 to 720 in critical reading, 650 to 750 in math, and 640 to 730 in writing. The ACT range was 28 to 33. Sixty-two percent graduated from public high school. Outside of the 55 percent who went to high school in California, other top states included Texas, Illinois, Hawaii, Washington, New York, and New Jersey. The largest number of students, one-quarter, entered the College of Letters, Arts & Sciences. Nineteen percent entered the School of Business, 16 percent were undecided/undeclared, 15 percent chose engineering, 13 percent one of the arts schools, 6 percent the School for Communication, 5 percent Cinematic Arts, and a very small number the other three schools.

USC offers no early-action or early-decision program. Unlike many universities, USC offers undergraduate majors across its many distinct schools and students can indicate which are of interest to them in the admission process. They must then be sure to address any particular school- or major-based requirements in their application and supporting materials.

The Ideal Student

USC is clearly not a school for passive students or social wallflowers. The university is diverse, fast-paced, and urban, with an active social life and a need for students to be hardworking, street-smart, socially aware, and assertive in order to be successful. Students need not enter USC with a clear academic plan, but those who do may be best able to take advantage of the mixing and matching that is available among USC's many liberal arts and preprofessional options. The university has had a conservative political reputation, but this doesn't seem to be the case today; rather, USC seems quite balanced politically and socially, with liberal, conservative, and mostly moderate students. Students from many social groups are welcome and have multiple outlets to connect with their fellow travelers.

Students say those who do best are "driven students; self-motivators; doers or action-oriented students with a desire to make an impact on their community and the world." And "students who are unafraid to ask questions and capable of breaking out of their comfort zone will succeed. It is important for students to be resilient and understand that failure builds character in the end." A senior says, "Students who are overachievers tend to do the best because they get involved in multiple clubs and organizations and learn to manage their time very well." And according to a sophomore, "Students who want to learn. The students that take time to get to know teachers and really participate in their education." Another student says, "What I love so much about USC is that there is a place for everyone here. No matter what major you are or what social groups you partake in, there is an opportunity for everyone to succeed. It definitely is a place with a lot of competition, but that just encourages students to do their best and work their hardest, which is what I believe gives USC students a great image."

Student Perspectives on Their Experience

Students at USC choose the university for reasons as diverse as the university itself. Some common themes include the sunny location, the city of Los Angeles, the school spirit and pride, family or local tradition, financial support, and specific academic or preprofessionally oriented programs. A senior says, "Aside from academics, the Trojan family, and its diversity, I simply wanted to go to USC because I knew I'd have fun here. The campus and Southern California weather are beautiful, and I loved how the campus is small enough to have a very 'college-y' feel to it: quads to play football/

Frisbee in, trees to read under, and sporting arenas within walking distance. USC is also very social, and I wanted that in a university. And finally, my one requirement for my college of choice was that it had to have a football team, and USC *definitely* has not let me down in that aspect." A sophomore says, "The energy and vibrancy of the campus, combined with the people I met while here, sold me." A senior says, "I have a wide variety of academic interests, and I wanted to explore all of them during my college education. USC encourages students to pursue diverse majors and minors, and even awards a distinction known as *Renaissance Scholar* to students who have successfully completed such course work." A senior says, "USC was the only college that I was considering that had the full package of everything that I wanted out of an undergraduate education. It had academics, athletics, ways to get involved, a good social scene, and a lot of school spirit." A freshman says, "I chose USC because of the variety of opportunities it bestowed upon me. First and foremost, USC offered me a chance to skip my last year of high school under the Resident Honors Program. Having exhausted most of academic resources in high school by my junior year, I jumped at the prospect of leaving high school a year early to attend one of the finest academic institutes on the West Coast."

Words used to describe the university include: *spirited, exciting, excellent, genuine, revolutionary, vibrant, buzzing, accepting, comfortable, driven, creating, changing, football, sporty, Greek life, diverse, sunny, landscaped, opportunities, connections, social, large, passionate, challenging, casual, research-oriented,* and *international.*

A senior says, "Academically, USC is superb. No other university has a communication program of the caliber that Annenberg (School for Communication) does." A sophomore comments on some other ways in which USC tries to make a large university smaller: "I like how the dean of the Marshall Business School takes the time to familiarize himself with business students by attending events and even teaching a course. I find this gesture meaningful, especially when USC is a fairly large school." On the other hand, "Some professors at USC have a hard time 'teaching' the material, causing many students to skip class and only show up for the exam. Another drawback is the large class sizes. Many of the lectures have over 100 students, creating a very distant relationship with the professor. Moreover, the counselors at USC do not devote enough attention to students." Says a senior, "Academically, USC offers many challenging opportunities such as the Thematic Option honors program, which replaces general education requirements with reading and writing intensive courses. USC also offers a freshman science honors program to challenge incoming students with a tough science curriculum."

Support for students on campus is generally viewed as strong, though there can be mixed reviews of some access to campus administration, larger class sizes particularly among the general education courses, and individual advisors. One senior says, "Really, what USC truly does best is cater to its students' needs and ensure that their experiences are of optimum quality. I have personally experienced this with everyday e-mails of internship/job postings, with awesome events to attend every week that cater to my interests and introduce me to new ones (e.g., it was open to the general public at USC to go hot-air ballooning!), through my incredible academic counselor, and through my classes and professors. Aside from the few general requirement classes I have had to take, I have not been in a class with more than thirty people for many semesters." Freshmen in particular seem to find good support, from the university, their RAs, and others, support which decreases over time, especially in terms of the residential life and dorm situation: "The USC freshman-year experience cannot be beat. Almost every freshman is immersed in a rich residential culture and given an experience can never be duplicated. I would say that USC could try to create this atmosphere for more than just the freshman year—becoming a residential campus for all four years." A sophomore says, "USC is the full package: brilliant professors, down-to-earth students, and an overall nurturing environment."

Of the famed Trojan alumni network, a senior says, "I was impressed with USC by its Trojan family: it's real. It exists."

Campus activities at USC abound. From the obvious (the football team and other major sports) to club sports, intramurals, arts and cultural events, and theater and music performances, students have much to do. A senior says, "I love the Visions and Voices initiative at USC. It is a program that offers students free events that pertain to arts and humanity issues." Another student says USC does a great job of "engaging students. Whether it be getting involved on campus or in Los Angeles, the culture at USC is one that pushes us to *do* something and do it exceptionally." A sophomore says, "Research as a freshman (getting published!) and having the opportunity to work so closely with a professor was a great experience. The Undergraduate Symposium for Scholarly and Creative Work allowed us to present our research." Another notes, "The Global Leadership Program was very beneficial to me since I

was able to travel abroad to Beijing, China, over spring break to witness the rapid growth and development. Overall, the Marshall School of Business emphasizes globalization by means of not only teaching material in the classroom, but also bringing students abroad to see what it is like in action." A senior says, "Attending USC has opened up the doors to do almost anything I could possibly imagine. I have had the opportunity to travel to both Shanghai and Taiwan, learn how to play Ultimate Frisbee, listen to speeches by Tom Brokaw and Wendy Kopp (founder of Teach for America), take a class taught by the university president, and watch a Heisman-winning running back in action, all through USC. Like nearly everyone else I know here, I have done things I would never have imagined before attending USC."

Diversity at USC is seen as plentiful and groups tend to mix fairly easily, though there are certainly cliques to be found. Says a senior, "USC is extremely diverse and international. I love that. When I was in Rome, for example, I was wearing my USC gear and someone shouted to me, 'Fight on!' But on a personal sense, I have made friends with so many different people from around the world—Pakistan, Hong Kong, Guatemala, London, the list goes on." Another says, "USC is diverse. We have many cultural clubs and organizations that are open to all students at USC. Resident advisors also attempt to foster a community of diversity in which its residents are tolerant of people of different backgrounds. The Lesbian, Gay, Bi, and Transgender Association is also heavily publicized and openly accepted on this campus. USC is also socioeconomically diverse, but I have to admit that a lot of the people I've met at this university come from similar, affluent backgrounds. I feel like people of similar cultural, ethnic, and racial backgrounds tend to group themselves together. Even though our university is extremely diverse, if people do not take advantages of opportunities to intermingle with others, then they might be stuck in their own social circle throughout all of college and never experience diversity." Says a senior, "Many students feel a pressure to match the 'image' of USC—which is mainly upper-class Southern California stereotypes based on shows like *Laguna Beach* and *The O.C.* I think this image does not correlate to what the students are really like, but to incoming students especially, it may feel like everyone fits this stereotype."

USC's surroundings get mixed reviews, and students are careful to mention crime issues to prospective applicants. The neighborhood around the campus has issues in terms of muggings, dilapidated houses, noise from hospitals and emergency services, and town-gown relations. Says a senior, "I wish USC were in a neighborhood that was more 'student-friendly.' It'd be nice to walk around and have shops and stores to hang out at, but we are not afforded that option." A sophomore says, "Safety and the ability to balance academics and social life are primary concerns." On the positive side, a senior says, "I think USC's location in downtown Los Angeles is unique and very beneficial to its students because it gives us a chance to interact with the community. Because of our location, I have been able to mentor and tutor inner-city children for the past four years. I have also been able to lead theater and therapy workshops with inner-city high school students. I have learned valuable life lessons through these experiences, and I think it is a very unique aspect of my college experience that most colleges cannot offer its students." Another student

notes, "At USC, the main concern usually revolves around safety. The university does provide several ways of staying safe—like the DPS (Department of Public Safety) and the Campus Cruiser (which provides students safe rides-home after hours)."

Finally, here is some great advice from a graduating senior for prospective freshmen: "USC students have a wide variety of interests, so take the time to find your niche. There is much more to USC socially than the Greek system. Take classes outside of your major and study what you really love and want to learn more about. Don't do activities for your resume; find activities you enjoy and that will enhance your college experience. Don't be afraid to branch out and talk to people you would not have spoken to in high school. Be accepting of others and enjoy yourself."

What Happens after College

According to the university, "USC alumni form a close and supportive community that spans the globe under the banner of the Trojan family. Our graduates aim high and challenge the status quo. They are leaders who take initiative, welcome the free exchange of ideas, and are continually testing out new ways of understanding the world. Trojans care for one another and treat all people with fairness and respect. Together with personal achievement and responsibility, our graduates embody a genuine team spirit that extends beyond the playing field to every aspect of life. The shared values of our alumni are reflected in the work they do, the friendships they make, and their approach to the large and small challenges they face every day. As a member of the Trojan family—now more than 200,000 strong—you will be part of a remarkable group who are shaping the future of the planet."

USC has an extensive Career Planning and Placement Center to help connect undergraduates with internships, jobs, and graduate admissions. A "Trojan network" allows students to search for alumni who have volunteered to serve as resources. A Trojans Hiring Trojans initiative with the alumni association promotes networking and alumni recruiting of USC graduates. According to the university, about three-quarters of USC alums live in California.

As you can imagine, students pursue a very wide range of graduate degrees and careers after USC. Some will stay right on campus through such guaranteed admission programs as those USC offers through its Schools of Medicine and Pharmacy.

Stanford University

Montag Hall
355 Galvez Street
Stanford, CA 94305-6106

(650) 723-6050

stanford.edu

admission@stanford.edu

[

Number of undergraduates: 6,520
Total number of students: 18,680
Ratio of male to female: 49/51
Tuition and fees for 2008–2009: $36,030
% of students receiving need-based financial aid: 44
% of students graduating within six years: 94
2007 endowment: $17,164,836,000
Endowment per student: $918,888

]

Overall Features

Many who know Stanford University today as one of the premier universities in the country (and the world) often assume that Stanford is a member of the Ivies, and are surprised to learn of Stanford's fairly recent origins and rapid rise in stature. Founded in 1891 through a gift of Leland and Jane Stanford in honor of their son, Leland Jr., who had died of typhoid fever at age fifteen, the university was intended to be different from the start. The assumption that Stanford University would be progressive and have impact on society underlay the development of the university and remains central to its mission today. As the university notes, "the Stanfords determined that, because they no longer could do anything for their own child, they would use their wealth to do something for 'other people's' children. They settled on creating a great university, one that, from the outset, was untraditional: coeducational in a time

when most private universities were all-male; nondenominational when most were associated with a religious organization; and avowedly practical, producing 'cultured and useful citizens' when most were concerned only with the former."

Thus today's policies of need-blind admission and generous financial aid; active recruitment of faculty and students of color and nontraditional students; and the pairing of forward-looking research and development across the arts, science, engineering, and technical fields with study in the arts, humanities, and social sciences are directly tied to Stanford's history. Leland Stanford bequeathed to the university more than 8,000 acres of livestock land in Palo Alto that eventually became the campus, still referred to today as "the Farm." Frederick Law Olmsted and Charles Allerton Coolidge designed a campus that is unrivaled in physical and environmental beauty. Few visitors can believe their eyes when they see the lush gardens, green lawns, sandstone buildings, and red-tiled roofs and realize they are the site of a midsize university serving some 6,000 undergraduate and 8,000 graduate students. Stanford's campus and its programs have a strong focus on sustainability and conservation. Stanford's most recognizable facilities have a traditional Californian and Spanish Mission feel, but the university also maintains modern and cutting-edge facilities. All of Stanford's schools are located contiguously on the same campus, which aids in the interaction between students, faculty, departments, and programs. The nearby cities of San Jose and San Francisco provide innumerable outlets off campus, and outdoor activities, whether on the beaches of Half Moon Bay and Santa Cruz or in the mountains of the Sierra Nevada and Yosemite, are plentiful within one to four hours drive.

In the words of a college communications official, "Stanford has evolved along with the nation and the region. Post World War II, Stanford president Wallace Sterling and Provost Frederick Terman conceived of steeples of excellence for Stanford in several key academic areas, including engineering. Terman's role in fostering close ties between Stanford students and the emerging-technology industries has led some to consider him the father of Silicon Valley. He fostered an entrepreneurial spirit that today extends to every academic discipline at Stanford. In addition, Terman's aggressive recruitment of the best graduate students and faculty worldwide, and his understanding of the essential role the creation of new knowledge plays in education, helped propel Stanford from a very good regional university into a world-renowned university. That said, what hasn't changed is Stanford's absolute commit-

ment to a broad liberal arts education as preparation for life. That commitment is as old as the institution itself."

Stanford's student body, like its home state, is as diverse as that of any university one can find. Says an administrator, "Stanford is very ethnically diverse, and that's important to us because we believe the leaders of tomorrow will need a global perspective that can be gained only through exposure to those who are from different parts of the world. It is a priority for the institution and a true pleasure to experience. This is how Provost John Etchemendy describes the importance of diversity: 'Diversity allows for new shapes, textures, and imaginings of knowledge; it encourages the innovation and insight that are essential to the creation of knowledge. A diverse community of scholars asks diverse questions and has diverse insights, and so pushes the forefront of knowledge further faster; providing in turn, a richer educational environment for our students.'"

Stanford's diversity includes not only a racial and ethnic mix. The university provides significant financial aid to about half its undergraduates, and notes that more than 70 percent receive some type of assistance. Students from families earning less than $60,000 pay no tuition or room-and-board expenses, and there are no loans in Stanford financial-aid packages, which are based only on need. The university expected to provide some $110 million in need-based scholarships in 2008–09. About 17 percent of Stanford's freshmen are first-generation college students. Some two-thirds attended public high school. Says a college official, this "is consistent with the vision of Jane and Leland Stanford, who wanted their university to 'resist the tendency to the stratification of society, by keeping open an avenue whereby the deserving and exceptional may rise through their own efforts from the lowest to the highest station in life. A spirit of equality must accordingly be maintained within the university.'"

The undergraduate student body at Stanford comes from fifty states and sixty foreign countries, and is 6 percent international. Just over a third of students are Californians. African Americans comprise 9 percent of students; American Indian or Alaska natives, 2 percent; Asian American or Pacific Islanders, 24 percent; whites, 42 percent; Mexican Americans, 8 percent; other Hispanics, 4 percent; and unidentified students, 5 percent. In addition to the undergraduates, Stanford's graduate student body is also highly diverse and international in background. To support its students, the university offers a number of community centers that provide specialized advising, lectures, educational programs, cultural activities, mentoring, resource libraries, volunteer opportunities, and social events. The centers include the Asian American Activities Center, the Black Community Services Center, El Centro Chicano, the LGBT Community Resources Center, the Native American Cultural Center, and the Women's Community Center.

Stanford guarantees housing for all four years, and 95 percent of students live on campus, with the remaining 5 percent studying abroad or taking part in Stanford's Washington program. The housing system is eclectic, with seventy-eight diverse options. Says an administrator, "Our residential system integrates intellectual pursuits into the dorms, encouraging students to appreciate even more broadly the life of the mind and to learn from their peers." Students may choose from traditional residence halls; houses focused on an academic area, language, and culture; or cross-cultural emphases. First-year student residence halls are staffed by faculty or administrators who serve as resident

fellows; these fellows also live in some of the affinity-based houses and upperclass residence halls. Seven of the seventeen fraternities and three of the eleven sororities offer housing, and about 13 percent of students overall join a Greek organization.

Stanford offers more than 600 student organizations to participate in, and the campus regularly hosts arts and cultural programs. Stanford Lively Arts sponsors theater, dance, and music performances and student programs. Community and public service are fostered by the Haas Center for Public Service, which provides opportunities for Stanford students to get involved during the year, as well as in the summers and through postgraduate fellowships. The center also facilitates service-learning courses and community-based research projects.

Athletics is a major feature of Stanford life, much like its Division I counterparts Northwestern, Duke, and Georgetown. As a college official notes, "Something that distinguishes us from our Ivy peers is that we retain high academic standards while competing successfully in Division I athletics. Stanford combines academic and athletic achievement in a way that is unique in higher education and among the Ivies. Our athletic facilities are extraordinary." All this is certainly true. Stanford's "Big Game" against the University of California, Berkeley's football team is only the tip of the iceberg. The Stanford Cardinal athletic program is nationally competitive in numerous sports, and Stanford has won the Director's Cup, recognizing the most successful Division I athletic program, for the last fourteen years. Stanford offers about 300 athletic scholarships, and 800 students participate in thirty-five varsity sports. Thousands of students participate in intramurals and twenty club sport programs. Stanford's facilities include premier-level indoor and outdoor fields, stadiums, and arenas: a 50,000-seat Stanford Stadium; the 7,000-seat Maples Pavilion; the 4,000-seat Sunken Diamond; a fourteen-court tennis stadium; an aquatic complex with four pools; and a 6,786-yard golf course.

Academically, Stanford offers topflight programs and faculty across many disciplines. The university employs 1,019 tenured and 295 tenure-track faculty, not including non-tenure track fellows and those affiliated with the medical center. Virtually all Stanford faculty hold the highest degree in their field. The faculty at Stanford includes sixteen Nobel laureates, twenty-three MacArthur fellows, and winners of the Pulitzer Prize, National Medal of Science, and Presidential Medal of Freedom, among other honors. With its significant endowment (424 professors hold endowed chairs), Stanford is able to maintain a 6.4 to 1 faculty-to-student ratio, which rivals most small colleges. Stanford offers undergraduates the bachelor of arts, bachelor of science, and bachelor of arts and sciences degrees. The university consists of seven schools: Business, Earth Sciences, Education, Engineering, Humanities & Sciences, Law, and Medicine. Undergraduates can major in the Schools of Humanities & Sciences, Earth Sciences, or Engineering, with about 80 percent choosing the first of these as their home base.

Stanford is in the midst of a major capital campaign begun in 2006, the Stanford Challenge, with a goal of raising more than $4 billion. The main components of the campaign are "multidisciplinary initiatives . . . designed to make groundbreaking advances in human health, environmental sustainability, and international peace and security. Other initiatives improve K-12 education; strengthen Stanford's undergraduate programs; reinvent and enhance graduate programs; and engage all students in the arts and the creative process through exhibitions, performances, and research. Core support and annual giving sustain Stanford's breadth of excellence in teaching and research."

What the College Stands For

According to a college official, "The Founding Grant, created by the Stanfords in 1885, says the objective of the university is 'to qualify its students for personal success, and direct usefulness in life; and its purposes, to promote the public welfare by exercising an influence in behalf of humanity and civilization, teaching the blessings of liberty regulated by law, and inculcating love and reverence for the great principles of government as derived from the inalienable rights of man to life, liberty, and the pursuit of happiness.' That's still the case."

"Stanford's unofficial motto," says an administrator, "is 'The Wind of Freedom Blows,' a quote from the sixteenth-century humanist Ulrich von Hutten, and has been interpreted over time as a call to unfettered research, service, and discovery." Stanford president John Hennessy says, "The pioneering spirit that inspired Jane and Leland Stanford to start this university more than a century ago and that helped build Silicon Valley at the doorstep of the campus encourages boldness in everything we do—whether those efforts occur in the library, in the classroom, in a laboratory, in a theater, or on an athletic field."

Curriculum, Academic Life, and Unique Programs of Study

Stanford provides undergraduates with an extensive and flexible set of options, but also requires a fair amount of academic breadth of study in order to graduate. Students will enter through the School of Humanities & Sciences, and eventually choose a major there or in the Schools of Engineering or Earth Sciences. Students must complete requirements in writing, rhetoric, foreign language, and three additional areas: Introduction to the Humanities (one course in each quarter of the first year); Disciplinary Breadth (one course each in engineering and applied sciences, humanities, mathematics, natural sciences, and social sciences); and Education for Citizenship (one course each in two areas, chosen from ethical reasoning, the global community, American cultures, and gender studies).

A college official shares these thoughts on Stanford programs: "Stanford is aggressively expanding and enhancing its multidisciplinary programs in recognition that solving the world's problems will require collaboration by people conversant in many disciplines. The leaders of tomorrow must be able to see the links among disciplines in ways traditional approaches to education do not necessarily promote. What makes Stanford different is that all of our seven schools are located within close proximity on the same campus. We are focusing in particular on making groundbreaking advances in human health, environmental sustainability, and international peace and security. Initiatives are also under way to improve K-12 education, reinvent and enhance graduate programs and engage all students in the arts and the creative process in recognition of the value of creativity to innovation."

Stanford allows double majors, minors, and individually designed majors. Well-known programs include biology, computer science, political science, engineering, and economics. Other interesting options include archaeology, film and media studies, Chinese and Japanese languages, classics, earth sciences (Stanford has its own Institute for Particle Astrophysics and Cosmology), numerous fields of engineering, feminist studies, public policy, and urban studies. Study-abroad opportunities abound, including programs in Kyoto, Australia, Beijing, Moscow, Oxford, Florence, Berlin, Madrid, Paris, and Santiago, Chile. A quarter of students study abroad before graduating.

Stanford's Washington program places sixty students at a time in the capital to study government, public service, environmental sciences and health, economic and education policy, and international relations. The Hopkins Marine Station, located ninety miles from campus in Pacific Grove, offers students the chance to conduct research and study in marine biology, conservation biology, and oceanography. A Diversity Exchange program facilitates exchanges between students at Stanford and students at three historically black colleges and universities: Howard, Morehouse, and Spelman. Native American studies students can participate in a similar exchange with Dartmouth College.

Major Admissions Criteria

According to a senior admissions officer, "Academic excellence is the primary criterion for admission, and the most important credential is the transcript. The recommendations we receive are crucially important, as are the personal qualities and character that students express through the essays they write us." She describes the successful Stanford student this way: "Students who derive pleasure from learning for its own sake thrive at Stanford. We look for distinctive students who exhibit energy, curiosity, and a love of learning in their classes and lives. We seek students who have selected a rigorous academic program and achieved distinction in a range of courses. We also take into consideration personal qualities—we want to know how students have taken advantage of available resources and their promise for contributing to the campus community and the world beyond Stanford. In some cases, exceptional ability in a particular area may be considered if an applicant is otherwise highly qualified."

Admissions to Stanford is among the most difficult of any U.S. college or university. For the class of 2012, just 2,400 students (9.5 percent) of 25,299 applicants were admitted. One thousand and seven hundred and three freshmen matriculated. Fifty-nine percent graduated from public high schools, and almost 9 percent were internationals. Ninety-two percent ranked in the top decile of their high school class. Fifty-seven percent scored between 700 and 800 in SAT critical reading, 63 percent did so on the writing, and 67 percent on the math.

Stanford offers a Restrictive Early Action (REA) policy, which is a single-choice, non-binding application plan. Students may apply early to Stanford, but not early action or early decision to any other college (public institutions with rolling admission or standard early deadlines like the University of California are excepted). Students do not gain a substantial benefit by applying REA.

The Ideal Student

A college administrator describes the Stanford environment and expectations: "Students learn best when given the opportunity to work closely with faculty members and to participate in the creation of new knowledge. What Stanford does best is combine the intimacy of a liberal arts education in the first several years of study with the research opportunities available only at a research university in the last several years of college. So, undergraduates at Stanford benefit greatly from the six to one student-to-faculty ratio; small classes; extraordinary research opportunities and grants; a phenomenal Stanford study-abroad program; excellent advising; wide-ranging academic support services; lots of off-campus study opportunities, including a marine station in Monterey; the honors programs; and many other offerings, including those in student affairs. Plus, the weather and gorgeous setting make Stanford a very enjoyable place to be. But, essentially, there is a culture of excellence and entrepreneurship throughout campus that encourages people to have high expectations of themselves and others."

According to the college, "Students who derive pleasure from learning for its own sake thrive at Stanford. We look for distinctive students who exhibit energy, curiosity and a love of learning in their classes and lives." Prospective students should be clear that Stanford is not a counterculture, alternative, or lackadaisical setting. Students here work as hard as anywhere; yet they tend to be health conscious, active, and relaxed. They work out physically, through intercollegiate and intramural athletics and on their own at the gym, in the pool, biking, hiking, windsurfing, and sailing. They also work at their academics quite seriously. The residential living and learning emphasis at Stanford provides good support for students who might be concerned about attending a huge university, which Stanford is not, and yet the university provides ambitious students with any number of social and academic outlets. "It is the people that make Stanford different," says an administrator. "They are gifted, high achieving, friendly, creative and idealistic risk takers committed to making a difference."

A student says, "The students who do best here are those who want to do more than just take tests and write essays; those students who use the resources of this world-class university to do more with their time at college. Whether a student wants to start a nonprofit group, intern with a company started by Stanford graduates, contribute to a research journal or express themselves in a student group, those students will always get more out of their time here than those who *only* eat, sleep and study." A recent graduate says, "Passionate, driven, independent students who possess a strong work ethic and a willingness to assume risks to achieve their goals will perform well at Stanford."

Student Perspectives on Their Experience

Students love Stanford, and the university's high graduation and professional and graduate school success rates attest to their satisfaction levels. On this "perennially sunny," "intellectually vital," "brilliant and innovative," safe, happy, and friendly campus, students find each other to be laid-back but hardworking, "collegial," supportive, and not cutthroat. Social life takes place in the dorms, at some frat parties, and on the sidelines of major athletic events. Students feel that they know one another and have a lot of contact, due to the residential-life programs of the university. Traditions like the Big Game and Full Moon on the Quad—when seniors kiss freshmen at midnight during the first full moon in the fall—are cherished. Students recognize a diverse, eclectic bunch of very impressive classmates, with their individual passions and accomplishments. They like the many housing and academic options, on and off campus, and appreciate the freshman and sophomore seminar program, which offers small seminars with strong faculty.

Says a recent graduate, "I chose Stanford because of the unparalleled freedom offered by the university. The quarter system, and the ability to declare a major any time before junior year, gave me that freedom." Having chosen Stanford, he notes, "I really appreciated the chance to study outside of my immediate discipline both by studying abroad in Florence for a quarter and also just by taking classes that caught my eye. Not being tied down to one concentration helped me make the most of my college experience." Another recent graduate praises "the Coterminal program at Stanford that allows undergrads who have made satisfactory progress toward their degree to begin pursuing a master's degree. I took advantage of this program and it shaped my academic path in unimaginable ways."

Students feel supported to try new things and implement their academic and extracurricular ideas. Says one, "Stanford provides students with the opportunity to explore their ideas fully. Having raised $18,000 to take a group of eight students to the Edinburgh Fringe Festival, I know that if a student can come up with a good reason to pursue their passion, there may be advice, faculty support, or grant aid to help them accomplish their goals."

"There are myriad resources to help students succeed both inside and outside the classroom," says a student. "Students meet their faculty advisor during orientation and keep that advisor until they declare a major by which time they have an advisor in their own department." However, "The advising system can produce disparate results. Some advisors tend to be very involved in their advisees' curriculum, while some can be more distant. I would have liked a more uniform advising system."

Students appreciate the administration and faculty's efforts to support them and provide them with freedom and flexibility. "The administration *trusts* the students; academically, socially, and in the way they take risks to achieve their ambitions."

Diversity on campus is recognizable and natural. One student comments, "Stanford is a melting pot of the world's cultures and peoples. Whether it be in residential life or in the classroom, I have met persons from all walks of life and different backgrounds. Almost all if not all ethnic groups have representation on campus through special housing, community centers, or extracurricular activities. More important, Stanford students embody an open spirit—one eager to meet new people and expand existing frontiers."

Advice for prospective students includes learning to balance academic, extracurricular, and social life. Says one graduate, "Ease yourself into freshman year. You won't be able to take on as many activities as you could in high school, so pace yourself with both academics and extracurriculars until you have learned some serious time-management skills."

What Happens after College

According to a senior administrator, "Stanford takes great pride in its graduates and believes they exemplify our approach to education. For instance, eighteen members of the astronaut corps hold Stanford degrees, including the deputy director of the Johnson Space Center. University graduates who have founded Yahoo!, Google, and other Silicon Valley high-tech companies are well known to the public. Other alumni have achieved equal distinction in humanities-related fields: poet Dana Gioia is chairman of the National Endowment for the Arts, and fellow poets Robert Hass and Robert Pinsky have both served as U.S. poet laureates. But the university is equally proud of those students whose careers do not necessarily lend themselves to public acknowledgement, including the vast number of Stanford graduates who are teachers."

Here are some postgraduate statistics from a campus dean: "About 40 percent of our graduates pursue advanced degrees: 1 percent in business, 10 percent in law, 16 percent in medicine, and 73 percent in arts and sciences. The most successful recruiters tend to be: Google, Goldman Sachs, McKinsey & Company, Apple, Microsoft, Lockheed Martin, Yahoo!, Intel, and Boeing. We also have many students who work for the Peace Corps and Teach for America." Many Stanford graduates have received top awards, including ninety-eight Rhodes scholars, seventy-eight Marshall Award winners, and fifty-three Truman scholars among Stanford's 73,007 undergraduate alumni.

Swarthmore College

500 College Avenue
Swarthmore, PA 19081-1397

(610) 328-8300

(800) 667-3110

swarthmore.edu

admissions@swarthmore.edu

Number of undergraduates: 1,477
Total number of students: 1,477
Ratio of male to female: 51/49
Tuition and fees for 2008–2009: $36,490
% of students receiving need-based financial aid: 47
% of students graduating within six years: 92
2007 endowment: $1,441,232,000
Endowment per student: $975,783

Overall Features

Swarthmore, one of the group of liberal arts colleges founded by the Society of Friends, commonly referred to as the Quakers, is distinguished both for its academic programs and quality of student body and for the spiritual and ethical values that define the community. Thanks to its reputation for academic excellence, the quality of its faculty and their commitment to teaching, and its enviable financial strength, Swarthmore has become one of the most selective institutions in the nation. Its standards for admission are equal to those of the Ivy colleges by every measure. The size of the endowment and the ratio of financial resources for educating each student is a story in itself. Swarthmore can afford to attract and retain academic stars in their individual disciplines who combine research and writing with teaching one of the brightest and most intellectually committed group of students to be found on any campus. The student-to-faculty ratio is eight to one, which allows for genuine personal interaction and intellectual exchange. The college is also able to meet the financial need of every student who qualifies for admission and has eliminated loans from its aid packages. Here is one of the handful of colleges that can still maintain a need-blind admission policy. Eugene Lang, a graduate of the college who used his earned wealth to create the "I Have a Dream" Foundation, donated $30 million for scholarships to ensure the continued presence of disadvantaged students on campus. Another famous former scholarship student, James Michener, also left millions to the college, part of which was designated to fund financial aid. This is how a very small college can attract and enroll a diverse and talented entering class each year. Students come from all fifty states and some sixty foreign countries (7 percent are internationals), with the mid-Atlantic states having the largest representation (45 percent). At present almost one-third of the student body is of color and one-half of all students receive financial support from the college.

Swarthmore students are renowned for their academic work ethic and genuine interest in intellectual pursuits. As with the other outstanding strictly undergraduate colleges in this study, all of the campus resources are directed toward the undergraduates, and the faculty teach a full course load and are available and helpful to students. At the same time, they place great demands upon their students, as their goal is to teach them to think for themselves in a critical, analytical way, in part by assigning a heavy reading load and requiring independent study and research assignments. Classes are small and depend on student participation at all times. Swarthmore is justly proud of its magnificent campus, which is a 399-acre, nationally designated arboretum. Philadelphia is a short thirty-minute train ride away from campus for a touch of reality when desired. Swarthmore's historical affiliations with its neighboring colleges, Haverford and Bryn Mawr, as well as the University of Pennsylvania, add to the intellectual and social opportunities for undergraduates. In a supposedly non-sports environment, Swarthmore fields twenty-two varsity athletic teams for men and women in Division III competition. There are numerous intramural and club activities for the remaining students. It is safe to say that the typical Swarthmorian puts academics well above athletics in importance but does enjoy the chance to get away from his or her studies through sports and the multitude of organizations available on campus.

A dean says of the college environment, "At Swarthmore, students are supported by a close-knit community of faculty, students, and staff united by a common interest in learning. Students are taught and advised by outstanding faculty who choose to teach at a liberal arts college because they want to work closely with bright, creative, hardworking young students. Faculty often go out of their way to make themselves available to students, and students reciprocate by seeking out faculty and others for ideas, constructive feedback, and advice about their interests. As a residential learning community where many members of the faculty and administration live within walking distance of the campus, Swarthmore is a place of frequent contact and interaction, which enables the development of open, collegial relationships."

How will Swarthmore develop in the future? As with other of the Hidden Ivies, the college's focus is on globalization and connections to applications of theoretical knowledge. A senior administrator says, "Swarthmore should and will become more global. We will educate more students from overseas and provide even more foreign study opportunities for U.S. students, whether in semester/year abroad programs or summers. We will maintain our academic excellence, but also develop closer engagement with the 'real' world, testing and applying what is learned in the classroom through community-based learning experiences."

What the College Stands For

Although Swarthmore considers itself a secular institution today, its mission statement clearly reveals its commitment to Quaker ideals: "We seek to illuminate the lives of our students with the spiritual principles of the Society of Friends." It adheres to the Quaker ideals of "hard work, simple living, and generous giving; personal integrity, social justice, and the peaceful settlement of disputes." Simplicity in all aspects of life is another significant tenet encouraged among students. This set of ethical values permeates the campus and the classroom. Students certainly adhere both to the work ethic and respect for one another's opinions and beliefs. A key element of the Quaker heritage is a commitment to community service, which all students are encouraged to participate in through the Swarthmore Foundation, which sponsors students in a wide range of service activities. There is a definite liberal attitude toward social and political topics but less of the aggressive activism and effort to effect conformity of thinking that prevails at other well-known liberal-thinking institutions.

What the college sees as important in itself encompasses "very committed professors who are as devoted to teaching as to scholarship. A huge endowment that lets low-income students concentrate on their studies and not their part-time jobs. . . . A campus 'culture' in which academic competition—believe it or not—barely exists. Grades are never posted. The emphasis is on one's personal best." As another administrator puts it, "A key objective here is to teach analytic ability—the ability to see beyond simplistic, black-and-white arguments—and to impart what our president, Alfred H. Bloom, calls ethical intelligence. . . ."

A senior administrator says the college's goals are "broadly, to educate students to realize their intellectual and personal potential and thereby equip them to assume leadership roles across society. We provide a rich curriculum of significant and relevant substance, solidly grounded in traditional fields of learning and more unusually, including educational studies and engineering, while also addressing the frontiers of expanding knowledge, such as Islamic studies and cognitive science. We want students to learn to think analytically, to construct persuasive and well-substantiated arguments, to express themselves effectively both orally and in writing. We want them to become creators of new knowledge. We teach them to test theory against practice, to use their intelligence and learning to advance knowledge and in turn improve their larger worlds." Another dean says, "Swarthmore is committed to academic rigor and excellence, and it is our goal to build each student's capacity for excellence, and each student's commitment to using their talents and abilities to address the complex issues and problems facing our communities and the world. What distinguishes Swarthmore is the degree to which the analytical, creative, problem solving, and communication skills developed in the classroom are also applied to students' activities outside of the classroom, whether through their involvement in clubs and organizations on the campus, through their volunteer work in neighboring communities, through national and global initiatives like War News Radio or genocide intervention initiatives, or through their ongoing service and professional activities as alumni. The habits of heart and mind that are essential to the Swarthmore experience are the habits that help alumni serve, contribute, and lead in many diverse activities and settings throughout their lifetimes."

"I want my students to develop habits of mind that make them open to ways of thinking about the world, themselves, and themselves as actors in the world," says a senior administrator and professor. "I want them to learn to be critical thinkers (not negative thinkers but thinkers who explore a range of ideas) who can tolerate the tensions, unanswerable questions, and dilemmas within any field of study—and within their own lives. I want them to develop (or develop further) a sense of responsibility to themselves, their communities, and the larger society—a responsibility that results at least in part from the privileges they have and/or gain by attending Swarthmore. And I want them to have the depth of understanding and skills they need to act effectively on that sense of responsibility."

Curriculum, Academic Life, and Unique Programs of Study

In keeping with its mission of exposing its students to a broad range of history, science, human behavior, creative expression, and languages, Swarthmore requires a number of courses across these disciplines. However, there is no core curriculum and the choice of subjects to meet the distribution requirement is extensive. Swarthmore's curriculum and academic opportunities have broadened over time to allow students many options on and off campus. First semester classes are graded on a credit/no credit basis to alleviate the academic pressure and potential for competition within the class. Serious students are encouraged to combine fields of study, assist faculty in research projects, study off campus in meaningful programs, and take advantage of the academic resources of Bryn Mawr, Haverford, and the University of Pennsylvania.

Programs in interdisciplinary fields cover a diverse range of topics, from environmental to French or German to peace and conflict to public policy or women's studies. For the highly motivated, the college has created an external examination program that lets students participate in eight honors seminars in their junior and senior years. They are examined at the end of their special studies by a board of external examiners from other colleges and universities.

An unusual policy that helps to explain the accomplishments of the faculty is the four-year cycle of sabbaticals in order to carry on teachers' research, travel, and writing. An administrator says, "What strikes me here is the integrity of the academics. Swarthmore has created a community where learning is very highly valued."

Swarthmore maintains a long-standing academic honors program "built on the idea of dialogue," which challenges students at an even higher level. Other programs of note include the Lang Center for Civic & Social Responsibility, War News Radio, and an externship program that helps students shadow alumni in their jobs for five days.

Academically, Swarthmore's engineering program is an uncommon offering for a small liberal arts college, allowing students to earn a bachelor of science in engineering while simultaneously pursuing the liberal arts. About half of B.S. recipients double major to earn a B.A. in another field outside of engineering. Engineering majors can take up to three-eighths of their courses in humanities and social science departments.

Swarthmore has a very diverse environment for a small college, and works to foster understanding and dialogue in its community. In addition to many student organizations and mentoring and advising opportunities, a dean says, "More important than facts and figures is how differences are appreciated, even welcomed. Each year, before the start of classes, members of the entering class are offered an opportunity to participate in the Tri-College Summer Institute: Seminars on Race, Gender, and Class. Additionally, during orientation week, students are provided with an opportunity to discuss diversity at Swarthmore as a first step in understanding the college's commitment to provide a liberal arts education that values understanding and engaging diversity, in addition to the importance of 'learning across difference' as part of the cocurricular educational experience."

Major Admissions Criteria

What does selectivity translate into, in terms of the credentials for admission to Swarthmore? The college takes seriously its policy of enrolling serious learners who demonstrate intellectual talent and the commitment and initiative to take advantage of the college's resources.

For the class of 2012, admission was extremely competitive. Just 963 of 6,121 applicants (16 percent) were admitted. Three hundred and seventy-four enrolled. For those who reported class rank (44 percent), 24 percent ranked one or two in their class; 21 percent were in the top 2 percent; and 42 percent were in the top decile. Middle 50 percent SAT scores were 680 to 760 in critical reading, 670 to 760 in math, and 660 to 760 in writing. The ACT range was 28 to 33.

A senior admissions dean says Swarthmore seeks "the most excellent and intellectually engaged students who possess tremendous capacity, love ideas and learning, and aim to direct their talents and gifts toward building a better world. While we have always brought excellent students to the college, our mission continues to inform our vision of what this intentional community should look like, and over the past fifteen-plus years, we have placed an emphasis on access and inclusion to bring the most excellent students with diverse backgrounds from all corners of the world. To accomplish this, the admissions office has historically employed a holistic review of each and every application, read in its entirety, by at least two readers, before rendering a final ad-

missions decision. In addition to the traditional measures of accomplishment such as grades, difficulty of curriculum, and test scores, the staff intentionally focuses on noncognitive abilities as evidence of fit and potential to thrive in Swarthmore's rigorous intellectual academic environment. We seek to build an intentional community with recognition of the role that students will play in contributing to each other's education inside and outside of the classroom."

Following from this, the most important admissions criteria, according to the dean, is that "students must possess the ability or potential to take on a rigorous course of study. We place the most emphasis on the high school record, strength of curriculum, and consistent or upward trends as academic difficulty increases. We want to see evidence of analytical ability or potential as evidenced in essays and recommendations, and we want to know that the student is open to different opinions, ideas, and that she or he has a willingness to challenge his or her assumptions inside and outside of the classroom. In addition, we hope the student will be an active contributor in one or several areas outside of the classroom. The campus is as active outside of the classroom as it is inside."

The Ideal Student

Swarthmore students put learning and intellectual exploration above all other priorities. They are attracted to this small, intense college for the quality of its faculty, its resources for learning, and probably most of all, the peer group they will interact with for four years. The ideals of respect for others and a determination to take advantage of their education lead high-performing high school students to choose

Swarthmore over the comparable leading colleges they would qualify for. The nonacademic features of the bigger, more traditional colleges do not appeal to them as much as do the premier teaching and learning that distinguish Swarthmore. So a potential student has to be clear about his or her own priorities and how hard he or she is willing to work to have a successful experience. Students are expected to be independent, to be self-starters open to new and differing ideas. Passion for an academic subject or the arts or athletics or helping others is a trademark of most of the students.

A senior administrator notes, "Swarthmore's distinctive intellectual and campus culture make it attractive to students who seek a powerful learning experience in and out of the classroom, so in many ways we start off predisposed to success in meeting our educational goals for students. One finds evidence of our success in the participation of so many students in the college's honors program and in the varied research and foreign study experiences that students pursue. It is also found in the impact that so many of our students have in business, education, health, government, and community and public service. Students thrive in the college's culture of inquiry, engagement and action."

Notes a professor, "Students at Swarthmore seem to do best if they are willing to ask for and use feedback. Learning new things in a new place surrounded by others who have always done well can be daunting for any student. Without some strength (and an accompanying humility) that allows them to recognize when they need support, how to get it, and how to use it, they can struggle on their own, assuming that they, alone, are incapable. I love first-year advisees who walk into my office overwhelmed with the opportunities, who

have a list of classes they want to take because they look interesting, who don't know for sure what they want to do during their four years—except try out a lot of different things." A graduating senior says those who do best are "intellectual students who have a deep passion for anything—from interests in lacrosse to track to science fiction to computer technology to politics to philosophy."

Student Perspectives on Their Experience

Location, academics, and the personality, philosophy, and feel of the college attract prospective students to Swarthmore. Says a junior, "The combination of strong academics and social life fits me, and the location is great. The campus is beautiful, being on the Scott Arboretum and with the Crum Woods adjacent to the campus (a great place to escape to, to go running in, and be 'off campus' while really not going far on a busy day). The suburban setting is also nice, because students are not limited to the college campus as a safe haven. The SEPTA station (local commuter rail) is literally on the edge of campus, there is a mall within walking distance, and the town of Swarthmore—'the Ville'—is all around." A recent graduate describes the differences she liked about Swarthmore when she visited campus. "First, I was struck by the college's incredible beauty; second, and much more important, I noticed the obvious intellectual atmosphere, different from that of other schools because it appeared to be one in which students did not take themselves too seriously. They seemed excited and genuinely interested in learning for its own sake. They wanted things to be done in their community or the world, not for personal rewards or recognition. The students felt their efforts were worthwhile and their goals should be attained whether they or someone else was responsible for them. It appeared to be a very sincere, grounded, yet exciting atmosphere (and since attending, I have not been disappointed)."

According to students, what Swarthmore does well is teach, encourage, challenge, and support. "Swarthmore is excellent in inculcating academic rigor and curiosity in its students. It therefore graduates very intellectual and maybe rather idealistic students," says a senior. The workload is rigorous, and students will spend a lot of time on their own and with others pursuing their intellectual interests. "Students who do best here must be prepared to study *all* the time. Grade inflation does not exist and so to get an A one has to spend about 90 percent of the semester in the library. That is the only way to stay on top of the curve. Therefore students tend to be very academic and have little time to do co-op programs to gain industrial experience alongside their course work during the semester. Students succeed here because of the elaborate support systems in place to aid in academic advising, tutoring, mentoring, etc. Students feel welcome to go to any source of support when they need it so it is very easy to find help here."

A junior says, "Swarthmore has a strong sense of community. There is an incredible amount of personal attention given to each student, and it isn't uncommon for a professor to become a friend as well as an educator. Swarthmore is a great place to get a solid broad-based liberal arts education—not just superficial in a lot of areas but an understanding of each field. Students have widespread talents and a lot of the time you will find people with very diverse interests or with interest combinations that are unexpected."

Students also pick up on and reflect the college's mission to teach more than just academic subject matter. "Swarthmore is special because of its close-knit nature. Many students feel a sense of community among about 1,450 other students. Students are frequently reminded to strive for 'ethical intelligence' and moral responsibility." Another student says, "One of the most important things to me at Swarthmore is the code of conduct that all students abide by. It's not a written code that every student signs like it is at some colleges, but more a way that everyone lives life at Swarthmore. It fits well into my Quaker heritage and that of the school (though it is nonsectarian now)."

Students at Swarthmore seem to defy generalization, as a junior points out, "There is no one type of student that does well at Swarthmore, and I can think of counterexamples for each adjective I am about to give you. Nevertheless, I would say that the majority of Swatties are intelligent, involved, self-motivated, hardworking, and need to know how to budget time. There are a huge number of college-funded activities and organizations (I believe the number is over a hundred), and you'd be hard-pressed to find a student who isn't a member of at least two or three." Another student says, "People who are internally driven to learn do best at Swarthmore. A competitive atmosphere, which some people use to propel themselves through high school and other colleges, is nonexistent here, so personal motivation is a requirement. Quickly, after getting over the feeling of being less intelligent, socially aware, and politically active than everyone else, students realize they're 'all in this together' and really support one another. Resident assistants, peer academic advisors, and upperclassmen are also instrumental in helping students to succeed at Swarthmore through their constant availability to dispense advice, wisdom, and support."

Students enjoy the diversity of the college and the financial-aid budget that helps to foster it. Says one recipient, "The school's policy of need-blind admission has been particularly important to me and I am happy that Swarthmore is doing all it can to maintain that policy." One student also comments on the travails of small-college life: "One thing that I don't like about being at a small school is how fast gossip spreads. The entire campus is active, and a small incident can spread into a big issue before any authority has time to get the real story out to the students. It's part of what comes with being at a small school, though—and I would say that the benefits more than outweigh the negatives." Some students would like Swarthmore to enjoy more of the national and international reputation they feel it deserves.

For a recent graduate reflecting on her experience, "The school is excellent at providing a rich and diverse array of activities and speakers outside the classroom through a constant influx of scholars, professors, and social and political figures from around the world. . . . The majority of students at Swarthmore (certainly by the time they leave) feel compelled to get out and change the world for the better, whether through the more typically thought-of avenues of medicine and law or through social justice work, innovative research, or grassroots organizing. . . . I feel I have *definitely* come out with an increased social awareness, armed with the tools to challenge and evaluate the opinions and issues around me and to elicit positive, constructive change."

Here are some other comments that reveal much about Swarthmore's students and community: "The school does best to support an atmosphere of intellectual curiosity by encouraging the formation of deep relationships between the faculty and students." A senior advises, "Late at night, after you have been writing your first twenty-page paper, when you have the choice to either sleep or have a philosophical discussion with your hall mates, don't always choose to sleep, you may miss out on the most intellectually stimulating conversation you may ever have had." "You will meet the most interesting, intellectual, and quirky people you have ever gotten to know. It is an environment that pushes a young student to grow quickly over the course of four years."

"Swarthmore's an amazingly supportive place," says another student. "In my three years here, I've always felt that there was someone (faculty, dean, staff, fellow students) to talk to, for everything from picking classes and majors to dealing with personal crises to listening and supporting your grand idea for helping save the world. I used a Summer Social Action Award to learn more about special needs education (something I'm very interested in, through a program I run here). My roommate, a Lang scholar, is currently in Romania doing a project with schoolchildren on the painted churches in Moldova."

"Swarthmore has a reputation for being, and is, a very liberal school. We love it, we celebrate it, we applaud it. However, there is a legitimate concern about the political diversity that exists on campus because of this. I've heard fellow students talk about how they sometimes feel marginalized because of their views on religion, abortion, Israel-Palestine, Republicans vs. Democrats, etc. It's clear that some students have felt uncomfortable because of their political/ideological differences. What is typically Swarthmore about this issue is that people have been talking about it, and I think there's been a lot of effort to hold a dialogue about this that fits with Swarthmore's goal of being a place for open and safe dialogue."

"The great thing about being a student at Swarthmore is that you can be absolutely confident that your voice will be heard. Any issues or problems that you might want to raise with the administration *will* be considered, no matter how casual or far-fetched you consider them to be."

"We Swatties are so spoiled! We get so much attention from professors, so many academic resources to exploit, and so many opportunities to expand our cerebral horizons—it's mind-boggling to think one would wish to be elsewhere. There are no graduate students here, so the entire focus is on you—the undergraduate. Ask yourself: Would you prefer to be a big fish in a small pond, or a small fish in a big pond? Swarthmore may be a small pond, but every droplet of its water shines with the luster of a fresh pearl, and no fish here can be called *ordinary*."

What Happens after College

Thanks to the caliber of teaching and the skills developed, Swarthmore graduates have an outstanding record of acceptance into graduate schools, with a particularly impressive record in the sciences and humanities. Swarthmore ranks among the very top colleges in the percentage of college graduates who attain the doctoral degree. Many enroll in medical school, engineering, education, business, and law. Their success in winning prestigious fellowships is impressive, including Fulbrights, Watsons, and National Science Foundation fellowships, plus many other graduate scholarships. As the college states it, referring to actual alumni, "Swarthmore graduates edit national magazines, run worldwide advertising agencies, write *New York Times* bestsellers, start successful software companies, build manufacturing empires, teach schoolchildren,

raise families, and run for president. Among other things." And they leave campus having established a very particular bond with their classmates—past, present, and future. As one administrator puts it, "Alumni administrators at our peer colleges often comment to me about the apparently unique and almost mystical bond that our alums seem to have for each other, regardless of age, geography, or politics. Our vice president has observed that at social gatherings off campus, our alums seem more interested in probing college policies and principles than in swapping gossip with each other. He's also noticed that alums are more likely to tell him about the ethical values they learned or reinforced here than courses they took or the big game they played in."

According to an administrator in the Career Services office, "Our students' solid foundation in the liberal arts—including their ability to think critically and analytically as well as communicate their ideas effectively to others—helps them succeed in careers as diverse as investment banking and founding a nonprofit to combating genocide in Sudan. Students graduate from Swarthmore with the desire to make a difference in the world—perhaps by bringing light to an injustice, improving the lives of others, starting a business or nonprofit in their home communities, or launching a microfinance initiative in the developing world. It is this combination of a deep ethical sensibility and intellectual brilliance that has led us to coin the term 'ethical intelligence' as the best descriptor of the unique identifying characteristic of our students. Surveyed immediately upon graduation, 65 percent of students pursue employment, 21 percent graduate or professional school, and 14 percent pursue other study or other options including postgraduate fellowships. Twenty-one percent of graduating seniors enter business careers, including consulting, banking, financial services, marketing and management. Additional career fields with high concentrations of graduates are scientific research (17 percent of students) and teaching in the United States or abroad (16 percent of students). Fifty-one percent of our graduating seniors secure employment in the for-profit world, 32 percent in the nonprofit sector, and 17 percent in government. Within five years of graduation, 87 percent of our students typically enroll in graduate or professional school. Of the seniors who enter graduate school immediately upon graduation, 47 percent enter Ph.D. programs, and 28 percent master's programs. Thirteen percent pursue law degrees and 10 percent medical degrees. The universities most frequently attended by our seniors who immediately enter graduate school are the University of Pennsylvania, Columbia University, Harvard University, The University of California at Berkeley, The University of Michigan, The Johns Hopkins University, and New York University. Twenty-five percent of students pursue graduate degrees in humanities, 24 percent in math and physical sciences, 19 percent in life sciences, and 8 percent in social sciences."

"The top career fields of our alumni are: business, healthcare, law, and teaching at the college or elementary and secondary school levels. Within business, the majority of our alumni are in management careers (35 percent), then consulting (20 percent), finance (17 percent) and sales/ marketing (11 percent)."

Here are some examples of graduates provided by a professor: "Students who take courses in educational studies, who do a special major in educational studies and another field, and who complete the teacher preparation program at the college, go on to be educators, doctors, nonprofit workers, lawyers, financial consultants, and journalists, among other things. JJ, a first-generation college student from the Bronx who was a special major in educational studies and sociology/anthropology, is completing the requirements for medical school in a postbaccalaureate program in New York City. JA, a linguistics major and educational studies minor, is an elementary school teacher in the school district of Philadelphia who is also pursuing his doctorate in literacy at the Graduate School of Education at the University of Pennsylvania."

Trinity College

300 Summit Street
Hartford, CT 06106

(860) 297-2287

trincoll.edu

admissions.office@trincoll.edu

[
Number of undergraduates: 2,243
Total number of students: 2,432
Ratio of male to female: 49/51
Tuition and fees for 2008–2009: $38,733
% of students receiving need-based financial aid: 38
% of students graduating within six years: 87
2007 endowment: $440,195,000
Endowment per student: $181,001
]

Overall Features

On a tree-lined, collegiate-looking quad in urban Hartford sits one of the few classic liberal arts college situated in a city environment. With its balance of academic and social life, friendly and happy atmosphere, and moderate small-college size, Trinity offers a wonderful balance for many types of students. Unlike many city schools, Trinity represents a fully residential college campus where most students spend the majority of their time with each other through their academic, social, and residential life. The second college in Connecticut, Trinity was founded in 1823 as Washington College by the Episcopal bishop of the state. The college changed its name in 1845 and moved to its present location in 1878. The college's charter "prohibits the imposition of religious standards" on members of its community and, despite its name, Trinity does not operate as a religious college. Trinity admitted women in

1969 and proceeded gradually to become more open and diverse, and national in reputation.

Whereas Hartford, the so-called insurance capital of the world, was once considered one of the most livable and desirable cities in the country, home to Mark Twain and Harriet Beecher Stowe, among other notables, the twentieth century saw a deterioration in city life and the rise of significant urban poverty and crime. In an attempt to turn what has been and could be a negative aspect of Trinity's campus and educational life into an asset and opportunity for the college, Trinity has in recent years striven to turn itself outward to the city. It has implemented extensive community-service and experiential learning programs; networked with city businesses, nonprofits, and government institutions; and promoted student and faculty engagement in the community.

So, from the 100-acre campus of well-maintained and recently renovated Gothic and modern buildings, Trinity students venture into the city through numerous programs. Trinity undertook a $175 million community revitalization initiative, building out from the campus a "Learning Corridor" that includes numerous public schools, a resource center for students and teachers, a Boys & Girls Club (the first located on a college campus), and a health and technology center. In addition to the wide variety of outreach efforts in the community, Trinity has also sought to change its campus composition and culture.

A senior administrator says, "When I arrived at this institution, over ten years ago, there was little or no oversight regarding student behavior. The reputation was that of a party school, and the current administration was very interested in changing the culture. At that time, many resources and policies were put into place and slowly the culture has changed to accept these changes. We still struggle with that history and the students now do not realize how much has changed in the past decade." A dean echoes these thoughts: "Trinity continues to attract students from diverse backgrounds, which makes for a less homogenous but more interesting group of students. More students now are interested in public policy; community service; and global, nonprofit organizations, as opposed to the more traditional topics such as history and literature that characterized many Trinity students twenty to thirty years ago. I expect Trinity will continue to be less preppy and more cosmopolitan. Emphasis will continue to be placed on keeping the academic standards high, with emphasis on independent work and options for internships and credit for community service activities."

A dean points out "more public school admissions, faculty members from diverse backgrounds with more social science and globally oriented interests, and Trinity's own emphasis on Hartford and its surroundings." In its curriculum, Trinity has implemented "academic distribution requirements that force students to explore areas of interest outside of their 'known world.'" To promote and serve its student body, and facilitate dialogue among students, "Trinity has an active Multicultural Affairs office, alumni of color groups, a Muslim chaplain, a strong Hillel program, and many lectures and arts events that highlight and celebrate cultural differences. We have a first-year reading initiative which selects a book for all first-year students to read that spotlights cultural differences to enable students and faculty to have a common ground for such discussions early on in the academic year. We have advising programs that focus on students at risk, especially students of color, and several supplementary instruction programs for first-year students in calculus, chemistry, and biology—areas where students of color or students from less rigorous high school backgrounds sometimes struggle with courses that are required for pre-health programs. Our academic mentors, RAs, and PRIDE leaders are trained to create dormitory- or seminar-related programs that address issues of difference and community."

About 15 percent of Trinity students still come from Connecticut, and sizable numbers are from Massachusetts and other New England states, as well as New York, New Jersey, and California. About 4 percent are internationals from some thirty countries. About 22 percent are students of color. Financial aid is generous, but Trinity is not totally need-blind. The college will guarantee to meet the full need of admitted students. Some merit scholarships are offered, as well as a limited amount of assistance for international students. The college is seeking to increase its funds for all types of financial-aid awards. About 45 percent of Trinity students graduated from public schools, indicating the large proportion of private school graduates (and thus offering some substantiation for Trinity's "preppy" image).

Despite its urban location, almost all students live on campus in Trinity's coed dorms, and housing is guaranteed for all four years. Upperclass students are awarded their rooms by lottery; affinity-based theme housing is available in the residence halls and students may propose new themes to the college for approval. In 2008, Trinity completed a $32.9 million restoration of its central campus complex, called the Long Walk, renovating the offices, classrooms, and suite-style residences in these historic buildings.

College officials describe a supportive and accessible environment that is quite student-oriented, with 183 full-time faculty and a student-faculty ratio of 10.7 to 1. In the words of one dean, what helps students succeed at Trinity are: "A close relationship between students and faculty, many academic support programs such as the Writing Center, the Quantitative Center, the First-Year Program and its academic mentors; alternative social programs such as the Fred, and Gallows Hill. Trinity also has an excellent library staff who are very student friendly; a strong Computing Center supports student research projects; and there are seriously conscientious deans of students, a counseling center, and the chaplain's office."

Academics at Trinity are challenging, and classes are small in the large majority of cases. Students regularly engage in research projects with professors, and faculty-student mentoring and advising is seen as a strength. In addition to a core liberal arts and sciences program, Trinity offers one of the few accredited undergraduate engineering programs in a small-

college environment. Significant research facilities are present, including recently renovated Raether Library and Information Technology Center, which holds some 1 million books, and hundreds of thousands of additional audiovisual materials. The Trinity College Field Station, in Ashford, Connecticut, is a research facility for hands-on work in the environmental and natural sciences.

After class, social life is active on Trinity's campus. The college and various student organizations facilitate all manner of performances, parties, volunteer projects, and excursions. On campus a coffeehouse and bistro offer food, drink, and entertainment. Transportation is provided to help students access Hartford-area events, from museum exhibitions to indoor arena and outdoor amphitheater concerts. Students can take trains, buses, or their own cars into New Haven, Boston, or New York City for additional social options.

Greek life has diminished in recent years, but about 20 percent of men and 16 percent of women join one of the seven coed houses, which are open to sophomores and other upperclass students. Drinking is prevalent on campus, in both the fraternities and dormitories, where the majority of socializing continues to occur, but the college has been at pains to reduce underage binge drinking and to offer more social options on and off campus.

Athletic competition is a major element of Trinity College life, and the school has long been known for its spirit and athletic participation. Trinity offers fourteen men's and thirteen women's varsity sports. Some 41 percent of students will take part in an intercollegiate sport, and most students take part in intramurals or one of the numerous club opportunities, as well. Though Trinity is a Division III NESCAC college, it competes against highly competitive colleges and universities in and out of the league, being well known for its squash, golf, baseball, and rowing programs. The Koeppel Community Sports Center opened in 2006 and contains a topflight ice rink, recreational space, and other athletic facilities.

Of Trinity's location, an administrator notes, "This institution has the benefit of being a small private liberal arts college that resides in a metropolitan area—a state capital. There are many opportunities our students have for civic and cultural engagement that others in a more rural campus would not." Nevertheless, don't forget that there exists beautiful surrounding countryside and quaint towns in Connecticut and Massachusetts, with skiing and boarding, hiking, and mountain biking nearby.

What the College Stands For

"Trinity College is a community united in a quest for excellence in liberal arts education. Our purpose is to foster critical thinking, free the mind of parochialism and prejudice, and prepare students to lead examined lives that are personally satisfying, civically responsible, and socially useful." Thus goes the college's mission statement. Trinity identifies four key elements that serve as a foundation for its ability to reach its goals:

"An outstanding and diverse faculty who excel in their roles as teachers and scholars, bringing to the classroom the insight and enthusiasm of people actively engaged in intellectual inquiry. Working closely with students in relationships of mutual respect, they share a vision of teaching as discussion—a face-to-face exchange linking professor and student in the search for knowledge and understanding.

"A rigorous curriculum firmly rooted in the traditional liberal arts, but one that also inte-

grates new fields of study and interdisciplinary approaches to learning. Trinity encourages a blend of general education and specialized areas of study, and takes imaginative advantage of the many educational resources inherent in Trinity's urban location and international ties.

"A talented, motivated, and diverse body of students who are challenged to the limits of their abilities and are fully engaged with their studies, their professors, and one another. Our students take increasing responsibility for shaping their education as they progress through the curriculum, and recognize that becoming liberally educated is a lifelong process of learning and discovery.

"An attractive, secure, and supportive campus community that provides students with myriad opportunities for interaction with their peers as well as with the faculty. The college sustains a full array of cultural, recreational, and volunteer activities, and embodies the philosophy that students' experiences in the dormitories, dining halls, and extracurricular organizations are an important and powerful complement to their formal learning in the classroom."

Curriculum, Academic Life, and Unique Programs of Study

Trinity's curriculum is broad and balanced, and students will fulfill general education requirements on their way toward concentrating in one or more traditional departmental or flexible majors. Students must complete one course in each of five different categories: the arts, humanities, natural sciences, numerical and symbolic reasoning, and social sciences.

Trinity is proud of its innovative First-Year Program. A campus dean says, "The goals of the First-Year Program are to foster critical analysis, writing, discussion and debate, research and information literacy. The program is designed to provide incoming first-year students with an intellectually challenging experience, a successful transition from high school to college, and an introduction to a lifelong habit of learning." In addition to this seminar and mentoring-based freshman program, Trinity has also strengthened other curricular requirements for graduation. A dean says, "Trinity was a leader in developing its First-Year Program from the initial realization that students needed more support in the transition to college (especially as more students came who were first-generation college students) to the design of the seminars and an upperclass student mentoring program, to our recent faculty vote making a first-year seminar a graduation requirement that also fulfilled the first of two writing-intensive requirements. Also important: a new academic requirement to take a course designated 'urban and global studies' and a second-language proficiency expectation."

Trinity is known for majors in social sciences like political science and economics favored by the preprofessional crowd, but also for biology, chemistry, computer science, and the physical sciences as preparation for graduate studies in medicine and other scientific and technical fields, and English and history for those inclined toward the humanities. Among its thirty-eight major areas of study, and 900 courses, Trinity also offers less common or newer options, such as public policy and law; environmental science; women, gender, and sexuality; Jewish studies; and educational studies. Interdisciplinary minors include options like architectural studies; cognitive science; community action; human rights studies; performing arts; and writing, rhetoric, and media arts. An Interdisciplinary Science Program is available, as are special curricular programs.

These are: a five-year program in engineering with Rensselaer Polytechnic Institute's Hartford campus that leads to a bachelor's degree from Trinity and a master's degree in engineering from RPI; the InterArts Program, a two-year option combining performing and liberal arts; the Health Fellows Program, which allows students to explore health sciences professions; the Guided Studies Program, focusing on humanities and the evolution of Western civilization; and the Human Rights Program, the first such program at an American liberal arts college.

Trinity's engineering program is certainly a key calling card for the college. The major sits within the college's overall curriculum, and students are encouraged to study not only other areas of the sciences, but also the humanities and social sciences offerings of the college. Associated with the engineering department, Trinity sponsors the Fire Fighting Home Robot Contest for entrants of many ages and abilities.

The Trinity College Community Learning Initiative promotes service learning and research throughout the college. Other connections to Trinity's home base include: the Cities Program, an interdisciplinary program offered to selected incoming students interested in urban environments; City Term, which places juniors and seniors in urban organizations; the Kellogg Initiative to study Trinity's efforts to help revitalize its neighborhood; and the Hartford Studies Project.

If all that wasn't enough, Trinity promotes numerous study-away programs through its Office of International Programs. About half of Trinity students will study off campus prior to graduating. Among seventy-five U.S. and international programs in forty countries, Trinity maintains its own programs in Barcelona, Cape Town, Paris, Rome, Santiago, Trinidad, Vienna, and other areas. Trinity is a member of the Twelve College Exchange program, as well as American University's Washington Semester program, Woods Hole's Sea Semester and environmental science programs, the Williams-Mystic program in maritime studies, the La MaMa Urban Arts program in New York City, and the O'Neill National Theater Institute in Waterford, Connecticut.

Major Admissions Criteria

According to Trinity's admissions materials, "Your transcript is, arguably, the most important piece of information that we consider. It not only tells us your grades and the courses in which you received those grades but it provides us with patterns. Have your grades been consistent? Have you done better in some courses than others? Have your grades been improving or declining? It also tells us whether you have sacrificed higher grades in order to tackle more demanding honors, Advanced Placement, or International Baccalaureate courses. Typically, we will see grades from all four of your high school years, including first-term marks in your senior year. Senior course work is considered, so ensure that you maintain a solid record throughout the year. . . . We look for students who have done well in the most challenging set of courses they could take within the curriculum offered by their school. We're really interested in how you have done at the school you attend and not how you, or anyone else, might be doing at another school. . . . Trinity, like most four-year colleges and universities, requires some form of standardized testing. This could be either the SAT I, any two SAT II subject tests, or the test of the American College Testing Program (ACT). If English is not your first language, the Test of English as a Foreign Language (TOEFL) might be the more appropriate test for you. We

consider contributions that you have made to your community—both in school and out of school. Quality is more important than quantity. We are more influenced by the depth of your commitments rather than by the number of your commitments. If you are an artist or a musician, you might want to consider sending slides or a tape. Teacher and counselor recommendations are very important and we take their observations very seriously. Hopefully, you will be able to develop relationships with teachers where they will feel comfortable making positive comments about you. . . . In addition to what others write on your behalf, what you write is a significant component of your application . . . we do read those essays! Take time with your writing, not just your essay, but also any short answer questions you are asked. A lot of emphasis is put on writing in the classroom at Trinity, so we look at your writing closely when evaluating your application for admission. We try to balance all of these factors in making admission decisions. Ultimately, the admissions committee tries to determine whether or not Trinity and you are a good match."

The admissions office reports that applications are up 80 percent over the previous five years. Of the 5,136 students who applied for the class of 2012, 592 enrolled. Their average SAT scores were 640 in critical reading, 647 in math, and 657 in writing.

The Ideal Student

The common elements of a successful, happy Trinity student seem to be self-motivation, balance, and engagement. Trinity students tend to be active, social, and interested in multiple areas of academic and extracurricular involvement.

"We have excellent extracurricular options for students to participate in and make their out-of-classroom experience as fulfilling as their course work," says a college official. "The range of activities, events, clubs, intramurals, and the like keeps our students busy and helps them to develop as both individuals and leaders."

"Trinity is a balance of 'types' here on campus," according to a dean. "It is a small residential college that urges its students to get out in the world and connect with urban and global issues. (We have a brand-new Center for Urban and Global Studies.) We have a wide variety of departmental majors with the flexibility for many different student-created majors or minors. Trinity has an engineering program that works within a liberal arts context. A high percentage of Trinity students are scholar-athletes. Trinity students often 'think out of both sides of their brains.' Students performing on dance stages may be neuroscience majors; football players are poets and film majors. The stereotypical Trinity student used to be a preppy economics major who wanted to go to work on Wall Street. Happily, that stereotype applies to a smaller and smaller percentage of students. But now more are exploring a wide variety of interests and fields, from education to learning Arabic, from environmental studies to anthropology. The good news is that defining the Trinity student is getting harder and harder."

Student Perspectives on Their Experience

Trinity students report a roughly similar set of experiences with the college. Most find a balance academically in various fields, plus a balance between social life and academic work. It is seen as ok and even promoted to be well-rounded. Students see each other as motivated

and capable, but more mature as upperclassmen than as freshmen ready for the work-hard, play-hard stereotype of college life. Of course, Trinity is not alone in this respect.

Students appreciate the city of Hartford as a source for learning and experience. There are lots of service activities that are easy to participate in. Town-gown issues aren't immaterial, and students see a divide between the college and its neighbors. There are some safety concerns among current and prospective students, especially after some high-profile assaults in recent years. The college is seen to be working on this area with security patrols, lighting, and other safety measures, in addition to its active engagement in trying to foster development in the city. Students value the business, nonprofit, and public service internships available to them, and the college's role in helping to facilitate them. The urban environment can be challenging, which students need to take into account, but it is a real asset to the college.

Faculty are viewed as brilliant, supportive, caring, and involved. Close relationships with professors and peers are developed. Students, faculty, and administrators are seen as friendly, and Trinity as a college where people, including your teachers, know your name.

Social life admittedly takes place mostly at more-or-less open frat parties, and in dorms, and alcohol is prevalent but not pressured in both venues. Students also find it helpful to have access to Hartford and other area options, a fun dorm life, and college- or student-sponsored comedy nights, films at Cinestudio, theater and music performances, the Fred for open-mike and theme nights, and so on.

The student body is viewed as more of a mix today. Certainly the preppy types are there, but students also report more of a political mix and social and economic diversity than meets the eye. Such diversity seems less "in your face" than at some other colleges. The old-boy stereotype and reputation are outdated but still present in terms of some rich students with preppy, expensive clothes and accessories. The attractive and outgoing "beautiful person" is readily apparent on campus, but so are many other kinds of students. The diversity and dialogue seem a work in progress.

There is a lot of school spirit and pride in the college among students, particularly in relation to athletics. Trinity students find more than enough academic challenge and social outlets, though according to some students, some community members don't fully exploit either. Though some see Trinity as not the most highly "intellectual" campus, most will agree the college is full of many smart students who are highly capable, involved in the community and a multiple college offerings, and headed for future success. The faculty and academic programs are strong enough to challenge any student.

What Happens after College

Trinity alumni are loyal and hardworking networkers and supporters of their college. Fifty-five percent participate in college fund-raising efforts, and many serve as mentors and recruiters of Trinity graduates. Trinity students are widely represented in the business world and do well in terms of graduate admissions. Some 60 percent of Trinity alumni will have received a graduate degree after five years, and 75 percent after ten years. Nineteen percent will enter graduate school immediately after graduation; 79 percent will work; 3 percent will travel or volunteer, or serve in the military. Popular career fields for alumni include business (34 percent), education (16 percent), science and engineering (12 percent), and legal services (8 percent).

Tufts University

Bendetson Hall
Medford, MA 02155

(617) 627-3170

tufts.edu

admissions.inquiry@ase.tufts.edu

[

Number of undergraduates: 5,029
Total number of students: 8,280
Ratio of male to female: 46/54
Tuition and fees for 2008–2009: $38,840
% of students receiving need-based financial aid: 49
% of students graduating within six years: 89
2007 endowment: $1,452,058,000
Endowment per student: $175,369

]

Overall Features

Tufts has a long history of progressive thought and education. Founded in 1852 by Unitarian Universalists, the church's first college initiative, Tufts has grown from a small liberal arts college led by a succession of Unitarian ministers into a nonsectarian blend of undergraduate and graduate education with significant advanced-degree programs in the sciences, international affairs, and other areas. Long struggling to move out from the shadows of its Boston-area brethren Harvard and MIT, as well as the other elite northeastern colleges and universities, Tufts has in the last decade found its own identity as the largest college or small university in New England (take your pick). Here you will find a top-notch academic environment full of independent and energized students and faculty.

Tufts is situated in suburban Boston, just five miles (and a bit of a world) away from the city, which is readily accessible by subway (the "T") and bus. Home to many small and large colleges and universities, the area is a mecca for students and offers innumerable social, academic, and preprofessional opportunities. Tufts's main undergraduate campus is located on a hill in Medford/Somerville, including 150 acres of trees and attractive buildings. The university also maintains campus facilities in Boston; Grafton; and Talloires, France. Affiliations with Boston's School of the Museum of Fine Arts and the New England Conservatory of Music provide additional academic outlets. Medford/Somerville is an historic area in and of itself, and in the neighborhood of many small revolutionary-era towns and the "living history" of the city of Boston.

Tufts is predominantly an undergraduate college, but also presents elements of a nationally respected graduate university. Its Fletcher School of law and diplomacy offers a unique package of graduate degrees in law, international business, and international affairs—oriented disciplines, as well as opportunities for undergraduates to experience some of Fletcher's courses, speakers, and programs. Tufts's strengths in the sciences are represented by its schools of veterinary medicine; engineering; dental medicine; nutrition science and policy; medicine; and biomedical sciences. Tufts also offers degree programs through its Graduate School of Arts & Sciences. The Tisch College of Citizenship and Public Service offers a wide variety of internship, outreach, service, and fellowship programs to students through the university.

Most undergraduate classes at Tufts are small, with a few large lectures (100 to 200 students) in some introductory courses. Graduate students interact with undergraduates sometimes as teaching assistants and often as members of social organizations, clubs, and other programs. The university is well known for engineering, health sciences, and international relations, but is strong in many fields across the liberal arts, including political science, languages, drama, English, and child development. Undergraduates will enter either the School of Arts and Sciences (the College of Liberal Arts) or School of Engineering. They may also apply for an early-assurance program as sophomores in order to gain admission to the graduate School of Medicine.

Though there are ten fraternities and four sororities, only about 10 percent of men and 4 percent of women belong to a Greek house. Residential life is mixed, with the university guaranteeing and requiring housing on the up-hill or downhill campus quads for only the first two years, after which many students will travel abroad and then return to live off campus. Tufts supports many arts-oriented student activities

on campus, including six a capella singing groups, several dance troupes, and numerous plays and musicals. They have access to the $10 million Aidekman Arts Center, which houses exhibition space, dance studios, private music rooms, and even a theater in the round. In 2007, the Granoff Music Center opened, offering more high-level performance facilities and practice spaces. The Leonard Carmichael Society (LCS) is the largest student organization, with more than 1,000 students involved. It serves as an umbrella for community service and volunteer activities, supporting more than forty such programs each year. There are also more than fifty religious, cultural, and political groups on campus.

Not historically known for its sports focus or rah-rah spirit, Tufts nevertheless competes in the New England Small College Athletic Conference group as the largest member. Athletic competition is thus serious and a real outlet for student-athletes who must devote a substantial amount of time to their endeavor. Many intramural and club sport options exist, as well.

Tufts maintains significant diversity in its student body. Geographically, one-quarter are from Massachusetts, but the rest hail from all fifty states and some sixty countries abroad. Tufts provides only need-based financial aid, but is need-blind in admissions and meets full need. Tufts provides a limited amount of financial assistance for international students. According to a college official, "We aim to bring students from all walks of life to campus. For instance, 15 percent of our students come from an international background, 27 percent are students of color. In the freshman class, 40 percent received need-based financial aid, including 133 Pell grant recipients. Overall, the university awarded nearly $13 million in need-based aid to the incoming class, a 32 percent increase since 2006. A significant way that Tufts is accessible to diversity is through the advising program as well and over 200 student-run clubs, including cultural clubs like the Asian American Alliance, the International Club, the Multiracial Organization of Students at Tufts, LGBT, and the Pan-African Alliance, to name a few."

Traditions at Tufts are quirky, and include the Naked Quad Run (self-explanatory) and the cannon, placed on campus to celebrate the first collegiate football game and painted for all manner of reasons on a regular basis. Tufts-fest and Spring Fling provide opportunities for everyone to blow off some steam in the spring.

According to an administrator, "Tufts University is a medium-size research university on an intimate scale. It is big enough to support the global resources and teaching of a major research one university, but small enough to promote a community that values dialogue between students, faculty, and staff. Tufts University is a place that is very open to change and willing to accommodate various constituencies on campus in their desire to try new things and improve existing programs. Students and faculty are encouraged to be creative and explore their ideas, and the academic experience is structured with this in mind."

What the College Stands For

Tufts sets out this philosophy in its vision statement: "As we shape our future, quality will be the pole star that guides us. We will seek quality in our teaching and research and in the services that support our academic enterprise. Our programs will be those that meet our own high standards, that augment each other, and that are worthy of the respect of our students and of scholars, educators, and the larger com-munity. For students, our search for quality will mean opportunities both in and beyond the classroom to become well-educated, well-rounded individuals, professionals, and scholars. For faculty and staff, it will mean opportunities to realize their talents in the service of Tufts's goals." The university goes on to elaborate its focus on lifelong learning and preparing students "to use historical perspective and to be receptive to new ideas" and to "be sensitive to ethical issues and able to confront them." Tufts describes itself as a "teaching university," saying, "We value research and scholarly activities independently from their contribution to teaching, but they will never become so important that we forget our commitment to educating our students." With a focus on "citizenship," Tufts says "We want to foster an attitude of 'giving back,' an understanding that active citizen participation is essential to freedom and democracy, and a desire to make the world a better place." Other goals include diversity, a global orientation, and fiscal responsibility.

An administrator notes these goals: "To educate future leaders who will actively look for innovative ways to go into the world and make a difference in the global community. Our hands-on curriculum creates engineers who love the arts, multilingual business leaders, doctors who relish politics, and policy makers who are technically and scientifically literate. Tufts students are doers: They are assertive, entrepreneurial, socially committed, and involved."

Curriculum, Academic Life, and Unique Programs of Study

Tufts requires the completion of a demanding set of requirements in order to graduate. Students in the Arts and Sciences must show competence in the "foundation requirements" of writing (two courses) and foreign language and culture (three semesters of a foreign language, and then either three more courses in the same language, three courses in a different language, three courses in a foreign culture, or a mix of advanced language and foreign-culture courses). Students must also fulfill a world-civilizations requirement (one course on a non-Western civilization); a quantitative-reasoning requirement (if they have low SAT or ACT scores); and distribution requirements demanding that students take two course credits each in the humanities, arts, social sciences, natural sciences, and mathematical sciences. Engineering students will have a different set of graduation requirements, geared toward a different educational emphasis, but no less demanding. All students will need to fulfill a departmental or program major or concentration, or a plan of study offering more flexibility and individualization. Many interdisciplinary study options are available, including engineering science studies, Asian studies, film studies, leadership studies, and urban studies.

The Experimental College at Tufts allows students to select from some thirty nontraditional course offerings each semester, for credit, which are taught by faculty, students, and visiting instructors. The outside instructors are often working professionals with diverse backgrounds who bring very different perspectives to the classroom environment in courses like "Going Green: A Practical Guide to Environmentalism" or "The AIDS Epidemic in Theatre and Film."

There is much study abroad going on at Tufts, with about 40 to 45 percent of students participating in a one-semester or full-year program. About one-third of these students utilize Tufts's own programs in places like Chile, China, Ghana, Hong Kong, Japan, London, Madrid,

Oxford, Paris, and Tübingen, Germany. Tufts offers a variety of off-campus study options in addition to traditional foreign study abroad. Students can participate in an exchange with Swarthmore College, or join the Williams-Mystic College Maritime Studies program in Connecticut. Tufts students can also cross-register on a limited basis at Boston College, Brandeis University, and Boston University.

The Tisch College of Citizenship and Public Service "empowers students, faculty, and alumni to be active, engaged leaders in their local and global communities through a university-wide interdisciplinary program." All students—graduate and undergraduate alike—have access to the Tisch College programs, including some 100 courses focused on citizenship and service, faculty research fellowships, summer research and internship opportunities for students, and support for student initiatives.

Five-year combined bachelor's and master's degree programs are available in a number of academic departments for talented students who wish to accelerate their graduate studies. A combined B.S. or B.A. and master of arts in law and diplomacy is offered as a coordinated six-year program that students apply to during their junior year of college and is very much an offering unique to Tufts. Prospective engineers may apply to a six-year B.S. in engineering/master of arts in law and diplomacy degree program as early-decision candidates during their senior year of high school.

Specialized dual-degree programs are also available with the New England Conservatory of Music and the School of the Museum of Fine Arts, leading to the B.A. or B.S. from Tufts and bachelor of music degree, or bachelor of fine arts, respectively, in five years of study. Finally, early admission to some of Tufts's professional schools, including the Schools of Medicine,

Veterinary Medicine, Dental Medicine, and Public Health, offer additional advanced-study options for focused Tufts undergraduates.

A college official says of these important programs for the university, "The Institute for Global Leadership, the Experimental College, and the Tisch College of Citizenship and Public Service are opportunity bankers of the university, which empower our students to pursue civic engagement and intellectual citizenship beyond the traditional curriculum. The low student-teacher ratio has raised the level of engagement in the classroom. The Summer Scholars Program is designed to match faculty mentors from each of the eight Tufts schools and teaching hospitals with an undergraduate student interested in doing paid research for the summer. This has increased the amount of student-faculty research and the amount of honors theses completed each year. Students interested in pursuing internships in public service can receive a stipend through Tufts University."

Major Admissions Criteria

Tufts's admissions process has become more competitive and more challenging in recent years. The university uses the Common Application but has several supplemental essay options for students to complete. Students may submit either the ACT with writing, or the SAT and two SAT subject tests. Tufts accepts a substantial proportion of the class in two rounds of early decision.

According to an admissions officer, the most important admissions criteria are "academic performance (transcript, testing and curriculum); recommendations (secondary school counselor and teacher); and essays." The college states, "In addition to evaluating an academic fit, the admissions committee looks for ways a student may contribute to the community as a whole. They will assess the level and type of involvement in each activity and may ask questions such as: Has the student been a significant contributor or leader? How has the involvement contributed to the school or larger community? Does the student have a special talent in a particular area? We do not expect that every student be captain, president or editor in chief; rather, we look for meaningful involvement with their school and/or community."

The class of 2012, 1,302 students, was selected from 15,642 applicants, of whom 3,988 were admitted (25.5 percent). 85 percent ranked in the top decile of their high school class. Mean SAT scores (and middle 50 percent ranges) were 719 (690 to 760) in critical reading, 722 (690 to 770) in math, and 723 (690 to 770) in writing. The average ACT was 31. 1,111 students enrolled in liberal arts, while 192 entered engineering. One-third applied early decision. More than a quarter were students of color, and 13 percent were internationals or Americans living abroad. 62 percent graduated from public high schools, and 79 percent came from outside Massachusetts. 69 countries and 44 states were represented in the class, with significant numbers coming from California, Connecticut, Florida, Illinois, Maine, Maryland, New Hampshire, New Jersey, New York, Pennsylvania, and Texas.

The Ideal Student

A successful and happy Tufts student is willing to work hard and get involved. This is generally a campus where students know each other, even though it is a midsize university. It is an active place, physically, socially, and intellectually. The tone of campus is primarily liberal and is likely to stay that way. One of the more politically correct environments, Tufts is seeing conservatives establishing more of a presence but still feeling like a minority and fighting an uphill battle. Overall, Tufts students are hard to typecast, representing many different groups of Americans and internationals. The common ethos of most students would be professional, global, activist (whatever the cause), academic, and open-minded.

A senior administrator says, "Tufts students are risk-takers and entrepreneurs with clear plans to use their education to solve complex problems and make a difference in their hometowns and countries. They value collaboration over competition and, while they are all quite the smarty-pants type, tend not to engage in much discussion about how smart they are. Rather, they show their intelligence through action."

Student Perspectives on Their Experience

Students tell us they feel satisfied socially and with their peers at Tufts, and engaged with their faculty and intellectual pursuits. There is more than enough academic challenge to be found on and off campus (sometimes too much, especially in the sciences) and a general appreciation for the university's efforts to diversify and go global.

High marks are given to accessible faculty who are experts in their field, and the outside speakers brought to campus, from national and international political and human rights leaders to celebrities. Students find a lot of hard work, but most classes are small and well taught. It is seen as a close campus community, with great facilities including library resources and good food with lots of options.

Students expect to get to know most everyone over time, or to be only a few degrees of separation away. Students from different groups hang out and relate to one another. There is certainly a monied crowd, both American and international, and on campus one will see grunge, neohippie, D.C. politico, and fashionista existing side by side. Some say the cultural affinity houses and groups can promote segregation and that students don't always communicate with one another. Others cite excessive PC and gay or women's movement issues, while others are pleased that the campus is so tolerant and that lifestyle differences are not a big deal. In general, one can say this is a politically correct campus, but there are some dissenting opinions and media outlets. Some students describe "intense" debates between viewpoints on a "split" campus.

Students socialize and party often, from the parties at fraternities to dorm parties and parties at various affinity houses. They appreciate that they are able to do a lot of different things on and off campus in Boston and at surrounding colleges.

Students like the international feel of Tufts and its programs, as well as its global focus and facilitation of study abroad. It seems a cosmopolitan place, especially to those not accustomed to a city environment. Students spend most of their time on campus. Boston is a T ride away but can take some effort to manage. Mostly the older students use the city, which does provide a pressure relief valve, and more diversity and action, including clubs, arts, sports, and restaurants, than the somewhat dull towns of Medford and Somerville. Relations with town residents are given mixed reviews, given the number of students living off campus (about a quarter of the upperclassmen), which generates numerous lifestyle issues.

One of the interesting things about Tufts is the diversity of student perspectives on their experiences. There seems to be much less unity of opinion than at some of the other Hidden

Ivies. Responses can range from those who say the university life is competitive and hardworking, to those who say the school is a friendly community and laid-back; from those who say they find the more anonymous experience of a research university, to those who feel know everyone at a small college. The respect and admiration students share for the faculty and academic programs certainly is universal.

Students relish the many community-service and campus-leadership opportunities they find at Tufts. They deny that the college is full of those who wanted to go to an Ivy and hold a grudge. A large proportion entered Tufts through early decision as a first choice, and absolutely do not regret their decision. The high graduation rate bears this out. As a senior summarizes on the Tufts Web site, "When I first visited the campus I knew that I would not need to visit any other schools. The campus has an enchanting collegiate atmosphere that allows you to dream big and believe that you can achieve anything. The professors and students that I met were in love with the school. Everyone I met was passionate about something; you could tell that these people were going to change the world."

What Happens after College

Tufts students are strongly focused on career and professional activities, with many choosing graduate study immediately or within a few years of graduation. This does not preclude their becoming involved in service activities like the Peace Corps, AmeriCorps, or Teach for America for a few years prior to career or graduate placement.

An administrator says, "Our students come to campus poised to learn from the world, and are able to use the classroom and research as a starting point to transform communities around the globe. A spirit of activism exists on our campus through groups such as Engineers Without Borders and the New Initiative for Middle East Peace. Tufts continues to be in the top twenty-five American colleges and universities with graduates serving in the Peace Corps. Academically, our students are also making great strides in maintaining an intellectual campus. Tufts University has had four Truman scholars in four years (two last year), there is an increase in the number of undergraduates who have published research, and the number of honors theses continues to rise. There is a tangible intellectual vibrancy on campus that transcends the classroom—even cafeteria conversations are laden with interesting tidbits from a cross section of disciplines. It's not uncommon to run into a group of people dissecting a Korean horror flick in the middle of the Residential Quad (at least, in the warmer months). Our students go all over the world to work in politics, finance, medicine, and the arts and the sciences. Through a strong Arts and Sciences and Engineering curriculum, students are prepared to use an interdisciplinary curriculum to apply creative solutions to any profession they choose. Over three-quarters of students will also pursue graduate degrees within five years of graduation."

The Tufts Career Services office offers substantial resources for students, including access to a network of some 7,000 alumni and parents who serve as mentors and advisors. The office maintains an extensive internship database, and internship grants are available for students who want to pursue community or public service internships in the summer. The Empower Program of Tufts's Institute for Global Leadership facilitates Tufts student involvement in experiential learning and social entrepreneurship. These kinds of experience will assist graduates in job placement and graduate school admission.

Tulane University

6823 St. Charles Avenue
New Orleans, LA 70118-5680

(504) 865-5731

(800) 873-9283

tulane.edu

undergrad.admission@tulane.edu

[

Number of undergraduates: 6,408
Total number of students: 8,696
Ratio of male to female: 46/54
Tuition and fees for 2008–2009: $38,664
% of students receiving need-based financial aid: 33
(Class of 2012)
% of students graduating within six years: 76
2007 endowment: $1,009,129,000
Endowment per student: $116,045

]

Overall Features

Tulane is a traditional, academic, and well-known national university founded in 1834, yet an institution that bristles with contradictions. An academically competitive and demanding program requires a great deal of work from students who wish to succeed, yet the "Big Easy" New Orleans environment offers all manner of distractions and social outlets. A major research university with renowned schools of law, medicine, architecture, and sciences and engineering, Tulane caters today to its undergraduates to provide them with a more structured, residential, unified, and attentive collegiate education. Tulane is laid-back and intense; friendly and socially demanding; southern and northeastern.

One must view Tulane and its modern history and development from the standpoint of pre- and post-Katrina, the devastating hurricane that leveled much of the city in 2005. Tulane and its campus withstood the hurricane with surprisingly little long-term physical damage, though many millions of dollars of cleanup expenses were incurred. Nevertheless, the university used the hurricane as an opportunity to restructure its programs and to reach out to the city of New Orleans and its residents, which were far more negatively impacted by the storm and floodwaters. Amazingly, Tulane has showcased its own resilience in the face of the huge challenges posed by Katrina—including having sent its student body across the country to colleges and universities that accepted them temporarily as visiting students—by continuing to build on and renovate the campus, and by using its faculty, students, and resources to support the rebuilding of the city. Tulane has seen record application numbers in recent years, and many students now desire to go to Tulane precisely because of the hands-on learning opportunities they will gain academically and in terms of social service by engaging with New

Orleans. As an administrator points out, "Being a top research institution in the Deep South and going through and recovering from the nation's worst disaster make Tulane unique. Being located in the country's most unique city also helps to make Tulane unique."

One post-Katrina change, and one that generated a large amount controversy, was the combination of Tulane and Newcomb College, the undergraduate men's and women's colleges, respectively, into the unified Newcomb-Tulane College. All undergraduates enroll here, and then can pursue programs and their degrees through five undergraduate schools: the School of Architecture; the Freeman School of Business; the School of Liberal Arts; the School of Public Health and Tropical Medicine; and the School of Science and Engineering. Additional schools offering graduate degrees are the schools of law, medicine, and social work.

Across its several campus locations, including on its main uptown campus of 110 acres, visible signs of Tulane's continuity and progress include a $44 million Lavin-Bernick Center for University Life encompassing 142,000 square feet of space, featuring a ballroom, bookstore, eating facilities, and student meeting areas. Tulane has outstanding academic facilities, from architecture design spaces to science research labs, and a downtown campus that includes medical facilities. The uptown campus is quiet and pretty, situated across the street from the Audubon Park and zoo and in the uptown residential neighborhood. The St. Charles streetcar line takes students into the French Quarter, Magazine Street, the Faubourg Marigny, and other centers of life in New Orleans.

An honors program provides additional challenge and opportunities for Tulane's top students. Sixty-two percent of Tulane classes enroll fewer than twenty students; just 8 per-

cent enroll fifty or more students. Students will encounter graduate-student teaching assistants in some introductory classes. Some 1,300 faculty support students, including 832 full-time faculty, 92 percent of whom hold a terminal degree.

Tulane awards significant need-based aid but has been very active in using merit-based financial scholarships to attract more academically oriented students to campus. As a member of a competitive Division I athletic conference, Conference USA, Tulane also awards athletic scholarships to a fair number of students. School spirit surrounding athletic competition is strong, especially for the men's and women's basketball teams. The university had seven women's and five men's teams in 2008–09 and plans to be back to full varsity athletic strength by 2010–11. It expects to support sixteen varsity sports by then, after having reduced athletic competition temporarily following Katrina. Tulane has a prominent recreation center on campus for active students, and the university supports many club and intramural athletic options.

Residential life on Tulane's campus is quite diverse. Just half of students live on campus, including all first- and second-year students. Others compete by lottery for upperclass spots in university housing, but many opt to live off campus in the many houses and apartments adjacent or close to campus. A small proportion of male students live in fraternity houses, but women are prohibited by New Orleans law from living in sorority housing. About a quarter of men and women belong to one of the fifteen fraternities and ten sororities. While a fair amount of social life does take place in the Greek system, much of the excitement and amusement happens among the off-campus houses and throughout the city of New Orleans with its many bars, clubs, cafes, restaurants, arts facilities, professional sports arenas, and live entertainment venues.

Looking toward the future, a senior administrator says, "I expect Tulane to become even more diverse in the representation of its student body and its faculty, to continue to strengthen its relationship with the city of New Orleans, and to keep its efficient operating philosophy which was developed out of necessity post-Katrina."

What the College Stands For

Announcing Tulane's post-Katrina renewal plan, Tulane's president, Scott Cowen, said, "Tulane University, now more than ever, is a powerful and positive force as New Orleans and the Gulf Coast region begin the monumental task of recovery. We are determined to find opportunity in the face of adversity. Tulane will do more than just survive; we will thrive and continue our role as a beacon of learning and research for the region and nation, as well as a dynamic engine of growth and change for New Orleans and its citizens." Tulane's mission statement says, "Tulane's purpose is to create, communicate, and conserve knowledge in order to enrich the capacity of individuals, organizations, and communities to think, to learn, and to act and lead with integrity and wisdom. Tulane pursues this mission by cultivating an environment that focuses on learning and the generation of new knowledge; by expecting and rewarding teaching and research of extraordinarily high quality and impact; and by fostering community-building initiatives as well as scientific, cultural, and social understanding that integrate with and strengthen learning and research. This mission is pursued in the context of the unique qualities of our location in

New Orleans and our continual aspiration to be a truly distinctive international university."

An administrator notes these goals for Tulane: "To graduate students who will have a lifelong love of learning and academic exploration, an understanding of the responsibility of engaging in community and nationwide service, and the ability to celebrate the joys and the challenges of life." She continues, "Students are aware that there is more to college life and to getting a good education than sitting in a classroom; New Orleans teaches its Tulane students many life lessons about maturity, responsibility for individual and collective action, living in a diverse environment, and giving to the community."

Curriculum, Academic Life, and Unique Programs of Study

Tulane has a broad and deep curriculum, with more than enough academic and professionally oriented programs to suit most any student. The university has been reemphasizing the undergraduate academic life of the Newcomb-Tulane College, and in addition to numerous specialized programs, require graduates to fulfill a broad set of general education requirements. All students apply to Newcomb-Tulane as first-year students, and remain part of the college throughout their university years. They also apply to a major by the beginning of their fourth semester at Tulane, and become simultaneously a member of the school that houses that particular major choice. In addition to whatever major and school-specific requirements they must fulfill, all students must also show competence in "writing, foreign language, scientific inquiry, cultural knowledge, and interdisciplinary scholarship."

A public-service requirement means all students must complete a service-learning course by the end of the sophomore year; they must participate in their junior or senior year in another service-learning course, an academic service-learning internship, a faculty-sponsored public-service research project, a public-service honors thesis project, a public service–based international study-abroad program, or a "capstone experience" with a service component. An administrator notes, "Curricular changes made possible by the public-service graduation requirement have brought faculty expertise and student ingenuity to help solve some of the issues facing our communities. Additionally, the university has expanded its direct outreach to the community through the expansion of its community-based health clinics and its advocacy for public education through its School of Medicine and Scott S. Cowen Institute for Public Education Initiatives, respectively."

All first-year students must complete a one-credit Tulane InterDisciplinary Experience Seminar (TIDES) focused on any number of themes. Tulane seniors need to complete a capstone experience requirement tied to their major. All students must complete courses in "Perspectives in the European Tradition" and "Perspectives Outside the European Tradition," as well as "Comparative Cultures and International Perspectives."

Students should have no trouble finding courses of interest to fulfill these various requirements. Interdisciplinary programs include an emphasis in political economy, and strong preparation for graduate studies in law, medicine, and engineering. Tulane offers early-decision programs into its well-regarded law and medical graduate schools for those with more certain future career plans. A number of other integrated degree programs allow students to earn a bachelor's and master's degree

from the university in just five years. Tulane's historical strengths have been in the natural and environmental sciences, humanities, and international and Latin American studies. The Stone Center for Latin American Studies has several hundred thousand volumes in its library and 150 courses taught by eighty faculty. Tulane maintains a Center for Bioenvironmental Research with nearby Xavier University, and offers course and program opportunities in such areas as communications with Loyola University, which is literally next door. Numerous other centers and institutes provide research and course opportunities in a wide variety of areas.

Options for interdisciplinary, joint-degree, and self-designed major programs abound. These include African and African diaspora studies; digital media production; environmental studies; Jewish studies; women's studies; and, of course, jazz studies. The Tulane Honors Program enrolls hundreds of Tulane scholars annually, and participants take advanced honors courses taught by faculty, can design their own major, and can spend their junior year studying abroad. They can also choose to live in special honors housing where they will be advised by more senior Tulane scholars.

Study-abroad opportunities abound and are administered by the Center for Global Education. Options include semester programs around the world, as well as generalized and specialized Newcomb-Tulane College Junior Year Abroad programs in France, Italy, Spain, and the United Kingdom.

Major Admissions Criteria

Tulane has become increasingly competitive since Katrina, surprising many applicants who have assumed that the university must be easier to get into today than it was prior to the hurricane. While classes just after 2005 didn't have quite the yield Tulane was looking for, applications and enrollment rebounded significantly in 2007 and 2008. The class of 2012 saw 1,550 enroll out of 34,117 who applied, 27 percent of whom were admitted. Data for the class entering in 2009 show more than 40,000 applications received!

The middle 50 percent SAT ranges for the class of 2012 were 620 to 720 in critical reading, 630 to 710 in math, and 630 to 720 in writing. The ACT composite score was 28 to 32. Sixty percent of students ranked in the top decile of their high school class, and 85 percent were in the top quintile. Geographically, the class came from forty-seven states and eleven different countries. More than a third came from the Northeast, 15 percent from Louisiana, 9 percent from the rest of the Southeast, 16 percent from the Midwest, 9 percent from the West, 6 percent from Texas, 4 percent from Florida, and 4 percent from the mid-South. Just 2 percent were internationals. Almost two-thirds graduated from public high schools.

All students apply to the Newcomb-Tulane College, as noted, but they may indicate school preferences in the admission process and present supporting materials for specialized programs like architecture or music. Tulane is certainly interested in reading students' essays and recommendations, and focuses most intently on the curriculum and grades that students present in their transcripts. However, given the application numbers the university is seeing these days, test scores have also become a more important element of the admissions decision process.

Tulane has no binding early-decision appli-cation plan, but its nonbinding early-action program is popular and competitive.

The Ideal Student

Tulane enrolls a student body that is both wealthy and lower-income, diverse and predominantly white, fashion-conscious and grungy: more contradictions. Certainly there are many preppy or city-sophisticated students on campus, with many quite focused on preprofessional career tracks. "Why would a student want to come to Tulane University?" asks an administrator. "Because he or she will receive a first-rate education in a city and on a campus that provides opportunities unlike any other school in the country. Where else can an architecture student actually design and build houses following a hurricane? Where else can a medical student work in a community clinic? Where else does a student have to engage in community service as part of his or her core curriculum?"

Students of diverse backgrounds are welcome on Tulane's campus and in New Orleans, but must feel comfortable mixing in an ethnically diverse and high-crime urban area, though one not immediately visible on the Tulane campus. The public service requirement will ensure that hiding out uptown is not an option for most students. Says a college official, "The goal of the requirement is to create students who are more civically engaged and more cognizant of issues being faced by New Orleans, the region, the nation, and the global community. The requirement enhances the academic learning of all students by providing them with opportunities to enrich and reinforce the knowledge and skills gained in the classroom with practical application in the community."

Students who do best at Tulane will be hard-working, socially outgoing, or willing to dig deep into one or more research projects, and excited about and ready for the challenges and opportunities presented by a midsize, urban campus in a unique city and subculture. One senior says, "The students who do best at Tulane are those who are able to best balance a social life with their academic and extracurricular commitments. New Orleans offers countless opportunities to have a good time, so it's very easy to be distracted from schoolwork if you aren't focused. Tulane students also tend to be involved in some form of extracurricular activity (besides partying), so students must balance that, too. Time-management is key."

A freshman says, "The type of person that succeeds at Tulane is someone who studies, seeks the easily accessible help from professors, and someone who isn't afraid to step out of their high school comfort zone. The workload at Tulane is definitely not small but with dedication and effort it can easily be mastered and accomplished by any student who has the desire. The temptation to party is always there, but your peers at Tulane will never pressure you to stop studying in order to party. Anyone from all backgrounds and all learning types can succeed at Tulane. All you need is the dedication and the will to succeed."

Student Perspectives on Their Experience

Students choose Tulane for its combinations and contradictions. It is one of the premier midsize private universities with a campus in a city, and it offers both a social and academic life with many options. Students also appreciate some of the specialized departments and preprofessional offerings beyond the general liberal arts program found in many of the smaller colleges. Says one student, "I chose Tulane because for me it was the best confluence of culture and education. The cultural experiences of New Orleans are boundless and I desired to attend an institution whose atmosphere would enable me to learn both inside and outside the classroom." A senior says, "I ultimately chose Tulane because it was the (academically) best school that I could go to for free (I got a scholarship to Tulane). However, money issues aside, I probably would've chosen to come to Tulane anyway. I had always wanted to live in New Orleans, and it was far enough from home to satisfy my need for independence but close enough to make trips home convenient." A freshman points out, "The people are friendly, the academics are demanding, but at the same time the teachers are very personable, and [they] want to get to know you and offer their assistance. Finally, the city of New Orleans allows endless opportunities."

A senior agrees, as do many students, that the faculty at Tulane are accessible and try to work individually with students as much as possible: "Out of all of my classes this semester, the largest one had about twenty-five students, and that size, if not smaller, is typical. You are able to get a personalized learning experience and you are able to really get to know your classmates and your professors." Another student says, "The academics are rigorous and time consuming but the teachers are very accessible. You are not treated like a number or another face in the crowd. The teachers provide one-on-one help and actually want every one of their students to succeed. It makes learning enjoyable. Along with easy aid from the professors, Tulane offers a free and easy-access Tutoring Center and writing-help center. At Tulane everyone is willing to tutor a friend or join group-study sessions with no hesitation." A senior agrees: "Tulane provides an exceptionally supportive learning environment by supplying their students with the necessary resources to succeed."

New Orleans has its pros and cons for students, but primarily favorable reviews: "Tulane's catchphrase is 'Only in New Orleans. Only at Tulane,' and anyone who goes here knows that it's true. Our location and student body make a unique combination that cannot be found anywhere else. How many other schools have events like Crawfest, our annual outdoor crawfish and music festival? Not many, I bet. Tulane doesn't have much school spirit when it comes to athletics, but everyone here knows that going to Tulane provides an experience much more rich than just attending a college, and there is a sense of pride and camaraderie among all students. Even with more than five other colleges in the city, Tulane stands out." From Mardi Gras to Jazz Fest, to out-of-the-way local eateries, New Orleans offers many options for students on a big or small scale. "From microeconomics to music and culture of Nola, classes are career-based and interesting. After all the hard work in class, the city of New Orleans is a great place to relax, whether at Audubon Park across the street, shopping on Magazine Street or eating at one of New Orleans's famous restaurants, it isn't hard to smile in Nola. The balance created naturally by the university makes the high-school-to-college transition and the college-to-real-world transition easy." Another says, "How can anyone ask for a better city to live in than New Orleans?! There is great food, great music, and exciting culture. New Orleans gives a Tulane student the opportunities to not only learn a subject in the classroom but to apply what he/she has learned in the real world. Business students study local businesses, and premed

students, such as myself, take part in studies and work in the local charity clinics. Helping to rebuild a broken city is so fulfilling." Another student says, "The Center for Public Service has been particularly instrumental in my life the past couple of years. I entered the university before the public service requirement was mandatory, so I didn't need to do anything to graduate, but I've been very involved anyway because of the amazing opportunities it offers. I wanted to teach English as a Second Language for a semester, and through the Center for Public Service, I was able to make it into an internship and get school credit for it. Because of my work with that program, a professor nominated me to be a Public Service fellow, which I am currently."

The main drawback of Tulane's location tends to be the crime rate, or fears thereof. Says one student, "Personal safety is something any New Orleanian, including Tulane students, needs to be aware of, but Tulane has gone above and beyond to ensure student safety including implementing the Safe Ride program so that students and their guests can travel safely off campus at night." Another reports, "The campus and surrounding area are very safe. The Tulane police department and New Orleans police department are on twenty-four-hour patrol on campus and in the areas surrounding campus. My peers and I have never felt unsafe at Tulane."

Many students tend to be ok with the level of diversity on campus, or not comment on it at all, while others would like to see more university support for less-affluent students and students of color, as well as better efforts to bring students together. Says one student, "First, I would change the price tag to make Tulane more affordable. Second, I would like Tulane to be more diverse socioeconomically, racially, and ethnically." Another says, "Tulane's diversity is most evident in the religious beliefs of its student body. . . . Tulane also has an Office of Multicultural Affairs which oversees all organizations supporting diverse student populations. While Tulane offers support for students of different ethnic and racial statuses, the racial minorities of our country are in the minority at Tulane as well." Students encourage prospective applicants to visit the campus and city to find out for themselves what the environment is really like.

A final comment from one happy student at Tulane: "I know everyone thinks that their school is special. But after having spent my first semester of college at another school that I considered going to (because of Tulane's closure after Hurricane Katrina), I know that Tulane is special. I am so happy here and I know that I am getting an experience and education unlike any other."

What Happens after College

Tulane graduates tend to focus on professional careers as well as careers involving some type of service. Speaking of some notable alums and Tulane graduates in general, an administrator says, "Imagine our pride when Lisa P. Jackson, a graduate at the top of her chemical engineering class at Tulane, is named the secretary-elect of the Environmental Protection Agency by President-elect Barack Obama. Or when we learn that Marine Corps Lt. Col. Douglas Hurley will serve as the pilot for the STS-127 shuttle *Endeavour* which NASA plans to launch in 2009. Or that internationally renowned cardiologist Dr. Michael E. DeBakey received a presidential medal for his pioneering work in heart transplants. Not to mention countless stories of Tulanians who come to school from other cities in the United States and choose to stay in New Orleans after graduation to continue the city's post-Katrina rebuilding. Or international students who choose to return to their native countries after graduation to put their education to use." Another official says of Tulane graduates, "They can go anywhere they want. Some stay in New Orleans, some return to their home cities or countries, some go to major urban areas. They are prepared to work in medicine, law, finance, public health, community service, education, the arts, architecture, business, engineering, social work, and public health and tropical medicine."

A group of Tulane graduates will stay put for a fifth year to pursue a combined master's degree program, or will complete law, business, or medical school through the university's accelerated or guaranteed enrollment programs. About 10 percent of Tulane graduating seniors will enroll in medical schools around the country, with an acceptance rate of 80 percent. Ninety percent of law school applicants who worked with the university's advising office gained admission to at least one school.

Vanderbilt University

2305 West End Avenue
Nashville, TN 37203-1727

(615) 322-2561

(800) 288-0432

vanderbilt.edu

admissions@vanderbilt.edu

[
Number of undergraduates: 6,496
Total number of students: 10,551
Ratio of male to female: 44/56
Tuition and fees for 2008–2009: $37,005
% of students receiving need-based financial aid: 39
% of students graduating within six years: 89
2007 endowment: $3,487,500,000
Endowment per student: $330,537
]

Overall Features

Vanderbilt is named for its founder, Commodore Cornelius Vanderbilt, who had in mind two goals in donating $1 million in 1873. The first was to create a university dedicated to the highest level of teaching and research that would benefit society like the top northern institutions, and the second was to contribute to strengthening the ties that should exist between all regional sections of our common country. Both were noble purposes as the nation worked toward healing its wounds following the Civil War and began its economic and technological expansion. A visit to the campus and a survey of the academic programs, teaching and research facilities, and graduate schools make it clear that this early vision has been fulfilled beyond what the generous commodore could have imagined. With a traditional undergraduate college of arts and sciences at the heart of the university and a

number of specialized schools, Vanderbilt is similar to its Ivy League cousins. It is more diverse than some of them, however, in that there are ten defined academic schools that provide thousands of courses at the undergraduate and graduate levels.

There are 838 faculty in the four schools that serve undergraduates: the College of Arts and Science, the Blair School of Music, the School of Engineering, and the Peabody College of Education and Human Development. Total full-time university faculty number 2,997, 97 percent of whom have terminal degrees. The undergraduate student–faculty ratio is nine to one. The undergraduate faculty are known for their commitment to teaching and advising students at the same time that they are expected to pursue research in their respective disciplines. Students come to Vanderbilt from all fifty states and ninety foreign countries. Of the student body, 45 percent are from the southern states and the rest of the American students are from every corner of the country, with very large mid-Atlantic and Midwest contingents. Eight percent of students are internationals. Ninety percent of undergrads live on campus, with all first-year students residing in one of ten houses in an area called the Commons. Sixty percent of students receive some type of financial aid, from significant need-based awards with no loan component to substantial merit-based scholarships ranging from several thousand dollars to full-coverage distinguished honors like the Ingram scholars.

Vanderbilt is situated on a beautiful campus of 300 acres on the edge of Nashville, a vibrant and growing city of more than 1 million residents. The Peabody College section of the campus is registered as a National Historic Landmark, having been designed after Thomas Jefferson's model for the University of Virginia. The main campus was designated a national arboretum in 1988. Thus students have the opportunity of living in a handsome, well-appointed environment with a city famous for its music, entertainment, and film industries, and good restaurants, performing arts centers, and shopping. Nashville is also the state capital and home to a dozen other colleges. While campus life is very active with arts events, athletics, student organizations, and its Greek system, students like the fact of having so many attractions available down the road as well.

Students choose to attend Vanderbilt because they want a balance of very strong academics, an active social life, a spirited community, a moderate political tone, an extensive range of curricula and majors or specialties, and excellent facilities. On the whole they are from cosmopolitan backgrounds and find the size of the university and the location very appealing. A majority are thinking ahead to graduate school and careers and believe that Vanderbilt's education and reputation will prepare them well. Most undergraduates choose to live on campus, which is a requirement for all first-year students. Approximately 35 percent of men and 50 percent of women join one of eighteen fraternities or twelve sororities, which are one of the active social outlets on campus. With the many school-sponsored activities and outlets in the city, there is not a great deal of pressure to join a Greek house.

Vanderbilt commits a good portion of its large endowment to generous scholarships, physical facilities, and very competitive faculty salaries. Ten years ago it received the single largest gift ever given to a university from one individual, over $340 million for academic and financial aid-programs. Through the large amount of financial aid Vanderbilt seeks to ensure that the meritocratic ideal of the opportunity for talented young men and women of all racial and socio-economic backgrounds to attend the university is fulfilled. There is a definite commitment at the trustee and administrative level to diversifying the student body. Vanderbilt awards 300 renewable merit scholarships each year and additional athletic scholarships to students of exceptional ability. Honor scholarships are awarded each year to outstanding students of color, those with musical talent, leadership, community service or teaching commitments, interest in engineering and technology, or demonstrated exceptional intellectual ability.

Students are quick to compliment the university on the quality and range of facilities available for residential, social, physical, and academic life. The student recreation center is one of the largest and most comprehensive in the country. It includes a full-size gym, squash and racquetball courts, pool, weight-training center, a cafe and social activities room, and an 8,000-square-foot wellness center with full-time trainers. This facility, in combination with acres of playing fields, enables students to participate in any of forty different intramural teams and some forty club-level sports. The Sarratt Student Center has a major theater and film wing, restaurant, and bistro. Under the leadership of its former chancellor of twenty years, Vanderbilt has refurbished virtually every residence hall and classroom on campus, in addition to constructing a number of new buildings for teaching and research. Vanderbilt is a member of the NCAA-sponsored Southeastern Conference, which means they field a large number of varsity men's and women's teams that add to the campus spirit. However, due to Vanderbilt's high academic standards the teams find the competition from the large state universities in the conference a major challenge, particularly in football.

What the College Stands For

Vanderbilt inculcates a spirit of community caring through the honor code that has long been in place. Students are governed in their academic behavior by the principles of the honor code and tenets of community values.

Vanderbilt publishes each year a statement called the Community Creed, which establishes the principles of behavior and values by which the community is guided. They include the tenets of scholarship, honesty, civility, accountability, caring, discovery, and celebration. These are the values the university pledges to foster for the well-being of the entire community. In the spirit of its founding, the university is committed to providing students with excellent teaching and preparation for assuming a responsible role as professional and community leaders. The honor code and community service programs, the residential life, and extracurricular outlets are all designed to build individual skills, cooperation, and concern for others. The theme that learning takes place through a variety of activities in a variety of settings is strongly articulated. The individual must become self-sufficient and honorable in his or her dealings with peers and instructors. Volunteer service to the community has long been a value that the university encourages; to this end, an Office of Active Citizenship & Service coordinates the forty-five student service groups who interface with 100-plus community agencies and local schools in myriad support roles. The Ingram Scholarship Program, established to honor the late chairman of the board of trustees, underwrites community service projects that students create and provides tuition scholarships on a selective basis to those who commit a good deal of their time to assisting others.

A senior administrator notes the educational goals of Vanderbilt: "1) to provide a strong liberal arts education which challenges students to develop their intellectual potential; 2) to provide depth of learning in one or more majors as preparation for subsequent graduate and professional work; 3) to prepare students generally for the 'offices of life' by giving them ample opportunity to learn how to work effectively with others, to lead and to follow; 4) to instill in them an ability to reason, morally and intellectually, to recognize and develop their strengths in solving problems, to learn to apply what they have learned to the task of recognizing what needs doing and the ability to do it." Vanderbilt focuses a great deal of energy on its undergraduate students, seeking through a wide variety of means to help them succeed. These include "a high level of expectation from faculty and peers, but in a friendly, open, and nurturing environment; a strong academic advising and residential counseling system; the availability of faculty and staff who make getting to know and serving individual students a priority in all aspects of their work responsibilities; a widespread sense of civility in the community as a whole, combined with expectations of high industry from all members of that community; and pleasant physical surroundings—a beautiful campus, comfortable residence halls, student recreation center, etc."

What distinguishes Vanderbilt among many of the Hidden Ivies—and colleges and universities in general—is its combination of strong undergraduate teaching and education with a research institution offering significant resources and programs. As an administrator describes it, Vanderbilt has a "commitment to the finest undergraduate experience available which is enriched by being embedded in a leading research university, surrounded by nationally ranked professional and graduate schools; a commitment to supporting undergraduates with the kind of attention that gives students the space to learn to make their own choices, and the assistance and services to make those choices well informed; and an inordinate commitment to offer undergraduates the right balance of challenges and support required to prepare them for active, productive lives."

Vanderbilt's mission statement succinctly captures the universities goals and values: "Vanderbilt University is a center for scholarly research, informed and creative teaching, and service to the community and society at large. Vanderbilt will uphold the highest standards and be a leader in the quest for new knowledge through scholarship, dissemination of knowledge through teaching and outreach, creative experimentation of ideas and concepts. In pursuit of these goals, Vanderbilt values most highly intellectual freedom that supports open inquiry, equality, compassion, and excellence in all endeavors."

Curriculum, Academic Life, and Unique Programs of Study

Vanderbilt provides a sweeping range of courses and concentrations. The undergraduate college consists of four distinct schools. Students are required to experience a range of liberal arts and science subjects regardless of which school they are enrolled in. This is in keeping with the overarching mission of graduating thoughtful, humanistically educated students who have the perspective and judgment, the critical thinking, writing, and communication skills necessary to cope with the complex and rapid changes they will confront. The academic year is divided into two semesters and an optional May term in which one course can be completed for credit in any department. B.A. degree recipients must complete a set of requirements known as "AXLE: Achieving Excellence in Liberal Education." Mandatory components include a "First Year Common Experience" focused on a writing seminar, a writing requirement, and a liberal arts requirement. The latter demands that students take thirteen courses in seven different "areas of inquiry." Students also major in at least one discipline. Each of the schools has its own set of core requirements that builds on the general curriculum.

The 3,500 students in the College of Arts and Science have access to forty-six different majors, as well as interdisciplinary concentrations, and the opportunity to design their own major. Peabody College specializes in education at all levels and human development. In addition to centers for teaching and research in the traditional area of education, a center for entrepreneurship studies is one of the many special programs Peabody supports. Peabody's 1,200 students can major in dozens of fields that lead to careers in administration and teaching in schools or colleges, business, and human service agencies. Peabody is considered one of the very top education schools in the nation at both the undergraduate and graduate level. The School of Engineering has an enrollment of 1,300 undergraduates and offers eight specialized majors. The faculty and facilities are rated among the strongest in their fields. The Blair School of Music is one of the jewels in the university's academic crown, with over seventy faculty, many of whom are preeminent in their specialty, who instruct only 200 students, all of them undergraduates. The physical facilities are exceptional in all respects: a superb library of relevant books and musical scores, dozens of practice rooms, and a performance auditorium. Blair has received significant gifts enabling it to triple the size of the facilities and secure state-of-the-art equipment. Blair is committed to training only undergraduates, who thus receive the benefit of intensive training and advisement thanks to a faculty-to-student ratio of two to one. Blair musicians can major in music performance, composition and theory, musical arts, or musical arts with education. All majors are encouraged to elect courses from the other divisions of the university to round out their education.

To carry on its mission of educating students across so wide a spectrum of disciplines, Vanderbilt has created nine libraries, holding over 3.3 million volumes. The computer and technology resources are also among the very best to be found in an independent university. As a university with great financial resources and preeminent graduate schools, Vanderbilt can sustain many specialized centers, ranging from literary studies and creative writing to a program in moral leadership for the professions to the largest archive and study center of television news to numerous technical and scientific research institutes. One-third of all students will participate in one of the campus-abroad programs Vanderbilt sponsors in Spain, Germany, France, England, Italy, and Jerusalem. Thanks to its professional graduate schools, ambitious students can combine bachelor and master degree programs in five years of study in music/ education, A.B./M.B.A., and B.S.E./M.S.E., among other 3/2 programs. The odds are small, indeed, that a student will not find a subject or field of concentration available to him.

College officials also note as important programs and policies: the student-run honor code, an Alternative Spring Break where students participate in a variety of community service projects, the faculty pre-major and major advising system, residential-life and education activities, and an office to assist students with disabilities. A college official, in identifying the importance of the honor code, also highlights some of the differences among the Hidden Ivies in how such codes or principles are defined and implemented: "The honor code is integral to the functioning of the university. It requires students to conduct themselves with the highest standards of honesty and integrity. Each student must sign the pledge certifying that he/she has not received nor given aid for every paper and test. In general, the honor code functions as a cornerstone of our residential community. Violations are handled by a student-elected honor council."

Major Admissions Criteria

In its search for students Vanderbilt focuses on academic preparation, involvement, and consistently strong performance and leadership. One administrator says that the university is looking for "students with proven records of academic and personal success in high school as evidenced by doing consistently well in a strong, college-preparation curriculum, and students who, by their extracurricular activities, indicate breadth and variety of interests, and abilities. Students do well who respond to a high level of faculty-student relations and who thrive in an environment of supportive, but competitive, peers of similar preparation, interests, and abilities."

Key admissions criteria are "a proven record of academic success in a challenging high school curriculum; consistency of performance at the highest level, academically and in extracurricular activities, or an upward trend in performance indicating growing ability and maturity; and an indication of interest in and special abilities to make a difference." Another administrator says, "Vanderbilt seeks students of academic curiosity, leadership, and personal character. The students who do well here are involved in many things beyond the classroom. All students study at Vanderbilt, but the most accomplished ones also are involved in community service, sports, student government. It's an active, friendly, all-encompassing campus." Important criteria for this admissions office representative are: "First, academic preparation. Second, level of involvement outside the classroom that speaks to a student's passion, leadership ability, and scholarly pursuits. Third

(or really, tied with second) is teacher and counselor recommendations."

As with many of the Hidden Ivies, Vanderbilt has become significantly more challenging in its admissions process in recent years. For the class of 2012, 16,944 applicants applied and only 4,292 were admitted, with 1,569 enrolling. Fifty-seven percent graduated from public high schools. Twenty-four percent were minorities. The middle 50 percent ACT range was 30 to 34, and SAT section ranges were 650 to 740 in critical reading, 680 to 760 in math, and 630 to 720 in writing. The class included 170 National Merit scholars and eight National Achievement scholars.

The Ideal Student

Thanks to its size, its great array of academic disciplines and extracurricular programs and opportunities, and its location, Vanderbilt can meet the needs and interests of most students. It holds special appeal for the smart and academically motivated individual who is determined to experience a blend of social, academic, and nonacademic activities.

The notion that Vanderbilt is "very southern," with all the meanings contained therein, is simply not true. It does a great job of casting a wide geographic net to ensure a good mixture of students in the community. There is a feel to the environment that differs from that of northern-tier colleges. Students define this as the friendly attitude of the students, faculty, and administrators; an interest in taking time to develop close social relationships; and a more traditional view of issues and events. The majority of undergraduates are outgoing, energetic, and committed to doing well in their studies while balancing them with an active extracurricular life. The workload is heavy and the grading competitive, but there is not a sense of cutthroat competition associated with many other midsize teaching/research universities. Students consider their personal values traditional to conservative, which in part translates into a respect for family, concern for the disadvantaged and for threats to the larger community. There are few radical or highly politicized students and not too many such faculty visible on campus. There is lots of room for many kinds of personalities in the community, but the majority of students will be working that balance between the academic demands and their other interests.

In terms of social life on campus, fraternities and sororities are a major, though decreasingly important, force, and Vanderbilt and Nashville offer the interested student many alternative sources of enjoyment and involvement. As one administrator notes, "The university community is unusually friendly and supportive. Undergraduate life is rich in social activities, many of which involve a large sorority and fraternity base. And while the numbers of students directly involved with Greek organizations is decreasing, their influence on the social patterns of undergraduates remains high. A Dean of Students office is charged with increasing the opportunities for all students, regardless of their social affiliations. The commitment is extremely high at all levels of the university community to focus on the entire student experience and on making that experience more inviting for a wider range of students from many different backgrounds."

Student Perspectives on Their Experience

Vanderbilt appeals to students because of its friendly personality, its multitude of academic programs and opportunities, faculty involve-

ment, and location. One first-year student describes "the welcoming feeling I got after being on the campus for fifteen minutes. The beautiful campus and friendly kids who went out of their way to make me feel at home for absolutely no reason other than the fact that that is the norm at Vanderbilt. . . . The overall attitude of the faculty, staff, and students is warm, friendly, and relaxed, which is important." A junior says, "I think students at Vanderbilt, unlike at many schools of its academic caliber, manage a perfect balance of academics and social life. We know when to work, but we also know when to stop and have fun. Directly related to that, I think, is the relaxed academic atmosphere. There's none of the cutthroat competitiveness among the student body that seems to characterize other schools. If I ask a classmate to help me understand a given subject, he won't give me the wrong answer because we both might be competing for the same spot in a med school. If I do badly on a test, I won't be made to feel like I need to commit suicide. Also, no one's interested in how well you might have done in a class. Everyone is expected to do well, but if you talk about how well you're doing, no one's going to like you very much."

Some students note that their classmates can find almost any kind of student to befriend on campus. Others state that "students are almost clones of each other appearancewise: 'Vandy girls,' preppy, blond, stereotypical 'rich' kids." And one student wishes the university would do more to "celebrate diversity, approach controversial issues, and become more supportive of minority programming . . . Our campus lacks socioeconomic diversity, but that is to be expected. Great strides have been made to recruit minorities, but little has been done to retain them. Vanderbilt has an atmosphere which is friendly and receptive on the surface.

What makes it special is that it draws people from all over the world. The different cultures are available for us to explore, we just have to take advantage." Other students say, "There is a good, not a great, amount of diversity on campus. More should be done to draw in students of different backgrounds. Geographically, however, Vanderbilt has it down perfectly. Although the school is located in the South, the majority of students are from elsewhere." And fraternities and sororities still hold sway too strongly for some students. "Greek life, while by no means dominant, is a major part of the social scene on campus. Sure that can be fun, but it gets old quickly. The Greek scene, I think, should somehow be restrained so that it isn't the only option. Of all the people I know who've left Vanderbilt, all but one left because of not wanting to join Greek organizations but feeling enormous pressure to do so."

Those who do best at Vanderbilt are "focused, goal-oriented, down-to-earth kids. The 'grade grubbers' do not make it here. Kids who are balanced academically, socially, and individually." A junior says, "I think generally every group does equally well. Obviously the student who works nonstop will do better academically than the one who parties every night, who will have more fun. The great majority of students, I feel, has learned to balance these two effectively. I think it is these people who benefit most from the college experience—those who work when they need to work and who are social when they need a break."

What helps students to succeed at Vanderbilt is, according to one, "The willingness of the community overall to help anyone who asks for it—academically, emotionally. Every professor I have encountered has been more than willing to help in any way they can." What

makes the school special is, "It is a top-twenty university minus the overly competitive, stressful, snobbish environment. I think the best thing about the school is that I feel I can approach virtually anyone on the campus and get a warm, responsive reaction. I am not intimidated to limit myself to my own clique or group." Another student says that "a relaxed academic atmosphere fosters a good work ethic. Without the constant pressure that other schools seem to pride themselves on, I feel more compelled to work hard and succeed." A senior says, "The abundant resources and the availability of the administration, and an atmosphere conducive to learning" help students to do well.

"I have been astounded at the number of programs and groups that Vanderbilt offers," says a freshman. "Students can find anything they want." A junior adds, "Vanderbilt really is unique in that it has an excellent academic reputation and Division I SEC sports. On this level, I would put it on par with more recognized schools like Northwestern and Duke. However, Vanderbilt also has a singular easygoing, non-pressured atmosphere—something most schools of high academic caliber cannot say. This is what makes the school stand out. Also, its student body is relatively small. I can say that there are days I can be walking around campus and see no one I know, and there are days I can walk around and see everyone I know. It's large enough not to be confining and boring, but small enough not to be impersonal. Also, for me personally, Vanderbilt's southern location adds something important to the school. Being from New York, I'm used to people being unfriendly and largely uncaring. Then I came to Nashville. The first time someone I didn't know said 'hi' to me, I thought they were either going to mug me or had some kind of mental imbalance. In that respect (hospitality and friendliness), the school is very southern."

What Happens after College

The majority of Vanderbilt graduates will eventually obtain advanced degrees, with some 70 percent attending professional and graduate schools within five years. Most students go directly into the workforce and about a third go directly to graduate school. Many Vanderbilt graduates find work immediately after graduation in national and international businesses. One administrator says, "Graduates are prepared and equipped to continue to learn—they know how to accept challenges, overcome obstacles, and succeed." Top industries for seniors entering the workforce include consulting, investment banking and finance, and engineering. Students are well prepared for graduate degrees, not only in the arts and sciences, but also in the specialized fields in which they can concentrate as undergraduates, from music to education to engineering.

Vassar College

Box 10
124 Raymond Avenue
Poughkeepsie, NY 12604-0077

(845) 437-7300

(800) 827-7270

vassar.edu

admissions@vassar.edu

> Number of undergraduates: 2,343
> Total number of students: 2,343
> Ratio of male to female: 43/57
> Tuition and fees for 2008–2009: $40,210
> % of students receiving need-based financial aid: 57
> (class of 2012)
> % of students graduating within six years: 88
> 2007 endowment: $869,122,000
> Endowment per student: $370,944

Overall Features

Vassar was founded in 1861 as a traditional residential liberal arts college by Matthew Vassar, who apparently wanted his name associated more with the education of women who had the capacity to become independent, spirited leaders than with the brewery products that earned him his fortune. Like women at its sister institutions founded in the same period—Smith, Wellesley, Mount Holyoke, and Bryn Mawr—Vassar women were encouraged to think for themselves and presume that they were the equals of the young men of their station being educated at the traditional Ivy League institutions. The graduates of these early women's colleges have had a significant impact in the civic, artistic, educational, and professional spheres from the earliest years to the present. Vassar first admitted men in 1968 in recognition of the changes occurring throughout higher education at the time, including the coeducational movements at most of their peer institutions. Given its high selectivity of students through its first century, the college was determined to maintain its ability to admit intelligent students who demonstrated a true passion for learning, questioning, and taking a role in vital contemporary social issues once they graduated.

Today's student body reflects these early goals of the college. Vassar continues to offer a rigorous curriculum taught by a distinguished faculty who expect their students to think for themselves, analyze critically the topics and issues at hand, and demonstrate strong communication skills in their writing and speaking. The great majority of classes are small and require plenty of student participation. Vassar classes average seventeen students, and the faculty-to-student ratio is nine to one. One advantage Vassar students enjoy over students of most colleges is that 70 percent of the 290 faculty members live on campus or nearby, so their involvement with students occurs as much out of the classroom as in it. Both the political and academic tone is definitely liberal, open-ended, questioning, and free-spirited. If there is any conformity to be found it tends toward acceptance of the concerns that spark campus debate, and activist causes typically have to do with gender issues, racism, politics, and sexism.

Matthew Vassar and the early leaders of the institution would no doubt be surprised to witness the changes in the viewpoints of the student body. The earlier women admitted to Vassar were predominantly from families of high social standing and wealth, mainly from the Northeast, who developed over time an attitude of noblesse oblige with regard to their role in society. Contemporary students clearly are caught up in the many concerns our society is trying to cope with. They are, by contrast, a highly diverse community of talented men and women in keeping with the college's celebration of the diversity of people, ideas, and opinions. One-quarter of the students are of color and 60 percent of all students receive financial aid. Vassar guarantees aid to all qualified candidates international or domestic; it meets the full need of students with generous aid packages. Vassar has eliminated the loan component of financial-aid packages for families earning below $60,000. There are no academic merit scholarships, as the college commits its resources to those who demonstrate financial need. Today, 60 percent of enrolled students graduated from public high schools, which also represents a major change from the college's earlier constituency. Thanks to the passionate feelings a great many earlier alumnae had for the education and values they gained as undergraduates, their financial generosity has built a robust endowment and superb physical facilities.

Although it is located in an old and tired industrial city, Vassar has 1,000 acres of beautiful grounds with 100 buildings for students to enjoy. There are multiple residential living options, from the older, elegant suites to newer dormitories and small houses. Virtually all students live on campus for their four years. There are no fraternities or sororities, so social life revolves around the residences, student center, and the many student organizations.

The library has extensive holdings of books and manuscripts. There are separate art and music library collections as well. The entire campus is networked for immediate access to libraries, faculty, labs, and the Internet. The concert hall, studio art center, and theater are first-rate.

The college is intent on building a stronger science program and facility as more students are interested in combining their arts and humanities training with the sciences. Student life is enhanced by the presence of a refurbished student center, a $10 million, multiple-use athletic facility and a nine-hole golf course. The recently completed Vogelstein Center for Drama and Film offers state-of-the-art facilities for film production and theater arts, and Kenyon Hall offers a dedicated dance theater. The college sponsors twenty-five varsity athletic teams for men and women that compete at the NCAA Division III level. Many club and intramural organized teams are also available. As a result, three-quarters of the student body participates in some level of organized sports—rather impressive for a college thought of as a progressive, intellectually oriented community with little traditional school spirit. Vassar students also participate in a great many extracurricular programs, from a very competitive debate team to theater and music performance. Some students devote their energy to theme or affinity groups to make certain that their interests and needs are recognized by the community at large.

What the College Stands For

Vassar's educational leaders believe passionately in the positive role their graduates should play in the professions, the arts, and the community at large. Concern for the future of our planet, the rights of others, and a standard of ethics and humaneness in the professions and commerce are some of the themes that are articulated through daily life on campus and in the classroom. Leadership, intellectual independence, learning how to think and react critically, and the breaking down of artificial barriers are the major attributes Vassar hopes to inculcate in its students. The nature of teaching and the opportunities for different forms of learning as described below are intended to fulfill these goals. The overriding theme is the quality of the faculty, their dedication to teaching and guiding students, and their passion for and expertise in their discipline. "Nothing is more important to a college than the quality of the faculty," says a senior dean. "We recruit the best minds of each generation and we support them when they are here. Our support includes aid to conduct serious scholarship and a strong program to strengthen teaching skills." The administration and trustees are committed to upgrading particular facilities in order to maintain the caliber of programs and learning opportunities beyond the classroom. Financial resources have been dedicated to upgrading the athletic facilities and expanding athletic activities, as well as refurbishing the theater building and classrooms. The development of information technology and the use of computers, both wired and wireless, and the Internet have been other changes at the college.

A senior administrator identifies some of Vassar's educational goals: "Through careful planning that combines independent action on the part of the student with wise counseling on the part of faculty advisors, Vassar offers a wide range of educational opportunities. Our liberal arts curriculum stresses depth of study within a discipline while providing breadth of exposure to many areas of inquiry. This depth-and-breadth approach results in intensive involvement in scholarly work, whether it be scientific research in a faculty member's lab, the writing of poetry with a member of the English department, or conducting archival research with the guidance of an eminent historian. Whatever the path chosen by a Vassar student, he or she will graduate knowing how to write and think, and will be open to and ready for the challenges of graduate or professional study or the demands of the business world." A dean notes these educational goals for students: "With the help of their professors and advisors to develop and carry out a program of study that balances curricular breadth and depth, engages them with different modes of thinking, has intellectual coherence, and meets their own educational goals; to learn to question assumptions and received ideas; to take a position and support it with an argument (which also involves learning the distinction/relationship between opinion or personal response and argument); to articulate arguments clearly, both orally and in writing; and to acquire general skills in critical thinking and quantitative analysis that can be applied in a wide range of contexts after graduation. Ideally, a college education gives you not only a certain body of knowledge but also an understanding of how new knowledge is acquired: We hope to teach our students how to teach themselves."

The college sees its history and philosophy as contributing to its current unique identity: "The origin, growth, and transformation of Vassar from the mid-nineteenth century to the present have created a liberal arts college of unique distinction. Our commitment to the highest standards of intellectual discourse and scholarly achievement coupled with a social setting that provides respect and support for all, women and men, whatever their race or ethnic origin, make Vassar different from and—we truly believe—better than other coeducational institutions." A dean continues, "With its incredibly strong heritage as a pioneer in the education of young women (at a time when most didn't even go to high school), Vassar has always stood for equal opportunity. . . . Underpinning this equality is the very fact of the institution's existence, and as we approach our thirtieth year of coeducation, we celebrate the fact that neither women nor men have had to 'move over' to a secondary place to allow for the other. I have heard our men and women graduates speak so eloquently of this equality and describe how it has prepared them for their professional and personal lives."

Curriculum, Academic Life, and Unique Programs of Study

Vassar has a very open curriculum. The college requires all students to show proficiency in a foreign language, and complete a Freshman Writing Seminar and one quantitative course. Students then concentrate in a department, pursue an independent program, or focus on an interdepartmental or multidisciplinary program. Vassar encourages students to ex-

plore a broad range of intellectual fields and to choose a field of concentration that they will delve into with great zeal. Well over forty majors and interdisciplinary concentrations (such as American culture, women's studies, and environmental sciences) are available, with over 1,000 courses to choose from. The admissions dean states that the array of options for students is truly remarkable. Conversations with various deans regarding the special features of the academic program revealed the common theme of students taking charge of their educational experience by virtue of the flexibility of the academic program. The key to realizing their goals is the commitment of the faculty to advising and supporting students, not just transferring information in a formal classroom setting. Faculty provide personal guidance to the many students who wish to design their own field of study.

Vassar is one of only a handful of major colleges where faculty members live in each of the nine student residences. They are committed to a three-year term so that they form strong relationships with their students. A positive reflection of the willingness to engage with undergraduates is the waiting list of professors who apply for house residencies. It is very easy to create an independent study program or an interdepartmental major or double major under the aegis of a professor. The primary method of education the faculty emphasizes is to teach students how to execute firsthand investigation using primary texts, analysis and synthesis of information, and on-site learning. The college states that "in a typical semester, seventy-five to one hundred students pursue independent study, in everything from political psychology to the physics of black holes. Five percent of each class creates independent majors, combining independent course work with available offerings in various disciplines. One faculty member, a biologist, put it this way: 'Our students engage in research at a level they would not be able to approach at a research university. So far this year, I've submitted three papers for publication, all of them with student coauthors.'"

All students must demonstrate proficiency in a foreign language either through placement testing or completion of intermediate-level course work. The college encourages students in all fields of study to spend a term or a semester on one of the multitude of approved foreign study programs that can find a Vassar student in Africa, the Middle East, the Far East, Australia and New Zealand, Russia, Eastern or Western Europe, or South America. College-sponsored opportunities abound for research and internships in the natural sciences, the performing arts, the humanities, and social sciences. Vassar is a member of the Twelve College Exchange program that lets students spend a year at any of the other colleges. Students confirm that they have a good deal of flexibility to develop a program of study that suits them and that the faculty is happy to assist them in formulating their ideas and structuring an appropriate set of courses and individual study and research. Strength in writing and critical thinking are hallmarks of the skills the faculty wishes to develop in students irrespective of the field of study a student pursues. Other interesting programs of note at Vassar, from the traditional to the more modern, include: Chinese, Japanese, and Asian studies; earth science and geography; Jewish studies; science, technology, and society; medieval and renaissance studies; and Victorian studies. English, psychology, and political sci-

ence are far and away the most popular majors at Vassar over the last ten years, with economics, art, and history next in line.

Major Admissions Criteria

Vassar is very selective in the admissions process, although it is not as fiercely competitive as the Ivy universities and some of the other Hidden Ivies. The great majority of applicants are bright, high-level performers who have succeeded in a demanding curriculum and have a variety of talents and interests. The dean of admissions describes Vassar students as being feisty, meaning independent in their attitudes and approach to learning. They are willing to take a personal stand on ideas presented to them. Thus, in addition to a challenging high school course of study covering all of the traditional subjects and strong grades, the admissions committee gives considerable weight to those who demonstrate clear evidence of a joy for learning; stretch their capabilities by taking the highest-level courses; and display independent thinking and leadership through clubs, sports, or student government. Relatively less weight is given to standardized testing as the committee believes test results do not reveal the personal qualities that predict success and add value to the Vassar community. Another administrator identifies "independence of thought, accomplishment in a range of the most challenging high school courses, and personal integrity" as the most important admissions criteria. How does the committee ascertain these traits? Through the courses a student has taken in high school; teacher and counselor recommendations; reports from employers, coaches, and advisors of clubs; interviews with alumni; and the applicants' personal statements in the application. We can attest to the personal attention and care the admissions committee gives to every applicant and their appreciation for the unique applicant.

The class entering in fall 2008 had middle 50 percent SAT scores of 2,040 to 2,210 combined, and 29 to 32 on the ACT. Ninety-five percent of students at Vassar today graduated in the top 20 percent of their high school class. Americans arrive from all fifty states, and the 8 percent who are internationals hail from fifty foreign countries.

The Ideal Student

Conventional thinkers and those in need of a highly structured academic format will not appreciate the more individualistic nature of a Vassar education. Nor will those students who flourish in a socially conventional or a highly competitive athletic community. The Vassar environment calls for an open mind on intellectual and social issues. Active learners are the most successful students. As one administrator puts it, "Open, inquiring, serious, and engaged (and engaging) young men and women flourish here. . . . High intelligence, therefore, is assumed, as is a capacity for active involvement and participation in the classroom and in extracurricular activities, including, increasingly, athletics. Vassar students are not passive learners, regurgitators of their instructors' lectures. Rather, they are taught to think for themselves and to express their thoughts clearly in writing. Much reading and writing are expected of our students, whatever the curricular area, and the students themselves are the fortunate beneficiaries."

Another dean identifies these qualities as important: "Self-motivation. Vassar students must learn to take responsibility for their own education and to know for themselves why it is important to them. Initiative. Vassar offers its students a rich variety of opportunities and a great deal of freedom to set their own academic course, but it also expects them to speak up for themselves, to pursue their own interests, and to ask for help when they need it. A tolerance for freedom (and the conflicts and uncertainties that go with it). Students who need a highly structured program, who want decisions to be made for them, or who prefer to live in a homogeneous community will find Vassar a difficult place. And a *passion for something*."

Vassar graduates will enroll in graduate schools at a rate equal to that of students in other top colleges, but they will take advantage of their undergraduate years to experiment with an eclectic range of courses or concentrations and explore unusual opportunities off campus. Grade competition and a tone of preprofessionalism are not in evidence in comparison to the more conventional colleges. Strong communication skills and an aptitude for independent learning are requisites for success. The individual who has little concern for the political agenda and social issues that flourish on the Vassar campus should look elsewhere.

Student Perspectives on Their Experience

Students choose and enjoy Vassar for its flexible educational program, academic strength, location (a beautiful campus two hours from New York City), and social climate. Vassar "gives each student the flexibility to explore their interests without putting restraints on them academically or socially," says a senior. "I like that Vassar pushes me to do more and always want more." Another sees as most important "the atmosphere. No two people are alike on campus. Since I came from a very cliquey high school that was all about fitting in with the status quo, Vassar gave me a chance to be myself. I feel as if I'm liked for who I am, and not where I came from, who my friends are, what clothes I wear, etc. I felt that atmosphere at Vassar during my first visit here. It definitely wasn't like the other colleges I had visited. The other colleges reminded me too much of high school." "For me," says another student, "I thought the best undergraduate education I could receive was at a small liberal arts school. Vassar has an excellent reputation among such schools and I knew I could play soccer here as well. The campus is gorgeous, the students are friendly, and I enjoy the suburban environment so close to New York City."

Students have difficulty identifying a type of student who does best at Vassar, but generally feel that independent students who like a non-competitive atmosphere will get the most out of their experience. "I think everyone excels each in their own way and in different things," a senior says. "The great thing is that there is no pressure to be number one in your class, like in high school. It's more important to do whatever it is you do to the best of your ability and not worry so much about everyone else. . . . You really have to be a strong individual in this strong environment, because while Vassar is a great place to be, it is very challenging academically, and you really have to be ready." Another senior says students do well who are "outgoing and active individuals who make it a point to participate in extracurricular activities and/or nourish their nonacademic interests."

Vassar students succeed because of "the friendships you make," according to one student. Another says, "I feel as if it's the professors' desires to want their students to excel. They make an effort to get to know their students and make themselves extremely approachable. They encourage questions and feedback and make an effort to tailor their courses to our wants and needs." And "Vassar provides an extensive advising system. Each freshman is assigned a student fellow who can advise them informally on personal and academic matters. Also each is assigned a pre-major advisor, then later a major advisor and thesis advisor."

Such offerings as the Junior Year Abroad program and the college's generous financial aid budget are identified by students as important elements of their Vassar experience. A senior says, "Through Vassar, I was able to study abroad in Madrid. My professors there were visiting Spanish professors from the local universities. I had the privilege of taking a course with a famous European theater critic. Every class was exciting because he spoke of actual experiences with working with actual contemporary artists. He knew and was known by everyone. It was an experience that I was able to have through Vassar, and for that I am grateful." Another senior notes, "The fact that there are over a hundred student groups on a campus of twenty-four hundred is amazing! There are so many options to be in whatever you want to be in, no matter what your major, your interests. It's so easy to try something new and there are so many things to keep you busy (maybe a little too busy at times!)." Vassar is seen as a diverse campus in many ways, with a supportive environment overall. As one student quips, "We were diverse before diversity was cool."

Students leave Vassar challenged and changed. "Vassar makes us think," says one. "It makes us start a dialogue with each other and encourages us to continue it wherever we go." Another senior says, "Vassar has done a fabulous job of helping me to think more critically, creatively, and constructively. I feel as if I am no longer looking at the world with rose-colored glasses. Instead of ignoring injustices, Vassar has taught me to confront them and to try and change them instead of going with the flow." Another adds, "The challenging classes prepare you to be responsible, attentive, well-read, and articulate." And for another senior, "Vassar is a residential college with a very open, socially aware community. The students here are independent, freethinking, and often very talented. The academic environment is intense, but there is *no* competition between students for grades. Students here are also given a great deal of academic flexibility, since there is no core curriculum. The most important factor that dictates to what extent a student has a positive experience at college is the people he or she meets. Vassar excels on the people front."

What Happens after College

Vassar believes that a strong indicator of the quality of the intellectual skills and substantial value system students acquire is their success in a wide range of careers and the high rate of admission to graduate and professional schools; also, the many competitive fellowships (including Fulbright, Rhodes, Watson, Goldwater, and Marshall) and top national awards they garner each year. A senior college official noted, "We intend to remain rooted in the liberal arts tradition, because our education has proved to be the sturdy foundation for flexibility in professional growth after college. The philosophy major may become a physician, the biology major a journalist, the English major a banker. Making such possibilities available to our students will continue to have high priority." While a good many will train for the professional fields of law, business, medicine, and education, others will undertake studies in the various arts fields, social work, journalism and communications, and public service. Prospective Vassar students are told of the power of the alumni network that helps undergraduates and recent graduates to land worthwhile internships and full-time positions in competitive fields. There is much truth to this promise, as many Vassar grads will attest to.

The college reports that alumni identify a number of factors as being particularly important to their development: "Vassar's small size and the fact that virtually all students and many faculty live on campus, which allows the development of a thriving sense of community; the strong emphasis on individuality—extending even to the development of personal academic programs by students under the guidance of professors; the emphasis on equality between the sexes, that women and men learn to work and study side by side; the astounding variety of extracurricular offerings that give students opportunities to explore new interests and learn the skills of administering organizations and working in groups; and the rigorous academic requirements, which leave alumni feeling that anything is possible if they have earned their Vassar degree."

According to the Career Development Office, about 20 percent of graduates will immediately pursue postgraduate studies, while 70 percent will do so within five years. Vassar reports a recent medical school admission rate of 85 percent, twice the national average, and a law school acceptance rate of 88 percent. Vassar ranks high as a producer of Ph.D. candidates, and provides a career mentoring network of some 3,000 alums.

Wake Forest University

PO Box 7305, Reynolds Station
Winston-Salem, NC 27109-7305

(336) 758-5201

wfu.edu

admissions@wfu.edu

Number of undergraduates: 4,405
Total number of students: 5,752
Ratio of male to female: 49/51
Tuition and fees for 2008–2009: $36,975
% of students receiving need-based financial aid: 36
% of students graduating within six years: 89
2007 endowment: $1,248,695,000
Endowment per student: $217,088

Overall Features

In 1834, the North Carolina Baptist Convention established Wake Forest to train young men in the region for the ministry or for lay leadership in the Southern Baptist Church. Wake Forest admitted women in 1942, and in 1956 the university relocated its campus from Wake Forest, North Carolina, to Winston-Salem. The Baptist sponsorship carried forward until recent years when tension between the goals and strict traditions of the Convention and those of academic freedom and intellectual inquiry came to a head. As the university grew in stature, thanks to the vision of the administrative leaders to place Wake Forest on the national map as a top small university, it attracted an increasingly more selective student body from outside the state and talented faculty who wanted the opportunity to carry on advanced research and teach a more diverse cohort of students. While teachers were frustrated by the limitations on the topics they could discuss and on their critical assessment of religious, philosophical, and scientific concepts, the students felt constrained by restrictions on drinking, dancing, and intimate dating that governed campus life.

After considerable discord and antagonism on both sides, the university dissolved its relationship with the Baptist Convention in 1986. The board and academic community voted to become an independent, nonsectarian institution. This is not to say that the ethical and moral teachings that had permeated the school for 150 years were also suddenly erased. Wake Forest continues to emphasize its traditional values that are based on the teachings of the Judeo-Christian tradition. Ethical behavior is infused throughout the community. Today, it is most pronounced in the emphasis assigned to the honor code, which governs the academic and social life of the student body. Wake Forest remains a conservative social and academic environment compared to the majority of colleges with which it compares itself today. However, it is highly successful in raising money from its loyal alumni to create professional advancement programs for its faculty and scholarships for talented students. It is also asserting its presence as a topflight liberal arts college that cares greatly about excellence in teaching and educating young men and women in the critical learning skills. The aim of the undergraduate experience is to bring together accomplished academics who love teaching and care about their students. Their mission is to challenge students to think critically and extend themselves to their fullest capacity. From the flattering comments of many students, the faculty is succeeding both in stimulating and challenging them. There are 393 full-time teachers available to students across a wide spectrum of disciplines. This leads to a ten-to-one student-to-faculty ratio. Every leader of the institution has expressed his or her admiration for the outstanding quality of the faculty and their commitment to teaching.

The heart of the university is the College of Arts and Sciences, or Wake Forest College, which represents the largest enrollment. There is one other undergraduate division, the Calloway School of Business and Accountancy, which enjoys an outstanding reputation for training students for the world of finance, accounting, and business administration. It has a proactive outreach program to attract minority students, scholar athletes, and students from other regions of the country. Until recently the percentage of students who receive financial support had not been as high as in a majority of the other Hidden Ivies even though the university now operates on a need-blind admissions policy. Today, Wake Forest meets 100 percent of students' need. It provides need-based aid to

36 percent of undergraduates, and families with income below $40,000 have their loan debt capped at $4,000 per year. The university offers merit-based scholarships to about 5 percent of incoming students each year. Some major awards require on-campus interviews as part of the process. At present 16 percent of students are of color and over one-half of all undergraduates are from the greater South including 25 percent from North Carolina. Wake Forest has much to offer a bright student who wants to experience a traditional educational and social environment that puts a premium on good teaching and faculty interaction with students.

The physical campus is very attractive and well equipped in all respects. The great majority of students who choose Wake Forest are attracted to the strong blend of academics, athletics, and social traditions that epitomize the school. Athletics are definitely a major component of student life. Wake Forest is a member of the Atlantic Coast Conference, which is one of the most competitive in NCAA Division I-A sports. As the smallest member of the conference, Wake Forest has to attract a high percentage of student athletes. Its success in intercollegiate competition among eighteen varsity sports is exceptional when the university's size is taken into account. School spirit is a defining feature of campus life, as it revolves around athletic competition. The athletic facilities for men and women who compete in intercollegiate and intramural athletics are outstanding and a major selling point for the university. The other element that defines the social environment is the large Greek system, which is comprised of fifteen fraternities and sororities. Affiliation with a Greek house is high— about 35 percent of men and 47 percent of women (45 percent total). The administration continues to evaluate the traditional role of the Greek presence and has moved first-year student rush to spring, as have many other selective residential colleges.

What the College Stands For

Wake Forest is clear about its purpose as an educational institution and about what it means to be well educated. "The undergraduate school of arts and sciences is the center of the academic life; through it, the university carries on the tradition of preparing men and women for personal enrichment, enlightened citizenship, and professional life. . . . It seeks to honor the ideals of liberal learning, which entail commitment to transferring cultural heritages; teaching the modes of learning in the basic disciplines of human knowledge; developing critical application of moral, aesthetic, and religious values; and advancing the frontiers of knowledge through in-depth study and research and applying and using knowledge in the service of humanity." This is a bold and ambitious mission statement that is implemented through a traditional arts and sciences curriculum that is informed by humanistic values and ethical issues and steered by a very committed faculty. "We have adhered to high academic standards and a liberal arts philosophy enriched by an emphasis on personal responsibility and *pro humanitate*," says a senior dean. Indeed, the college has asked all applicants to assess its *pro humanitate* (for the good of humanity) mission in an application essay.

The college sees student success as emanating in large part from student-faculty interaction and student support services and attention. Officials note the importance of "a careful admissions process, comprehensive orientation, advising, low student-teacher ratio and much faculty-student contact, and support services and resources"; "close relationships with faculty, a strong sense of community, leadership and training opportunities, the large number of clubs and organizations to become a part of, and great technology." "With its small class size and low student-to-faculty ratio," the college argues, "Wake Forest ensures that there is time and opportunity for all points of view to be shared and explored, and that students receive the attention and intellectual stimulation for which they chose this small, private university. Teacher-scholars encourage critical thinking and analysis, encouraging students to form and express their thoughts and opinions."

In the future, Wake Forest will likely advance in its educational delivery and academic programs, but do not expect this institution to lose sight of its traditions. Says one dean, "I think Wake Forest will change in degree but not in kind. We will stay abreast of innovative initiatives in teaching and scholarship but not abandon our purpose and heritage." The university tries to combine the best features of a research university and a small liberal arts college and the important aspects of tradition with educational changes. "Wake Forest has stayed faithful to its liberal arts curriculum and has continued to stress that students take courses in a planned and systematic way so they are exposed to the liberal arts in a sensible way. Wake Forest has also been innovative in regard to new initiatives, such as our use of technology to teach. We have remained small and flexible. We combine, I think, the best of the old and the new." Another dean adds, "The dedication of our faculty to both research and close interaction with students is remarkable and highly unusual. The resources made available to students and the attention given to them are also unusual."

Wake Forest's strategic plan, begun in 2006, identifies the Wake as the "Collegiate University": "We have integrated the intimacy of a college with the academic vitality of a research university. We have adhered to the teacher-scholar ideal in recruiting faculty. We have been shaped by a culture that is distinctly North Carolinian; and at the same time, we have emerged as a national university with international networks. A rich religious heritage is present within a climate of academic freedom and an unfettered search for truth. The size of the university ensures personal attention to individuals while fostering the bold ambitions of a major institution. Our model is rare in higher education."

Curriculum, Academic Life, and Unique Programs of Study

To achieve its educational goals, the university requires students to gain exposure to subjects across all of the academic disciplines. The requirements are fairly specific. There are thirty-seven majors to choose from that can meet the special interests of virtually all students. Honors programs are available in some departments, as are minor programs in such interdisciplinary areas as East Asian studies, environmental sciences, American ethnic studies, international studies, and women's and gender studies. The student with unusual or very specialized interests who desires more flexibility than the structure curriculum provides can opt for the Open Curriculum program, which lets him develop his own field of study with the close supervision of a faculty advisor. Another opportunity for the highly motivated student is the Research Fellowship program, which is designed to match an undergraduate with a professor who is carrying out advanced research in his area of study. In addition to providing invaluable experience in research and writing, the program also awards $4,000 financial scholarships to each research fellow.

The university encourages exposure to other cultures through its 400 study-abroad programs, including its own residential centers in London, Vienna, and Venice. Wake Forest faculty accompany students and direct a curriculum that for one semester focuses on the culture of the particular country. Over 60 percent of students participate in international study. Wake Forest has high expectations for its students and thus the academic load is heavy and challenging. Requirements to graduate are well-defined.

Wake Forest also encourages volunteerism and service, offering many opportunities for students to make a difference. "Rooted in its liberal arts tradition and 'Pro Humanitate' motto," says an administrator, "Wake Forest University strives to use knowledge for the benefit of humanity. In this spirit, students and faculty commit tens of thousands of hours to community service each year. Student volunteerism is reflected throughout clubs and organizations, including the Volunteer Service Corps, which serves as a clearinghouse for more than fifty-five agency and school volunteer programs. In addition, special programs include volunteer service trips to Calcutta, India, and Kayamandi, South Africa as well as domestic sites. Our faculty are also committed to these service ideals. Through the Academic and Community Engagement Fellowship Program, faculty integrate service-learning opportunities into their existing courses. As a result of this program, students are able to synthesize their learning experiences both in and outside the classroom while also providing valuable service to the greater community."

Major Admissions Criteria

Wake Forest seeks "high achievers, academically curious, motivated students with strong character, students who value community and service to others," says an admissions officer. Top admissions criteria are "academic achievement, character, and motivation." A senior administrator says, "We look for academically bright students who are well-rounded and have a sense of purpose. These types of students do well at Wake Forest." And, he stresses, "Wake Forest looks at each application individually." For fall 2008 enrollment, Wake Forest admitted 3,473 of 9,050 applicants, of whom 1,312 matriculated. Middle 50 percent SAT scores for critical reading plus math were 1280 to 1400. Beginning with the class entering in 2009, Wake has made standardized testing optional. Accompanying this change is a strong encouragement for personal interviews, preferably on campus. The college says in its materials, "Wake Forest has high academic standards and competitive admissions criteria. The university seeks intelligent, curious, and highly motivated students who are interested in a lifetime of learning and in using their knowledge in service to humanity."

A senior admissions officer says, "We seek students with intellectual curiosity, character, and energy. In reading applications to Wake Forest, selecting our class and then observing the student body, I am struck by how eager our students are to expand their worldview. They look outward, eager for new experiences and to master new skills and to use their skills in real-world problem solving. Instead of pigeonholing themselves into one subject area, they look for linkages between science and literature, economics and humanitarianism, music and mathematics, art and entrepreneurship. They read good books, keep up with current events, seek close academic relationships with faculty and approach 'school' with excitement. Our students enjoy Wake Forest's emphasis on

community, school spirit, and service." In the admissions process, she looks for "demonstrated intellectual and academic achievement, character and humanitarianism, and curiosity and love of learning."

The Ideal Student

Wake Forest is a distinctive institution in terms of its traditional social and intellectual culture and academic structure. Those who want to live with a group that is enthusiastic about athletics, a predominantly Greek social system, and a demanding, traditional approach to learning will find Wake Forest to their liking. Students should prepare for Wake by taking a broad and challenging high school course load, and should be interested in and excited about continuing to study a wide range of topics in college. The rewards of the Wake Forest experience are considerable: the accessibility of the faculty and the quality of its teaching, the camaraderie accruing from the residential-life programs and active Greek system, and the opportunity to participate in or cheer for topflight athletic teams for both genders. Do not consider Wake Forest if your priorities differ dramatically from these. Also, one will find only some, but not many, "alternative" students (students who are very far out of the social mainstream) at Wake.

Student Perspectives on Their Experience

Students are attracted to Wake Forest for its location and beautiful campus, small size and faculty attention, school spirit, and traditional values. "Wake Forest offered an incredible liberal arts education in a great environment," says a senior. "I knew the smaller classes would be beneficial and I loved the location. Wake

Forest was one of the most beautiful campuses I had ever seen, and the care they gave to the grounds demonstrated the attention Wake seemed to give to every aspect of college life. Additionally, I liked the idea that my classes would be taught by professors, not teaching assistants and grad students.

A graduate reflects on his first visit to campus: "I can still remember my September visit to Wake Forest. I will always remember attending an 11:00 a.m. religion class. The professor lectured on decisions and their effects on our lives, using one of C. S. Lewis's novels as his text. He described God's role in our decisions, and as I looked around the classroom, I noticed quickly that each student was just as intrigued as I was. The class piqued my interest, and I was disappointed only when the fifty minutes were at an end. I was so impressed by the class that I knew that Wake would be among my top choices for college. Yet what happened that afternoon made it my only choice. I was walking on the main quadrangle and I happened to pass the religion professor. I stopped him and started to reintroduce myself, but he cut me off. 'Tom [name changed here], how was your time on campus? I do hope you come here.' I was so surprised that he remembered my name, all I could do was respond with something like 'It was great' before thanking him and walking on. I thought a lot about that late-afternoon conversation as I made my college choice. I decided I wanted a place that felt like home, and at Wake Forest, I had found that place. I am still amazed at the familial atmosphere to be found at the university."

What Wake Forest does best, according to students, is foster connections between students and their faculty and peers. Says a senior, "I think Wake Forest has an incredible skill for getting to know each student and dis-covering his or her gifts. Rarely will you find a school at which you can walk around campus and see a half dozen faculty or staff members who have touched your life in such a way. Every one of my professors has gotten to know me as an individual, has discovered the things I have to offer in class, and has helped me to grow as a person as a result of the class. My professors have supported me during tough times and I have been taught incredibly valuable lessons as well." Another senior says, "I would say that Wake Forest does three things incredibly well: First, it emphasizes technological competence to its students; second, it emphasizes collegiate community; and third, it provides opportunities for its students that expand their horizons and challenge them to be better students and better people."

Diversity could use some work at Wake, where students feel the environment is friendly, but not as culturally mixed as it could be. Says one, "I think diversity is a problem at Wake Forest, although efforts are being made to remedy this problem. After coming from an incredibly diverse high school in South Florida, in which students spoke myriad languages and shared their diverse cultural backgrounds, it was off-putting to see the lack of differences at WFU." Another critique centers on the university's "overregulation" of student life. Nevertheless, the university is reported to be fair in its dealings with students and open to and encouraging of student involvement in and discussion to resolve issues. Finally, students should be aware of the very strong role the Greek system plays on campus, for "on the weekend it frequently feels like 70 percent Greek to 30 percent independent students." And as a student notes, "Though I definitely wouldn't describe Wake as segregated or racist, we do need to work on intermingling to appreciate the diverse cultural and religious backgrounds represented. People tend to be politically correct; one rarely sees a student with crazy-colored hair or extremely alternatively dressed. The very small lesbian and gay population is supported by 'straight' students in the Gay-Straight Student Alliance." Social life outside the Greeks (which one student calls "more like clubs than they are cliques") takes place at weekend football games, church events, retreats to the beaches or mountains, and the bars and concerts of Winston-Salem.

"Students who sit in the corner and don't say much may have done fabulously in high school, but at Wake Forest, participation counts," says a senior. "Many professors count participation somewhere between 10 and 15 percent of a student's final grade, so it helps to speak up. The students I see succeeding are those who go beyond the call of duty; they perform well, they visit their professors when they need assistance, they build relationships with classmates, advisors, and professors, and they make an effort to be seen and heard. By taking that extra step, the opportunities I have had at Wake Forest have been endless." Another student says, "There are three common characteristics of students who do well. Those who work hard, frequently spending up to six or seven hours on homework a night, will do well academically at Wake Forest. Students who value honor and a tradition of integrity will do well personally at Wake Forest, and students who are willing to volunteer of themselves and their time will do well socially and emotionally at Wake."

Says one student, "In general, the people who are happiest tend to be well-rounded. The typical Wake student could be described as motivated academically, involved extracurricularly, interested in others, works out fairly often, has a hidden talent or two (guitar, skiing, flying, etc.), and enjoys spending time with friends. Students also tend to be friendly—it's not unusual to be greeted by the smiling hello of a stranger as you walk across the quad. Socioeconomic status varies greatly, but I am continually impressed and grateful that, in general, money and status do not influence attitudes or relationships. Though many students place a priority on their studies, they are generally willing to help another classmate better understand course material, and would gladly take a study break to catch up with a friend in Shorty's."

The honor code at Wake is a real part of daily college life, where, one student notes, "From freshman orientation to graduation day, students are reminded of the strong and heavily enforced code. They sign tests to ensure honesty and individual work, and questionable events are placed before a strict judiciary system. This strong code has helped Wake become the reputable school it is today." Wake supports its students in many ways, including the Writing Center, learning disabilities support, and support for multicultural students.

One student notes the impact of one of Wake Forest's many student-faculty and internship programs on her educational experience. "The STARS [Student Technology Advisors] program has been incredibly beneficial to me because it is a paid job, it has given me technological training, I have had the chance to work with faculty and administrators, and I have acquired internship opportunities, all of which have better prepared me for the workforce. STARS pairs technically experienced students with professors to teach all levels of computers so that the professors can integrate more technology into their classrooms. I have had the chance to work with several professors, as well as the vice president of finance and administration; I was placed in an awesome summer internship last summer; and I received training, paid for by the program. As a senior eloquently puts it, "Charles Kuralt, the renowned recorder of Americana, reminds us that unique does not mean rare or uncommon, but in his words, 'alone in the universe.' I think that Wake Forest is unique in higher education. With its strong emphasis on technology, community, and opportunity, Wake Forest provides students with the tools and the resources to achieve great things, for Wake Forest educates students; it does not indoctrinate them."

A sophomore says that "scholarship programs are excellent; every scholar is taken care of and the scholar director is incredible—guides scholars from day one and begins to offer advice for graduate scholarship programs after undergraduate studies. Wake Forest also emphasizes unique, independent research projects (Richter) that require traveling and promote being a more global, well-cultured individual."

This young woman says the student who does best is "a Wake Forest student [who] doesn't accept mediocrity and is willing to work, as known in the nickname *work forest*."

A junior notes, "The motto of Pro Humanitate is very important to the identity of the university as well as that of the student body. Students try to embody this motto in many of their activities as well as within their course of study." Also, "a lot more kids attend Wake on financial aid than the average student would think. There is a perception that all Wake kids are rich and come from affluent families, but the reality is that many Wake kids come from disadvantaged backgrounds."

This student points out that "the communal feel at Wake is unrivaled. While individual success is important, we share in the success of one another. People do not compete against each other, but rather compete against themselves and push themselves to be the best that they can be." He advises students, "Be who you want to be. Wake may seem homogeneous, and you may think that you have to be a certain way to fit in. The truth is that the Wake student body is rife with diversity, and you will eventually find your niche if you decide to maintain your individuality."

"I love class change at Wake Forest," says a sophomore. "During these ten to fifteen minutes, you will not see stressed students watching the ground intently as they focus on the material for next class's quiz. Instead, you will see smiling students waving at each other across the quad and stopping to catch up on the past weekend, laugh about the sometimes questionable food selections in the Pit, and wish each other luck on an upcoming test. This is the spirit of Wake Forest: happy students and faculty encouraging each other to succeed academically and socially on the most beautiful campus in the country."

Finally, a senior says, "Wake Forest is best at creating a sense of community among each of its constituencies and instilling the importance of community in its students. This concept of community is addressed on multiple levels from the campus community, to the local community, to the national and international communities. Students are provided unbelievable opportunities to interact and establish relationships with faculty and staff members. There are also many opportunities to enjoy the campus community socially with campuswide events such as the annual Springfest Week which includes concerts, a carnival, and the main event, 'Shag on the Mag,' in which the campus community comes together to enjoy an outdoor dance hosted in a large tent on Manchester Plaza featuring lots of dancing to some of the Carolinas's best shag bands. Initiatives like Project Pumpkin and service trips during academic breaks allow us to understand the importance of giving back to and interacting with the local community. And finally, a focus on the international community is established through the university's encouraging of study abroad and through the offering of international service trips."

What Happens after College

Students enter Wake Forest with postcollege goals very much in mind. One-third of every graduating class immediately enters professional or academic graduate schools, including about 22 percent to academic graduate programs, 6 percent to law schools, and 4 percent to medical schools. Many more will earn an advanced degree over time. Not only business school graduates but also a majority of arts and sciences majors move directly into the business world. Career choices include accounting, financial services, consulting, education, technology, management, sales and marketing, social services, and others. About 7 percent pursue education and teaching immediately after Wake. About half of graduates find work in the Southeast, and a quarter in the Northeast.

Washington and Lee University

204 West Washington Street
Lexington, VA 24450-2116

(540) 458-8062

wlu.edu

admissions@wlu.edu

[
Number of undergraduates: 1,774
Total number of students: 1,779
Ratio of male to female: 50/50
Tuition and fees for 2008–2009: $37,412
% of students receiving need-based financial aid: 36
% of students graduating within six years: 89
2007 endowment: $692,797,000
Endowment per student: $389,430
]

Overall Features

What most defines Washington and Lee is its sense of history and enduring traditions. The college was founded in 1749 as an independent, nonsectarian liberal arts college to educate young men of social standing from the South for positions of leadership. Like so many of the eighteenth-century independent colleges established more on the ideals of the enlightenment than on sound financial footing, Washington and Lee was faced with financial disaster early on. The college authorities appealed to George Washington, who made a generous gift in 1786 that saved their tenuous institution from extinction. A century later the college suffered the ravages of the Civil War and badly needed a significant figure to lead its rebuilding. What more inspiring figure could be found than General Robert E. Lee who, as president, led the college to a position of strength and preeminence until his death in 1870.

Washington and Lee is committed to remaining a small college in order to cultivate strong personal relationships between students and faculty and among the students. There is a distinct effort to foster a culture of friendliness and respect for one another. Several of the distinctive features that influence the life of the school are the honor code, which governs all campus academic and social behavior, and the "speaking tradition," an invention of General Lee that requires that students, faculty, and administrators greet and speak to one another whenever they cross paths on the campus. Students express considerable respect for both the speaking tradition and the honor code as the guiding rules of expected behavior for the entire community. Students take exams unsupervised and their word is respected in all matters. The honor system is run completely by an elected student committee. All cases of violation are heard by the committee, which makes determinations on the rightness of the case and the punishment to be handed out. This is a major responsibility since the college operates on a single violation rule. The student panel is therefore required to vote for expulsion if a student is found guilty.

The trustees decided to open the gates to women only in 1985 after much deliberation. Conservators of the institution's traditions and sense of community, they wanted to be certain that this would not lead to radical changes. The consensus of students, faculty, and alumni is that Washington and Lee has maintained its special ambiance while improving the academic qualifications of the student body and the quality of life. The women who enroll seem to be as positively influenced by the cultural codes that affect the school as the men are.

Washington and Lee's rural location was an intentional decision by its founders, who wanted to create an intimate community of students and faculty and protect students from the distractions of urban life. Little has changed in this regard over two and a half centuries. Faculty are hired both to teach and support their students through small class sections and interaction on the campus and in the dormitories. Faculty are evaluated on the basis of their teaching more than on their research and publishing. The location of Lexington definitely engenders communal spirit and strong personal relationships.

Washington and Lee is a highly selective institution, with rigorous academic standards imposed by the faculty. A strong emphasis is placed on developing superior writing and critical-thinking skills throughout the curriculum. Unusual for a small liberal arts college, Washington and Lee supports both an undergraduate school of commerce offering concentrations in economics, accounting, management, public policy, and political science, as well as a graduate school of law along with the College (of arts and sciences). The Williams School of Commerce, Economics, and Politics has its own full-time faculty of forty-two professors and enrolls about 250 undergraduate students in one of its majors each year (about a third of the student body). All W&L students begin their freshman year in the College, the arts and sciences school, beginning to fulfill the university's general education requirements. Overall, W&L supports almost 200 full-time faculty for undergraduates, 95 percent of whom have terminal degrees in their field. Forty-one majors and more than a thousand courses offer huge opportunities for study. The student-to-faculty ratio is eight to one, and all freshmen and sophomores live in campus housing, creating an intimate and supportive community. The student body represents forty-six states, thirty-three foreign countries, and 10 percent to 12 percent of students are of color. Four to 5 percent are internationals. The College is committed to broadening its geographic, cultural, and socioeconomic student body to achieve national and social diversity. Financial aid is generous and the needs of almost all candidates who demonstrate need is met. W&L has eliminated the loan component of all need-based financial-aid packages. One hundred fifty merit scholarships are available to students who apply for them, and about 10 percent of incoming students receive the Johnson scholarship, which covers tuition, room, and board. W&L has increased its aid funding substantially in recent years. However, the percentage of students on scholarship does not yet match that of the more socially diverse peer institutions in the northern and western regions of the country. The southern ambiance, with all its appealing traits, for many still characterizes the college. There is not a highly charged engagement with the political, gender, or social issues that one finds on many of the more liberal campuses. Students are more focused on their social lives, studies, and athletics.

A student with an appreciation for the beauty of a collegiate campus and surrounding environs could not do better than Washington and Lee. It is a pristine, beautifully designed and maintained campus. The Greek system is central to the social life of the school with about 80 percent of men and 70 percent of women being members of one of fifteen fraternities or six sororities. There is little interest, it seems, in changing the role they play in the community, which is not the case at many of the other residential colleges of comparable academic quality. Varsity athletes compete at the Division III level in twenty-three sports, and the great majority of undergraduates participate in intramural and club sports. Washington and Lee has a longstanding reputation in intercollegiate debate and sponsors a number of newspapers and literary magazines. For a small college there is much to do on campus.

What the College Stands For

Few institutions are clearer or more direct in declaring their educational mission and philosophy: "Washington and Lee emphasizes personal honor and integrity. No one attends the university without becoming aware of new dimensions of honor and integrity. Accordingly, students are given a large measure of freedom in governing their own affairs and are represented by active membership on faculty committees. . . . The same code of honor that governs academic life guides personal life. Washington and Lee strives for intellectual distinction. Its steady purpose is to be one of the nation's great 'teaching' colleges. Research is encouraged as part of the learning and teaching process, not as a substitute for it, or as a way of determining promotion or tenure of its faculty. Ideally, the university believes teaching and research cannot, and should not, be separated. Washington and Lee fosters an academic environment in which both teachers and students constantly learn—in classrooms and laboratories, in private research, in the give and take of extemporaneous discussions. All students are taught by professors, not by teaching assistants or graduate students." This is as forceful a pronouncement of institutional commitment to a first-rate faculty and outstanding teaching with ongoing research as a means to better instruction as any institution has promulgated. The university also defends its historic independence to influence young men and women in its own fashion. "Free of any control of church or state, the university is dedicated to the democratic form of social organization, to the dignity of the individual, and to the ancient freedoms, particularly to liberty of the mind with its attendant right of inquiry. Hence, the university is free to chart its own course, consistent with the highest educational standards, its traditions, and its aim of service to mankind."

College officials are proud of Washington and Lee's history and reputation. Curricular development, strengthening of instructional resources, improvement of facilities, continued increase in financial aid based on gifts and endowment (rather than tuition), and increasing diversity are institutional priorities. Washington and Lee's uniqueness, according to one administrator, is found in "our honor society, our curricular breadth for a medium-size undergraduate institution, a law school as our only graduate program, top rank in all programs, and two hundred and fifty years of educational leadership and service."

W&L's mission statement captures well the nature of its environment and educational program: "Washington and Lee University provides a liberal arts education that develops students' capacity to think freely, critically, and humanely and to conduct themselves with honor, integrity, and civility. Graduates will be prepared for lifelong learning, personal achievement, responsible leadership, service to others, and engaged citizenship in a global and diverse society."

According to a college official, "In May 2007, the university adopted a strategic plan titled 'A Liberal Arts Education for the Twenty-first Century.' Within its overall tenets are several specific issues that the university will address. They include attracting need-based financial aid to make the university accessible to greater numbers of qualified students, including minority students, from low- and middle-income families; providing faculty compensation consistent with peer institutions; launching new undergraduate curricular initiatives while establishing an innovative third-year law curriculum; globalizing the curriculum and enhancing leadership and ethics programming; transforming the undergraduate Leyburn Library into a twenty-first-century learning venue; and renovating the historic Colonnade."

Curriculum, Academic Life, and Unique Programs of Study

The educational theme of Washington and Lee is to foster students' exposure to the broadest range of knowledge and human understanding beyond the limits of any particular special interest. "The aim of the curriculum is to free the mind, lead to understanding, create humility and tolerance, and afford a basis for continuing study and learning." The emphasis on a true liberal arts education is similar to that of all of the Hidden Ivies. The curriculum stresses studies in the humanities, and the social and natural sciences, especially in the first two years of study, which will give students the exposure to then choose a concentrated field of study and a preparation for graduate school. To meet the goals and purposes of the university, students are required to fulfill a number of distribution areas. However, there is a large number of available courses in all areas to meet their interests while also fulfilling requirements. Although there is less freedom to create a highly personalized program of study than at some other Hidden Ivies, there are formalized programs of interdisciplinary concentrations and independent study and research opportunities that professors help a student to design and manage. A college administrator says W&L's "distinctive curriculum blends traditional liberal arts and sciences with preprofessional programs in business, journalism, and law, giving students a contemporary perspective necessary to flourish in a complex world. These professional programs benefit because they exist in a liberal arts setting, and the liberal arts programs benefit because they exist alongside areas of inquiry attuned to the problems facing society. W&L has developed increasingly strong interdisciplinary programs in areas such as the environment, poverty, and women's studies. In addition, Washington and Lee respects its students' ability to govern themselves and fosters a community based on trust. As a consequence, students develop as leaders in all areas of campus life and go on to become leaders in their communities and professions." The faculty is determined to help students develop the vital skills of independent thinking, clear analysis and problem solving, and the capability to communicate their ideas in their writing. The journalism and commerce programs make it possible for undergraduates to combine the advantages of a liberal arts education with more defined studies in one of these two specialized fields. The university sponsors a good number of off-campus internships and study-abroad programs as a function of its mission to provide a breadth of intellectual and cultural experiences for students.

Major Admissions Criteria

The admissions office stresses that the single most important factor in evaluating candidates is the quality of their high school course load and performance in these courses. Indications of stretching oneself intellectually by undertaking the more advanced-level courses, and taking key subjects like English, mathematics, history, sciences, and language for three or four years are significant factors in evaluating worthy candidates. The committee prefers students who have taken the highest level of a subject its high school offers, be it honors or advanced placement, to those who breeze through with good grades in less challenging sections. As one senior administrator identifies the key admissions criteria, they are "academic achievement and ambition, demonstrated leadership, characteristics of honor and integrity, and energy." Another points to "academic achievement, intellectual ability, and honesty." Leadership as demonstrated through student organizations, social service organizations, and athletics are other significant factors in determining who is offered a place in the incoming class. Test scores are also taken into account with the mean of those admitted at the high end of the scale, but not as narrow as the range among the Ivy institutions.

"There is no formula for a successful application to Washington and Lee," says an admissions officer. "That is, we consider each piece of every application in our admissions decisions and strive to be thoughtful and careful as we individually review every application we receive. Each application is read by at least three admissions committee members, and most applications are read by the entire admissions committee. The most important part of the application is the daily record—what courses students have taken and how well they have done in them. We are looking for students who have a superior record of achievement in the most challenging curriculum available to them. This does not necessarily mean students should take every AP class their schools offer, but it does mean they should challenge themselves in every subject if they wish to be competitive applicants to W&L. . . . In a typical admissions cycle, we receive applications from a few thousand applicants who are well qualified to succeed academically at Washington and Lee. In order to select the thousand or so students whom we will admit, we consider a wide range of nonacademic factors in addition to the academic record. These include extracurricular involvement (quality is much more important than quantity), character (revealed by letters of recommendation, essays, and accomplishments), their interest in attending and their fit with W&L, and other personal qualities. We encourage all applicants to be honest, be themselves, and schedule an admissions interview."

The class of 2012, 454 enrollees, was selected from 6,386 applicants, of whom 1,074 were admitted. Middle 50 percent SAT ranges were 660 to 740 in critical reading, 660 to 740 in math, and 660 to 730 in writing, and the ACT range was 28 to 31. Fifty-seven percent attended public schools, and thirty-four were valedictorians or salutatorians. Twenty-three were National Merit finalists or scholars. thirty-four were first-generation college students. One hundred ninety-two were varsity captains, and fifty-six edited major campus publications. Because enrollment is so small, entrance into Washington and Lee is very selective, but also very personalized.

The Ideal Student

Any prospective candidate to Washington and Lee should review its mission statement and the essential features governing the life on campus. The school's location, size, and most popular social and extracurricular options should determine if this is the right match. The conservative nature of the student body and the approach to learning and living in this small environment will suit the traditionalist who wants a very challenging educational experience, a close-knit community, a spirited athletic program, and the social life that the fraternities and sororities provide. Washington and Lee cares as much about the values it imbues in students as it does for the quality of the academic education. One must buy into this system to grow and prosper over the four-year experience.

The school offers "a *total* commitment by faculty and administrators to support students," and maintains "a balance between high academic standards/superb teaching, and a spirited social environment." While not the most diverse of the Hidden Ivies, Washington and Lee continues its efforts to open up the university to more types of students. A senior administrator says, "The high level of dedication of W&L faculty to teaching, the individual attention they are able to give students, and the extent to which the faculty are able to integrate outstanding scholarship and creative accomplishment into their work with undergraduates is legendary. Many student services and activities, including a Division III athletic program where academics truly take priority, provide

formative experiences outside the classroom. The sense of community and personal trust, mainly due to the honor system, is much stronger than at most other schools and is key to the character development that is part of W&L's mission." Another notes, "For each incoming class, we look for an exceptional group of 455 students who are passionate about learning and will enliven our classrooms; who will embrace our community as their own and contribute of themselves to ensure its vitality; who will excel on the playing fields and in the arts; who will produce high-quality newspapers and magazines; and who will participate in the over 160 clubs and student organizations on our campus. The best indicators of success and happiness at W&L are demonstrated engagement in one's school and home community, intellectual creativity and curiosity, an interest in issues and ideas beyond one's own experience, and a willingness to be challenged socially and academically."

Student Perspectives on Their Experience

"I chose Washington and Lee because of the beautiful campus, excellent academics, social atmosphere, and sense of community," says one, characterizing the general sentiments of his fellow students. A classmate adds "the honor code and academic reputation" to the list. The university does a good job of "making each student feel like much more than just a number," and "puts on a helluva show when it needs to. W&L knows how to throw a party, official or unofficial."

The university could definitely attract a more diverse student population and expand the social scene beyond the Greek system, say some students. But others praise "the fraternity life. I've made the best friends of my life." And the honor code "is like no other anywhere in the world."

"While it's not perfect," says a senior, "one of the most important things to me in my college career has been sitting alone in a deserted academic building at 4:00 a.m., taking a test with my books across the room, *on my honor.*"

Those who do well are "extremely intelligent as well as socially compatible." And the faculty members are clearly the primary support structure for students. Says a senior, "The fact that all professors are so easily accessible helps students to succeed. You become very close with your professors and have the feeling that they can help you with many things." Another adds, "The professors many times go the extra mile to help out. They know their students. Ultimately, however, it comes down to God-given genius or work ethic."

Students who come to Washington and Lee should be prepared for a social scene dominated by fraternities, sororities, and athletics, a small college where "everybody knows everything," and a beautiful, challenging, and supportive campus and academic environment. "During weeks in which you sleep ten hours in order to prepare for the barrage of tests, quizzes, projects, and papers," says a senior, "sometimes you find yourself alone on the majestic front lawn at sunrise, and you find peace with what you're doing."

A senior says, "I chose Washington and Lee because of its reputation and prestige. I also liked the sense of community and the opportunity to participate in many organizations. Of course the scholarship money was an added bonus and sealed the deal." She describes W&L as "conservative, prestigious, proud, challenging, and supportive."

This student appreciates that "our speaking tradition (whether or not it is actually upheld) really reminds me of the type of person I strive to be, along with our honor system. In the 'real world' there isn't someone always watching over your shoulder, but you are expected to produce your own work, so I really appreciate our honor system."

Another senior "chose Washington and Lee for three reasons. One, its reputation for high academic excellence. I knew that I would have outstanding professors, as well as a very small student-to-teacher ratio that would enable me to access them. Two, the community atmosphere at W&L is unparalleled. The honor system, speaking tradition, and small town of Lexington contribute to a very unified, friendly, and hospitable school environment. Third, Lexington, Virginia is a beautiful atmosphere in which to live and study. The school itself is beautiful as well as the surrounding countryside. Activities such as the Outing Club make use of W&L's beautiful surroundings."

She notes, though, that "Washington and Lee's enclosed, even isolated atmosphere does not inspire diversity or dynamism within its students. The population tends to be homogenous though actions over recent years have tried to diminish this somewhat. Because it is isolated students are unable to take part in the opportunities that accompany big cities. The closest large metropolitan area is Washington, D.C., which is at least 2.5 hours away. Sometimes, Washington and Lee's small size is a blessing, other times it can feel claustrophobic or suffocating. The social atmosphere can be suffocating as well, as there is no place for uniqueness in the homogeneity. Every year this stereotype is somewhat diminished, but there is still limited diversity on campus. I think the campus is still open and friendly to people of different backgrounds and experiences, but diverse students may feel isolated because the campus isn't diverse."

According to this student, "The biggest help to Washington and Lee students are their advisors.

Many students have very close relationships with their advisors. Advisors help students in many activities from choosing classes to deciding what to do after college. They can be a guiding force and a source of advice. The administration is always willing to help W&L students. The administration loves student initiatives and is willing to help out in any way it can to see them to fruition. W&L has a lot to offer, but if a student doesn't see something they like, they can always start the club. Since W&L does not have that many students, each student is a priority for the administration. The social network among students also helps them to succeed. W&L isn't a scholastically competitive atmosphere. Students in general want their classmates to succeed and help each other. Friendships and close bonds, combined with a large Greek system, create a friendly atmosphere within the school."

What Happens after College

"Washington and Lee graduates leave the university as broadly and deeply educated men and women, capable of thinking critically and communicating their ideas and opinions with confidence," says a college administrator. "They are prepared to occupy positions of leadership in virtually every area of society, where they contribute through both their knowledge and their strength of character. About 25 percent of graduating W&L seniors go directly to graduate or professional schools, while 95 percent of the rest have jobs within six months of graduation. In the last few years, W&L graduates have won such major postgraduate awards as Truman, Goldwater, Luce, Watson, and Beinecke scholarships and Fulbright fellowships." About a third of the seniors heading to graduate school pursue academic graduate programs, a third choose law school, and a fifth medical school. A survey of one recent class's employment plans shows of 301 employed students, 74 were involved in finance, 39 in education, 22 in business, 22 in consulting, 20 in nonprofits, 17 each in accounting and government, 13 in advertising, 11 as paralegals, 10 in science, and others in a variety of fields.

Washington University in St. Louis

Campus Box 1089
One Brookings Drive
St. Louis, MO 63130-4899

(314) 935-6000

(800) 638-0700

wustl.edu

admissions@wustl.edu

[
Number of undergraduates: 6,339
Total number of students: 11,390
Ratio of male to female: 49/51
Tuition and fees for 2008–2009: $37,248
% of students receiving need-based financial aid: 39
(class of 2012)
% of students graduating within six years: 92
2007 endowment: $5,567,843,000
Endowment per student: $488,836
]

Overall Features

Washington University was founded in 1853 as an independent, nonsectarian institution to train men and women in the liberal arts and professions following the model of other major universities of middle size. The university is a combination of an undergraduate liberal arts college and four specialized schools and a graduate complex that offers degree programs in the professional fields of business, law, medicine, art, architecture, engineering, social work, and a host of academic disciplines at the doctoral level. Unlike the traditional small, strictly undergraduate colleges of note, Washington provides an extraordinary range of academic studies for undergraduates because of its size and the presence of the special schools. Within the undergraduate population of 6,300 full-time students, over one-half study in the College of Arts and Sciences. Another thousand are enrolled in the engineering school,

and the rest concentrate in architecture (200), art (340), or business (750).

The university has an astonishingly large endowment that places it in the top ten universities in the nation. As a consequence of its wealth, Washington can function as a midsize university with first-rate professional schools and a traditional college of arts and sciences that is the heart of the educational enterprise. Students have access to rich academic offerings, multidisciplinary concentrations, and independent study and research opportunities under the aegis of teacher/scholars. Fifteen hundred individual courses are available in ninety undergraduate programs that make it easy to meet the distribution requirements. More specialized majors must undertake a number of courses in the arts and sciences. Washington is a research university, and faculty members are expected to carry on their own scholarship, but they are also expected to take their teaching responsibilities seriously. More than 3,000 faculty keep 80 percent of classes with fewer than twenty-five students, and there are very few classes taught by teaching assistants. The overall environment is moderately traditional since the great majority of students are focused on their studies and future career plans. There is good variety in the backgrounds of the student body, who come from all fifty states and over 200 foreign countries. Of undergraduates, 86 percent are from out of state, 25 percent are of color, and 6 percent are of international background. Thanks to $65 million allocated to undergraduate scholarship funds, the university provides financial-aid packages to 60 percent of its students. Washington is able to fulfill its goal of attracting deserving students of all cultural, social, and economic backgrounds who qualify for admission. A large number of merit-based scholarships are also available to attract outstanding students to the various undergraduate schools. Students who apply for and receive such merit-based assistance may find a scholarship package worth $2,500 per year or one that covers the full cost of their education. Washington has also eliminated loans from the need-based financial-aid packages of families earning less than $60,000.

The campus, located in a suburban section of St. Louis abutting a 1,300-acre park and the famed Saint Louis Zoo, is handsome and well equipped. The majority of students live on campus in one of thirty residencies, while a small percentage live in fraternities on fraternity row. Only one-quarter of men and women join a fraternity or sorority, so their influence on the social life of the campus does not overwhelm the many other social organizations and student activities supported by the university. New students live in one of Wash U's residential colleges, which offer "living/learning communities" consisting of several houses, social life, peer mentors, activity programming, and resident advisors. There is a good sense of a campus community considering the size of the school and the urban location. Students are fortunate to have a wealth of activities available to them both on campus—from special interest clubs to myriad performing arts groups—and in the cosmopolitan city of St. Louis. Music, theater, art museums, restaurants, and professional sports teams are easily accessible to students. There is a strong Division III program of competition in nineteen sports for men and women. Washington helped to found the University Athletic Association of nine institutions similar in size, academic quality, and student body. Thanks to excellent athletic facilities, 75 percent of undergraduates participate in intramural sports. Many students who are attracted to Washington for its diverse academic programs, student body, and urban location like the opportunities to get involved in community service (two-thirds do so). This is clearly reflected in the large number of social and service organizations that students can join. The diversity of the students is also reflected in the many affinity groups that represent virtually all of the gender, racial, ethnic, and political concerns alive on most campuses. Yet the atmosphere is not charged with intense activist attitudes or behavior. One undergraduate commented that his peers have personal convictions and passions on various issues, but do not press them on others in the in-your-face style popular on some other campuses. It is rare to hear a student lament that there are not enough things to do or academic offerings to choose from. The more pertinent issue is how to choose from such a cornucopia of options in all respects while staying on top of the demanding workload. Washington students are concerned about their grades since most are headed to graduate studies but they also value downtime and engagement in clubs, sports, or the arts.

What the College Stands For

As a liberal arts college and research university, Washington has a dual objective of educating the beginning learner through training in the arts and sciences, and at the same time, adding to the store of human knowledge through extensive research programs in the humanities, sciences, medicine, and technology. For undergraduates, the goal is to "prepare students with the attitudes, skills, and habits of lifelong learning and with leadership skills, enabling them to be useful members of a global society." The model graduates will be bold, independent, and creative thinkers who will give back to the larger community.

Washington's identity is wrapped up in its midwestern location and ethos, undergraduate and graduate balance, and residential environment. Says an administrator, "During the late 1950s and the early 1960s, Washington University decided that it was going to become a national—indeed international—university. Our leaders envisioned a world-class university located in the Midwest as an advantage for our students and for the region in which we are located—a place where collaboration, interdisciplinary learning, excellence, hard work, and success are basic attributes. Add to that our midwestern qualities of friendliness, inquisitiveness, and respect for individual differences. The educational process here operates within the framework of these qualities and attributes. For our undergraduates we provide an experience that touches on virtually every aspect of a young adult's transformation from high school and home to becoming a mature individual who

has developed a strong grounding in world cultures, literature, the sciences, the humanities, and our socioeconomic frameworks. That transition occurs well beyond the classroom, laboratory, and studio. It happens as part of a student's residential-life experience, philosophical and spiritual growth, and a sense of purpose that distinguishes a well-rounded learner."

A detailed set of educational goals for the university's students, prepared in 1996 by William Gass, a professor in the humanities, begins by stating, "Education should be lifelong and lifewide. You may leave college, but you ought not to leave learning." Gass goes on to set out Washington University's specific goals in two major areas, illustrating the breadth of the liberal arts education that the university seeks to provide: "personal development (set high standards, know thyself, solidify good character, be ready to accept social and environmental responsibility, and recognize that the arc of learning is lifelong); and intellectual development (reasoning, communication, aesthetic appreciation, mathematical skills, historical perspective, scientific understanding, literature, international and cultural awareness, and information acquisition and research)." Washington's mission statement reads, in part, "Washington University's educational mission is the promotion of learning—learning by students and by faculty. Teaching, or the transmission of knowledge, is central to our mission, as is research, or the creation of new knowledge. The faculty, composed of scholars, scientists, artists, and members of the learned professions, serves society by teaching; by adding to the store of human art, understanding, and wisdom; and by providing direct services, such as health care."

Washington University's endowment has allowed it to develop its Frederick Law

Olmsted–inspired campus infrastructure. "We are still creating our campus," says an official. Several hundred million dollars were spent in the last decade on new buildings and refurbishments of older ones. This includes an athletic complex, a new business school facility, science and psychology buildings, and, most recently, the Danforth University Center and the Mallinckrodt Student Center, with lounges, eating spaces, meeting rooms, bookstore, and theater space.

Curriculum, Academic Life, and Unique Programs of Study

The presence of a large teaching faculty, an extensive curriculum in the arts and sciences, and the professional schools of visual arts, architecture, engineering, and business lets students develop almost any course of study of their liking. The university encourages students to combine studies in different fields within the arts and sciences and between the liberal arts and the professional programs. A few examples are programs in the biological sciences and medicine, architecture and engineering, and performing arts and business. Washington is an excellent place to explore potential long-term intellectual and career interests before making a final commitment. As an administrator says, "We want to remain a medium-size campus with the resources and services of a much larger university, but with the supportive, comfortable atmosphere of a small college. We also want our students to prepare for the twenty-first century by designing programs that draw from more than one discipline and also provide excellent, in-depth preparation for careers and for life."

Almost two-thirds of students earn both a major and minor, or more than one major. More than one-third complete requirements for two majors and some students earn two degrees from two different schools at the university. "These choices," says an administrator, "allow students an intensive focus on one or more subjects, uniquely combining such areas as engineering and art, business and Russian, architecture and history, computer science and visual communications, history and French, or English literature and management, etc. It is even possible for students to earn a bachelor's degree and master's degree in one or more disciplines under several combined programs." The dual undergraduate/graduate degree options in the University Scholars Program allow students to get a jump start on their advanced degree and career. Possibilities include law, business, medicine, and social work. Students must apply to both the undergraduate and graduate programs before they enter the university. Undergraduates do have specific distribution areas to fulfill but have great flexibility in doing so. The university encourages undergraduates to take advantage of its sponsored study-abroad programs that involve study in more than just languages and the arts. The Olin Library and departmental libraries provide undergraduates with outstanding facilities for study and research. The faculty incorporates a good deal of research and self-directed study for students within a formal course curriculum, and students are encouraged to partner with their professors on research projects. The theater, art museum and studios, and science facilities are considered outstanding. There are numerous specialized centers for research and study on campus that can appeal to a student who has a highly focused interest in one of the fields. Some additional programs to note include the Skandalaris Center for Entrepreneurial Studies, which helps students learn about and launch a new business or venture; an extensive domestic and international internship program; a Summer Scholars research program; the McDonnell International Scholars Academy; and Summer Language Institutes.

Major Admissions Criteria

Admission has become significantly more selective in the past ten years. In 1998, 16,000 high school seniors applied from all regions of the country for a first-year class of 1,475 men and women. For fall 2007, 22,428 applied for a class of 1,338 enrollees. The growing popularity of the university due to wider recognition of its academic excellence, considerable resources, and quality of life has attracted increasingly high-performing students. Says the college, "The word is out among high school students that Washington University is a great place to go to college. This is the result of genuine, word-of-mouth communications traveling from satisfied graduates to high school students and their parents around the country." Washington University has an excellent admissions and enrollment process, involving personal letters to prospective students, merit awards to talented admitted students, and early identification of potential applicants, which has also contributed to the university's higher visibility outside the Midwest. Almost all of those who enroll graduate in the top quintile of their high school class, take the most enriched courses available to them, and present high test scores. The generous financial-aid policy helps greatly in attracting many outstanding candidates of diverse backgrounds.

What is the admissions committee looking for as it sorts out such a large pool of qualified applicants? Beyond the statistical profile of grade point average, test scores, and class rank, it is looking for evidence of a serious commitment to learning and a particular talent or activity that indicates a student will make a special contribution to the community and take advantage of the many resources the university has to offer. Officials say, "Washington University seeks students with a range of qualities that we believe will contribute to their success in life:

- The student should pursue a challenging curriculum in high school where it shows he or she has received strong academic preparation as well as demonstrating strong performance.

- It is important that students have special talents and character—factors that can be revealed through the application for admission, the required essay, recommendations from teachers and others, or through the optional interviews offered with our alumni in many areas of the nation and the world.

- Successful applicants also show a commitment or passion over time for one or more extracurricular activities and/or a job or other responsibilities.

Strengths in each of these three areas are important for the successful applicant." Requirements and the importance of particular admissions factors, such as test scores, vary significantly among Wash U's undergraduate programs. Art and architecture, for example, focus significantly on a student's portfolio. As the university notes, "While there are university admission standards common to all of the schools, we look for the right preparation for each curriculum. For example, students interested in engineering or premed should have completed course work in physics, chemistry, and math at least through the precalculus level. Students applying to the Sam Fox School of Design & Visual Art should demonstrate interest, skill, or background in art and design by highlighting relevant course work or extracurricular activity, or by submitting a portfolio."

The Ideal Student

Men and women who are serious about their studies and who want to grow in other dimensions as well will enjoy their experience at Washington University. The range of opportunities both intellectual and social and the campus-centered urban setting works well for the adventurous individual who is looking to stretch his or her experiences beyond the home and school environment. Those with a natural curiosity to explore new activities and make new and differing friendships while balancing their academic obligations will gain immensely from their four years of living at Washington University. While it is not a huge institution, it does take individual initiative to profit from what the community has to offer. One cannot sit back and expect teachers and advisors or other students to automatically reach out. It helps to have some idea of interest areas in order to select from the comprehensive range of academic opportunities.

As an administrator notes, "We seek students who can thrive here, who are self-starters, and who look forward to sharing their college experience with people from abroad and different backgrounds. We want students who can explore their own potential to learn more about themselves and their talents. Our professors make themselves accessible both inside and outside the classroom—in fact, virtually all

courses are taught by regular faculty. These faculty are dedicated to the success of their students and at the same time lead in their research fields, write definitive books, create works of art, and win major awards. They embrace teaching as an integral part of their commitment to the university and to its students." Classes at Washington are larger than at most of the smaller colleges in the Hidden Ivies, but are still small relative to most universities. Students who thrive at Washington enjoy these more intimate class settings and the opportunities for research work available with faculty, but again, have more personal motivation to seek out support and contacts. "Most classes range between one and twenty-four students," say college officials. "Where classes sometimes may be larger at first, they generally become smaller as the student progresses into his or her chosen field or fields of study. And for those students who seek direct interaction, many have found that they can work shoulder-to-shoulder with faculty on research and scholarly projects outside the traditional classroom." Students are assigned advisors to help them along their path, and they find a wide range of athletic, social, and residential-life experiences to choose from. What students should not expect at Washington is too much hand-holding or an environment where everyone is studying the same subjects or is interested in the same social outlets.

Student Perspectives on Their Experience

Students choose Washington University for its diverse academic offerings, academic reputation, friendly urban/suburban community, and sense of balance. Says a senior, "I made my final decision when I came to visit the school—

the community here is so friendly and helpful. You feel like yourself here, not a number. The deans, professors, and faculty are all very accessible and they want to help you succeed. It felt comfortable when I visited and I knew this was where I wanted to be." A sophomore says, "I chose my college because of the people who work, teach, and go to school here. Sure, it's easy to read magazines and reports that give statistics and rankings of schools, but once you get to college none of that stuff really matters. It's going to be the teachers, the students, and the people that you come into contact with that impact you the most. You have to be able to see yourself as a part of them in order to feel comfortable and receive the best education you can."

An architecture student adds, "Washington University has a diverse selection of attributes and educational opportunities to complement its successful architecture program. A large part of Washington University's immediate appeal is the fact that the institution sends out a lot of mail and is very interested in all applicants. I chose WU because of its larger urban area location and its beautiful campus (in old Gothic style with pink granite). WU also had the educational opportunities and a higher caliber of students and faculty that I was looking for." A sophomore says, "I chose Washington University because it provided big-university resources while maintaining small-university friendliness and faculty involvement. On top of having incredible resources, the campus is gorgeous and conveniently located."

The university "offers millions of opportunities: internships, study abroad, extracurricular involvement, sports, work, majors/minors, everything. The possibilities are endless. If you want to do something, the school will do its

best to help you. The school also does well with keeping up buildings and staying on top of technology." Another student says, "There are so many resources for students, whether you need help with classes or are looking for a job or want to get research experience, the people at Wash U are there to help you. All you have to learn to do is ask. I was taken aback at just how willing to help everyone was. I wasn't used to that kind of treatment, but I realized that by taking full advantage of the opportunities presented to me I would become a better person."

Those who do best "keep up with their studies, have an interest in the subject matter, develop relationships with their professors, and seek help when needed," says a senior. A sophomore echoes this: "Everyone is smart at this school. It's how hard you work, how you use resources, and how much you concentrate on priorities. Slackers do not get very far at Wash U. Many students who breezed through high school by cramming for tests the night before have to change their study habits to survive here."

Students feel supported by faculty, friends, and the extra resources available on campus. "The staff at Cornerstone (the learning center) is wonderful; they provide tutors, organize study groups, help with time management, some of everything. They don't just get tutors who are knowledgeable about the subject matter, they hire tutors who are knowledgeable and who want to help others understand. The tutors are excellent in explaining difficult concepts in easier terms. Also, the support from academic advisors and peer advisors helps us to succeed, too." Another student adds, "Students have a lot of help in succeeding here. Though it's a very challenging school, the attitude is not one of competition but of camaraderie and group learning. Those who need a little extra help can still succeed when they take advantage of TAs, office hours of the professors and tutors, as well as their fellow students. Our motto is 'Play hard after you work hard.' Everyone here likes to have fun, but everyone is also very academically oriented. It helps a person to work hard when everyone else around them is doing the same."

The university is seen as containing a diverse mix of people. Says a senior, "Our campus is very diversified in accordance with culture, ethnicity, race, socioeconomic status, and other factors. There are various cultural organizations that are very much involved in making the campus community a comfortable one for all students. The university also provides diverse programming, whether music, lectures, or theater performances that touch on many different groups of people." Another student reflects, "WU is a multicultural melting pot. It is probably one of the only schools in the nation where students of all races, ethnicity, backgrounds, etc. get along like friends. By living in the residential college, students get to experience different cultures every day. For example, my floor contained fifty students—twenty-five male, twenty-five female; three of these were African Americans, eleven Asians, three Hispanics, six Indians—all from five different countries including the United States. Now, I just had to think hard about this, because we all became such good friends. By living together, we learned to put behind us the colors of our skin and we were able to see the true individuals that we are. That experience alone was almost as valuable as the education."

The only critiques of Washington seem to center on the need to improve communication between the various parts of this university, among the different schools and programs, and connecting academic and residential life. This is something the university sees as positive, reflecting the increasing number of students wanting to live on campus after their first year and through their tenure at Wash U.

"Washington U is a great place to be," a senior states. "It offers so many opportunities and the wealth of knowledge contained here is phenomenal. Don't let the price tag of attending turn you away: Need-based and merit-based financial aid is available and financial services will work with you to help find a price your family can afford."

What Happens after College

As already noted, a majority of graduates continue their academic career in graduate school either immediately upon graduation or after gaining work experience for several years. Some 84 percent pursue advanced studies within seven years of graduation. There is great diversity in the fields of study graduates pursue—they include not only the traditional professional schools but also the arts, architecture, sciences, technology, education, and social work, among other disciplines. If a student performs well in his or her undergraduate studies, he or she will have many opportunities for graduate work thanks to the university's reputation for rigorous academic preparation. The university, commenting on a recent graduating class, says, "Students have been recognized for achievements in areas as diverse as genetic engineering and fashion design, and have received such prestigious graduate study awards as two Rhodes scholarships in 2006, as well as Fulbright, Marshall, Beinecke, and Truman scholarships, and Goldwater, Mellon, Putnam, National Science Foundation, National Graduate, and Howard Hughes fellowships for undergraduate research." The university maintains a strong Career Center with many resources for students and alumni. Acceptance rates into the professional schools, particularly medicine, science, business, and engineering, are very impressive. Alums are increasingly loyal and, according to the university, report an "extraordinarily high satisfaction rate with their experience here . . . In the last two decades, participation rates in the annual fund have more than doubled."

Wellesley College

106 Central Street
Wellesley, MA 02481-8203

(781) 283-2270

wellesley.edu

admission@wellesley.edu

[
Number of undergraduates: 2,247
Total number of students: 2,247
Ratio of male to female: 100% female
Tuition and fees for 2008–2009: $36,640
% of students receiving financial aid: 59
% of students graduating within six years: 92
2007 endowment: $1,656,565,000
Endowment per student: $737,234
]

Overall Features

Wellesley's administrators, faculty, and trustees are quite clear on the priorities that define the nature and purpose of the institution. The college was founded in 1876 by Henry Fowle Durant, a prominent educator who believed strongly in the right of women to have access to a liberal arts education comparable to that which men received. There is no questioning the quality of the education Wellesley women have obtained over the course of the college's long history. After considerable discussion during the 1960s and 1970s, when all of the great New England men's colleges began to admit women and several of Wellesley's sister institutions decided to admit men, the trustees decided to remain faithful to Wellesley's original purpose of providing young women of talent and intellect with a distinguished education in an environment that put women's interests and issues at the forefront. Wellesley has maintained its excellence through these latter years of change and foment in higher education. No other liberal arts college in the nation can claim to offer stronger intellectual training or a more complete and beautiful campus on which to live and learn.

Having determined to remain in the forefront of educating only women, Wellesley's leaders and faculty include in their teaching and campus activities a women's agenda of significant issues. They promote the fact that every dollar spent on teaching, facilities, research, residential life, student organizations, and athletics is for the purpose of advancing the schooling of the women under their tutelage. There is one teacher for every nine students and all of them are engaged in teaching and advising. The average class size is seventeen to twenty. The administration takes pride in the equal balance of men and women on the faculty. In the opinion of the dean of admissions,

Wellesley has retained what has always been the best features of its education while developing major new directions and approaches in curriculum and teaching. Wellesley's leaders are convinced that the experience of associating with talented, curious, independent peers of many different backgrounds and a preeminent faculty is at the heart of the educational experience. Thus the residential college that is very selective in creating its community each year is considered a matchless experience for personal development and intellectual growth.

Wellesley is better able to carry out its goals than most liberal arts colleges today, thanks to its robust endowment, magnificent campus and facilities, and the passionate support of a majority of alumnae. Recent examples of their generosity include the Knapp Media and Technology Center, described as the pulse of the campus with its many advanced technological features for learning. All college buildings are wired to the center, which makes possible easy communication among all members of the community, as well as advanced levels of research and teaching. Some 1.5 million items are available through the college's computerized library system. Wellesley believes that its graduates must be literate in technology if they are to excel and lead in their chosen endeavors. The Pforzheimer Learning and Teaching Center is also a nice addition, providing academic support to students and training for faculty to learn how to teach their subject in the best possible manner. Wellesley's Science Center, Jewett Art Center, and Davis Museum and Cultural Center offer additional fine facilities for students. The college's wealth allows the admissions committee to admit any qualified applicant without concern for her financial need, which results in a highly diversified student body of high academic quality. Over 50 percent of stu-

dents receive financial aid today. The trustees, again in keeping with the goals of the college, support over $38 million in the financial-aid budget so that students' financial-aid packages will include more grant money and fewer loan obligations. Wellesley recently eliminated loans for families earning less than $60,000 and reduced loans for other aid recipients.

What has changed most in the past several decades is the composition of the student body. Wellesley women were, for most of the college's history, advantaged white women from economically comfortable, traditional backgrounds. At present 45 percent of the women are from multicultural backgrounds. Eight percent are internationals from thirty-three different countries in the class of 2012 alone. This is what gives the community its energy, diversity of opinions and cultures, and thus provides a growth experience for every woman in residence. The twenty-one residential halls are attractive, comfortable, and self-sustaining. Students are assigned randomly to a residential house for their four years to ensure exposure to a mix of students. There are dining rooms in the houses, apartments for faculty masters who are responsible for the advisory system, and cultural activities that have generous support from the college. Wellesley operates on a system of democratic government for the community; thus students have many opportunities to participate on committees and advisory boards that determine school policies and programs. Extracurricular activities take many shapes and forms. Theater, music, art, community service, political clubs, newspaper and literary magazines, and special interest groups make up the 150 organizations that are recognized by the college. Wellesley sponsors thirteen intercollegiate teams in all of the major sports in addition to dozens of intramural teams that compete among the residential houses. The sports center is an impressive modern complex that includes all of the facilities necessary for an outstanding athletic and personal training program. The crew program has a long and special tradition at Wellesley, with excellent facilities for training.

What the College Stands For

Wellesley's leaders sum up their belief in the importance and value of a liberal arts education in this way: "In a time when technology is widely accepted as a substitute for ideas, when action all too often preempts thought, and when instant analysis masquerades as understanding, a liberal arts education has never been more crucial. In such a time, Wellesley College may well offer more to its students than ever before in its history." What lies behind this pronouncement is the belief that the ethical and moral values developed over the history of mankind are essential guideposts in making critical judgments in our complex society. Wellesley faculty and advisors stress this concept throughout the curriculum and the residential life in which all students participate. The college uses its financial resources to admit outstanding women of all backgrounds, because they believe that exposure to a highly diverse community is one of the most significant elements of the learning experience at Wellesley. Another mission of the college is to play a leading role in advancing the interests and needs of women in contemporary society as well as to instill a general sense of social responsibility. These concerns permeate the academic and social life so that students will feel they can express their beliefs and will have the confidence to achieve their goals in the future. Self-governance also weaves its way into the fabric of the Wellesley philosophy. The

more opportunity for young women to make daily decisions for themselves and govern their own behavior, the better prepared they will be to lead independent and fruitful lives.

Another element of college life that takes on particular importance is Wellesley's honor code: "Throughout its history Wellesley College has based its student life policies on the concepts of personal integrity, respect for individual rights, and self-government. Established in 1919, the honor code rests on the assumption that individual integrity is of fundamental value to each member of the community. The honor code requires all students to display respect, responsibility, and honesty in all aspects of college life. The honor code provides students with privileges that are unique to the college. They include self-scheduled exams, open-stack libraries, guests in the residence halls, and full use of college resources. The fact that these privileges have been instituted and continued without major incident for eighty years is a testament to the respect Wellesley women have for the honor code and for one another." An administrator describes these goals for Wellesley's students: "Exploring the breadth of the liberal arts curriculum; gaining knowledge and skills in depth in pursuing a major; learning strong critical-thinking, quantitative, and communication skills through undertakings inside and outside the classroom; learning how to work effectively with others in a community of learners."

In the future, Wellesley aims to continue its academic strengths and traditions while ensuring its relevance for its students in a changing society. As an administrator notes, "Wellesley College will continue to adapt as needed to respond to the rapid globalization of the world's economic, political, and cultural systems. The people who will effectively manage these new global realities will be those who can move from culture to culture, who can collaborate and communicate with fluency across national, racial, religious, and socioeconomic lines, with knowledge, sophistication, sensitivity, and ease. The work environment of the future will demand both intellectual flexibility and task-specific disciplinary knowledge. There is no better preparation for this environment than a Wellesley liberal arts education." Another official notes, "The changes that I have seen in my time here that have been most significant in my view have been recent efforts to enhance the strong faculty-student connections here by developing stronger advising programs, the efforts to expand and enhance the range of academic support services to better meet student needs, and the efforts to support learning within a very diverse community."

Curriculum, Academic Life, and Unique Programs of Study

Wellesley is committed to the tradition of the liberal arts education, which means that students will gain both breadth in the major fields of the arts and sciences and depth in a chosen discipline. As an administrator puts it, "Wellesley's faculty members cherish the opportunity to work closely with their students; in fact, more than 400 students each year conduct research with their professors. Our strong liberal arts curriculum is broad, with more than 900 course offerings, and deep, with offerings from introductory classes to advanced seminars. Wellesley's distribution requirements ensure that each student's liberal arts education is built on a strong foundation of core knowledge. The goal of these requirements is to expose students to a wide range of issues and ideas that will broaden

their knowledge and worldview while providing the flexibility to pursue personal interests and curiosities."

The faculty undertook an intensive internal review to determine what skills are necessary to become an educated person and a leader in the twenty-first century. The results of this study serve as road signs for the college to navigate the academic curriculum and to plan for the constantly fluid world that awaits its graduates. As one senior administrator expressed it, the end goals are to have students seek the major truths of life and develop the tools to continue their search for the rest of their lives; to become good writers, avid readers and researchers for informed decision-making, and critical thinkers; to know how to solve complex problems; and to be able to articulate their ideas with fluency.

There are no core courses that must be taken, but there are distribution requirements to be certain that students choose widely from the exceptional number of courses available within the thirty-two departments and twenty-four interdepartmental majors Wellesley supports. It is safe to say that the breadth of subject areas and interdisciplinary programs offered are second to those of no undergraduate liberal arts college in the country. Wellesley has always encouraged its students to survey the sciences, which are considered among the strongest departments. The faculty encourages students to design a double major or opt for an interdepartmental major to fulfill their special intellectual interests. In order to broaden still further the learning opportunities for students, Wellesley has a full cross-registration policy with MIT. Typical courses taken by Wellesley students at MIT are technical sciences, architectural design, computer science, and business finance. Wellesley is also a member of the Twelve College Exchange program, the National Theater Institute, and the Maritime Studies program, among other cooperative arrangements. Students are encouraged to gain experience in other cultures through the many study-abroad programs that are sponsored or cosponsored by Wellesley. The school calendar includes a winter session in January that takes off the pressure of the regular semester academic obligations and allows students to undertake a special interest course, to work or intern off campus, to create and complete a special project, fine-tune certain skills, or do volunteer work.

Major Admissions Criteria

The admissions committee is quite clear on the most useful information that determines if a candidate is likely to succeed in this heady intellectual and social community: the high school transcript. The nature and quality level of the courses is the best indicator of intellectual curiosity, motivation, and the skills to meet the demands of the Wellesley education. The admissions committee is concerned that students do not overextend themselves, either. They look for signs of challenging oneself in advanced courses of major interest, but it is not critical to take nothing but Advanced Placement courses in the junior and senior year. Teacher recommendations that describe a student in depth are of great value to the committee since they are witness to the degree of intellectual curiosity, engagement in the learning process, maturity in thinking, and enjoyment of studies. Test scores do matter but there is no specific cutoff for consideration. As an official says, "The board of admissions seeks students who have intellectual curiosity, the

ability to think, to ask questions, and to grapple with the answers. Wellesley looks for a student who is excited about learning. This comes through in the board's review of her academic record, her recommendations from teachers, and in what she writes about herself in her application. The college also looks for signs of leadership skills, motivation, and commitment."

While Wellesley attracts its share of top academic performers each year, it can be somewhat flexible in its decision process because all of its available places go to women. Consequently there is great interest in the candidate who has achieved in a particular talent arena, be it music, drama, art, dance, athletics, or leadership. The dean of admissions summed up their policy this way: "Every decision made by our committee is an individual decision and the total person, not just grades or test scores, is taken into consideration." Wellesley is somewhat unique among the major colleges as it includes both elected faculty and junior and senior peer readers in the review and recommendation process for admission. Because its leaders believe so strongly that the heart of the Wellesley experience is the opportunity to live among talented, diverse, curious peers and outstanding faculty, the admissions community gives considerable weight to the kind of person who will enhance the residential community.

And again, needy students should take note that, as the college makes clear, "Wellesley College is one of the few most selective colleges and universities that has a need-blind admission policy and meets 100 percent of a student's demonstrated financial need. Applicants are considered for admission only on the basis of their talents and personal qualities, not on their ability to pay. If a student has financial needs, the college works with the student to establish a financial-aid package that meets 100 percent of the demonstrated financial need, based on a financial-aid formula. These policies mean that Wellesley is able to attract the best students, regardless of their financial resources. Keeping a Wellesley education accessible and affordable is one of the college's top priorities. We are fortunate to be in such a healthy financial position to be able to maintain these admissions and aid policies."

Statistics for the class of 2012 show a very competitive and diverse group of enrollees. Average SATs were 689 in critical reading, 689 in math, and 677 in writing. The average ACT was 30. Almost two-thirds of students were public school graduates. Twenty-three percent came from New England, while 21 percent came from the Pacific and Mountain states, 18 percent mid-Atlantic, 17 percent South, and 12 percent Central.

The Ideal Student

Wellesley has high expectations for its students not only in the classroom but also in the life of the community. While students have a great deal of autonomy in choosing the courses they find interesting and creating a major field of study that meets their intellectual goals, they will be confronted with a very demanding level of work. Sloppiness in thinking, writing, and presentation will not be excused by the faculty. A premium is put on individual interpretation and the ability to defend one's position on any topic. Wellesley women are also expected to carve out one or more activities from the huge block of opportunities and to excel in their elected endeavors. An administrator, speaking of the strengths of Wellesley students, notes, "Their intelligence and resilience, their ability and willingness to support each other as learn-

ers, and the wide array of support services available to them." He also points out that "Wellesley is a very diverse community, with a commitment to fostering open dialogues among students and other community members of different backgrounds and perspectives. Diversity issues are a very important part of new student orientation programming, sending the message that everyone has an active role to play and everyone has things to learn about living and learning in a diverse community."

Wellesley is definitely a women's environment. A concern for women's interests and challenges is also part and parcel of a student's commitment as far as the college leaders would have it. But for all of the advantages of the college's location in the Boston/Cambridge area and the cross-registration opportunities with other major institutions, many Wellesley women lament the lack of a natural and readily available social life. One has to reach out to the larger world beyond the gates of this semi-secluded campus to develop relationships with members of the other sex. While most women choose to attend Wellesley primarily because of the excellence of its education and facilities, a distinct cohort of feminists want little to do with men in the traditional dating scene. They find at Wellesley the opportunity to engage in active debate on the issues of feminism and the barriers to equal access to careers and leadership roles. For all of this, the prospective student should be more concerned for the distinctive educational and community programs Wellesley fosters than for the more traditional model of the coeducational collegiate experience that strikes more of a balance between intellectual and social components. Wellesley is also an appropriate place for the intelligent and inspired young woman who will feel more comfortable in a less competitive social environment, one that provides a good deal of support from the faculty and administration.

Student Perspectives on Their Experience

It would be rare to hear a Wellesley student complain about the quality of the education she has received. From the outstanding faculty and its interest in and support of a student to the curricular opportunities to the physical facilities and campus, Wellesley is as good as it gets. Most students express their appreciation for the many advantages Wellesley provides. They find that concern for their welfare is demonstrated by the many layers of personal and academic counsel and support, the financial aid that is generously dispensed, and the multitude of choices for learning on and off campus. As a senior describes, "Wellesley seemed to offer me the best of two worlds. First, it was an excellent small liberal arts college. Second, it is affiliated with a top research institution (MIT). There was something about Wellesley that clicked with my personality. I knew I had found an environment I could thrive and grow in. Wellesley is a student-focused institution. There is something for everyone. Wellesley provides its students with a wealth of resources to aid in their education process. These resources are not only financial but include the people (faculty, staff, students, and alums). We have great relationships with our alums, who are always ready to help."

The main reasons some women are unhappy center on the lack of a coeducational environment; they lament the absence of men in class discussions and hang-out time. Discontent seems to be less about formal dating than about having men around as an option for social and intellectual engagement. As one woman puts it, "Women at Wellesley, in general, do very well in life. Wellesley is able to maintain its reputation as a place that breeds great women, women who will make a difference in this world. But this achievement comes at a price. Women at Wellesley give up a normal college experience—complete with men, partying, and intellectual/social activity—for one that resembles the experience of a women's boarding high school. Most women accept their life at Wellesley for what it is, and instead focus their attention on their studies and the dream of life after college. The knowledge that Wellesley is not 'normal' is very stressful and difficult to accept for many students, especially those who spend time with friends at other schools." Some students feel there is too prevalent an edge to feminism or diversity for them on campus and that this puts pressure on individuals to conform to an agenda or to be part of a particular group; other students say there is too much emphasis on work at the college. One first-year student reflected on her desire to transfer to a coed institution where her friends were enjoying themselves. Her dilemma, she told us, is knowing what she would be giving up in terms of programs and faculty if she were to leave Wellesley. Yet other students want the college to be even more involved in social change: "I would like my school to be less isolationist and more radical," says one. "I would love for Wellesley to be a school for rebels and activists." Wellesley applicants are advised to measure the campus environment carefully, to spend time at the college and talk to current students to make sure this school is the right match for them, because, if it is, it offers an amazing and empowering undergraduate college experience.

A junior points out these aspects of her experience at Wellesley: "Whether for academic, financial, or personal matters, there are always people that students can turn to for help, including student organizations, professors, and the administration. Faculty are easy to reach and helpful, making the classroom experience that much more intimate. Wellesley also does an excellent job of keeping a check on us, especially in our first year, when we are required to find an advisor and are encouraged to seek help and develop networks. Speaking of the network, the Center for Work & Service is especially helpful in assisting you in finding an internship in unique places, and in introducing you to powerful figures." She describes her classmates this way: "Most students at Wellesley are independent-minded women who are among the best and brightest females to come from their high schools around the country and the world. Women at Wellesley are competitive, but more so with themselves than others. Because success is something that we must and will achieve, there is a lot of motivation to do well. Women who can handle that pressure while being able to remain confident do best at Wellesley." She says of the social-life issues, "If you want connections outside of the college, you will have to work for it. Wellesley's proximity to Boston and available transportation makes this an easy and fun experience. The question is if one is willing to utilize such resources. My social life at Wellesley has been quite an experience, one that has been both fulfilling and fun. I definitely make an effort to get to know people on and off campus and search out opportunities on my own." Of Wellesley's financial support, she says, "This financial aid has helped me to be where I am, as it not only meets my financial needs, but also continues to assist me in other ways. For example, during my sophomore year, Wellesley helped fund a Wintersession trip to Morocco, where I studied Moroccan history and culture. Wellesley also offers stipends for internships

and projects for students throughout your four years."

"I am also grateful for self-scheduled exams," this student says. "These allow us to take final exams when we are most comfortable after the reading period, thereby lowering the stress level and making it more convenient for students."

A sophomore had this to say about Wellesley: "Wellesley provides one of the best, if not the best, support networks for their students. There are so many academic and social resources, from the Davis Museum (our very own art museum), to (free) peer group and individualized tutors, to research guides, an expansive alumnae network, and daily transportation into Boston and Cambridge." What could the college do better? "Wellesley can improve at encouraging students to become more involved in humanity-oriented/volunteer programs. I think we should be told more often that it is ok to step away from our work and go outside. With that said, I know many students who are involved with volunteer work and are at the same time dedicated to their studies. Our college's motto certainly says a lot about how important helping others is, which I honestly see every day—in our student resident advisors, our academic peer tutors (students volunteer for these positions), and students simply helping one another on an internship application, or a problem set. I think that this type of benevolence is internal to our college, and I would love to see more ways to make it more external while remaining intact with the college's structure." Those who do best "are those who are open-minded enough to try a variety of course and social events, those who aren't afraid to ask questions or to ask for help, those who are passionate, those who think independently but work cooperatively, those who are engaged with their studies and extracurricular activities, and those who are proactive." She points out that "the active learning that takes place in classrooms makes Wellesley stand out from other undergraduate institutions. The professors certainly help initiate discussion, but it's the students who help keep it going. We learn a lot from each other."

A junior says, "Just in the past two years, I have connected with alums at the on-campus Learning from Lawyers conference and through the online W Network in order to get internship advice, and I have had meals with groups of alums of all ages in Boston and Washington, D.C. Wellesley alumna lawyers in Boston have even included me in their lunch group and connected me with their contacts in my field of interest."

What Happens after College

Any alumna will attest to the value of her Wellesley degree in the larger marketplace and to the doors it opens to prove her abilities and drive. The Wellesley alumnae network has long been a powerful resource for jobs in a wide range of professional careers. The college maintains a well-organized information and advisement center for its students to connect with potential employers and meet the requirements for graduate schools of all kinds. As with the other top colleges today, a majority of graduates will explore jobs and new parts of the world before continuing their formal education in advanced programs. Here, too, the reputation for academic rigor is such that Wellesley women are very successful in gaining admission to top graduate programs. A college official says, "Throughout our history, Wellesley has prepared women to be engaged citizens and leaders. . . . Among the college's 35,000 alumnae are a pioneering environmentalist, a construction engineer, a NASA astronaut, the first female U.S. secretary of state, journalists, artists, community volunteers, and government officials around the globe, to name only a few. Our alumnae recognize the value of their Wellesley education and remain tremendously supportive of and loyal to the college."

According to the college, "Wellesley's rigorous and well-rounded liberal arts education prepares its students for achievement in a wide variety of fields and a changing work environment. Approximately 20 percent of recent Wellesley graduating classes have entered graduate schools following commencement. Popular fields of study include education; medicine and other health-related fields; humanities and law; science, math, and computer science; and the fine or performing arts. Wellesley's acceptance rate to medical schools is significantly higher than the national average. The majority of Wellesley graduates each year pursue employment upon graduation. More than 130 top businesses and 40 nonprofit organizations recruit on campus each year.

Wesleyan University

70 Wyllys Avenue
Middletown, CT 06459-0260

(860) 685-3000

wesleyan.edu

admissions@wesleyan.edu

[
Number of undergraduates: 2,787
Total number of students: 3,192
Ratio of male to female: 49/51
Tuition and fees for 2008–2009: $38,934
% of students receiving need-based financial aid: 47
% of students graduating within six years: 93
2007 endowment: $710,800,000
Endowment per student: $222,681
]

Overall Features

Not exactly a small college, nor precisely a large university, Wesleyan is one of those few institutions occupying a niche between the two. A large college/small university, Wesleyan is predominantly focused on its 2,700 undergraduates, but offers a select few programs to some 200 graduate students, and an additional several hundred part-time students concentrating in Wesleyan's groundbreaking Graduate Liberal Studies Program. Wesleyan's resources—from more than 300 faculty providing a nine-to-one student-to-faculty ratio and 900 courses in thirty-nine departments and forty-six major fields of study; to its libraries holding some 2 million volumes; to its 340 campus buildings—substantiate the university as a national leader and rival to Amherst and Williams, the other "little Ivies."

Yet many compare Wesleyan more closely to Swarthmore, its small cousin, with which Wesleyan shares a decidedly intellectual, progressive, and liberal bent. Wesleyan is a diverse, cosmopolitan environment. Its students are 26 percent of color, hail from forty-eight states and fifty-three foreign countries, join more than 200 organizations ranging from ethnic affinity clubs to a cappella groups to community service programs, and represent some of the most academically talented young men and women from across the United States and abroad. Seven percent of Wesleyan students are internationals, and the college provides special financial-aid programs, such as the Freeman Asian Scholarship Program offering full scholarships to twenty-two top Asian students in each class, in order to attract strong foreign students to the university. Wesleyan is completely need-blind for American students, and meets full need. Those with income below $40,000 will not incur any loans in their aid package.

Wesleyan is situated in the small, central Connecticut town of Middlebury, which provides a safe campus environment, a fair amount of off-campus social and commercial opportunities, and easy access to New Haven, Hartford, and New York City. Most students spend the majority of their time on campus, taking advantage of student-run and college-promoted sports and arts activities. A member of the New England Small College Athletic Conference, Wesleyan competes against its Division III small-college brethren throughout New England and the Northeast in twenty-nine varsity sports. The college also is home to eleven club and nine intramural sports. Some 700 students participate in intercollegiate athletic competition, belying Wesleyan's reputation as an artsy, non-active institution. In fact, in recent years Wesleyan has been actively seeking to rebuild its athletic reputation and showcase its balance in traditional areas like sports. An impressive modern athletic complex offers a good outlet for those wanting more physical activity.

That said, Wesleyan does indeed have stellar facilities and programs in the arts, especially music. A new university center built with many sustainable design features houses a café, lounge, and theater and dance facilities. The campus' historic Memorial Chapel and '92 Theater have also been renovated. There is a center expressly devoted to film studies, another one of Wesleyan's strengths, which includes the college's Cinema Archives. For those pursuing Wesleyan's well-regarded programs in astronomy as well as earth and environmental sciences, there is the Van Vleck Observatory. The Davison Art Center houses an extensive collection, including works by Dürer, Manet, and Goya. There is also the Zilkha Gallery providing space for exhibitions by students and others.

Most students live on campus at Wesleyan, but the definition of campus is a broad one, as the university offers an extensive and eclectic mix of housing choices. These include larger residence halls with single or double rooms, wood-frame houses, apartments, and affinity-based "program houses." Starting out in more supportive residence halls housing resident advisors, students can choose in their later years between a few fraternity houses (just 3 percent of men and 1 percent of women belong to them), the Outhouse for members of the Outing Club, the Open House for LGBT (and beyond) students, the Science Hall, the Film Hall, numerous language/cultural houses, and other options. Thus many housing options are available, and the college recently renovated a number of residence halls in an effort to continue to improve on the residential-life experience for Wesleyan students.

A senior administrator sums up the college well: "Wesleyan is a dynamic community and educational setting. Faculty, staff and students are creative and innovative. This desire to be on the cutting edge both educationally and with respect to social issues drives continual change at the institution."

What the College Stands For

Founded by Methodists in 1831 and named for John Wesley, the founder of the Methodist faith, Wesleyan's original emphasis was on the liberal arts and social service, rather than training for the ministry. In 1937 Wesleyan fully separated from the church, and continued to advance its curriculum in the balanced liberal arts and sciences. While the university experimented with coeducation in the late 1800s and early 1900s, the college moved away from women's education through most of the twentieth century until it went fully coed in 1970. The 1960s and 70s, and more recent decades, saw Wesleyan continue its efforts to bring more diversity to campus, concentrate on its intellectual depth and breadth, and develop a balanced social and academic environment.

"Wesleyan is a community of active learners," says an administrator. "Our faculty bring their scholarship and creative work to their teaching and engage students as creators of knowledge. We enhance capacities for critical and imaginative thinking that can address unfamiliar and changing circumstances; engender a moral sensibility that can weigh consequence beyond the self; and establish an enduring love of learning for its own sake. Wesleyan articulates general education expectations to ensure that students experience the breadth of our curricular offerings. The faculty have also articulated 'essential capabilities' that students should hone during their time at Wesleyan."

These are the capabilities Wesleyan seeks to improve in students during their time at the university: "Writing; speaking; interpretation; quantitative reasoning; logical reasoning; designing, creating, and realizing; ethical reasoning; intercultural literacy; information literacy; effective citizenship." Wesleyan's most recent strategic plan, titled "Engaged with the World," suggests these goals for the university: "At Wesleyan we prepare our students to face a rapidly changing world with confidence and the sense of responsibility to want to make the world a better place. Our students gain that confidence from a strong education in the liberal arts and sciences—an education that engenders the ability to engage in critical thinking, to communicate effectively, to find creative solutions to problems, to develop the imagination to see the world as others see it,

and to make ethical judgments that come from deep within one's character."

Curriculum, Academic Life, and Unique Programs of Study

The college describes its curriculum as "open," but Wesleyan has not abandoned a set of distribution requirements designed to expose students to a wide array of topics across the spectrum of the liberal arts and sciences. Awarding only the bachelor of arts degree to undergraduates, Wesleyan requires students to fulfill the program for a departmental or interdepartmental major, or one of two special "collegiate programs": the College of Letters or the College of Social Studies. Students must apply for these latter two selective programs by the end of the first year. Interdepartmental majors include such options as African American studies; Russian and East European studies; feminist, gender, and sexuality studies; and archaeological studies. Students may also propose "university major" programs involving more than one department.

General education requirements at Wesleyan cover three broad areas: natural sciences and mathematics; social and behavioral sciences; and humanities and arts. Students are expected in their first and second years to take at least two courses in each of these three areas. Undergraduates are then expected to take one more course in each of the areas during their third or fourth year of residence. Throughout the curriculum, Wesleyan offers students the opportunity for independent study and tutorials. The First Year Matters program offers freshmen a shared experience and common reading component.

Wesleyan offers many study-abroad options, including college-run programs in France,

Germany, Israel, Italy, and Spain. A teaching apprentice program offers students the opportunity to teach a course with a Wesleyan faculty member. Wesleyan participates in the Twelve College Exchange program, as well as the Williams-Mystic College Maritime Studies program and the National Theater Institute sponsored by Connecticut College. Other programs and opportunities of note include: participation in the Venture Consortium, which promotes experiential learning and social responsibility; a college consortium with Trinity and Connecticut colleges; and 3/2 science and engineering programs with Columbia and Cal Tech.

To take advantage of its special graduate programs, Wesleyan offers undergraduates some interesting options. Upperclass students can take advantage of some course work in the university's Graduate Liberal Studies Program. They may also pursue combined B.A./M.A. programs in anthropology or the sciences in five years.

Major Admissions Criteria

"No specific high school program is prescribed for admission," notes the university's admission profile, "but admitted students have taken advantage of the specific curricular offerings of their high school to get the most out of their education, whether through AP courses, International Baccalaureate diploma, local college courses, or simply the most challenging teachers. Their curiosity and love of learning travels with them. Rank in class and standardized test scores are generally quite strong, but more telling in our assessment are the quality of academic program and teacher recommendations. The admitted students are diverse in experience, nationality, region, first language, race, ethnicity, and socioeconomic status. The vari-

ety of talents and commitments to school groups, civic and religious organizations, politics, and the arts makes successful candidates more likely to contribute to the life of this vibrant community called Wesleyan."

An administrator says this about successful applicants: "First, they must be academically qualified to perform at the highest levels. They should also be well-rounded in their interests and involvements, be creative and innovative thinkers, and represent diversity of identity, thought, and opinion. Students should also understand the responsibility associated with the freedoms they have on campus."

Wesleyan's class of 2012 was one of the most competitive ever. Applications were up 6 percent from the previous year, and 44 percent from 1998. 2,245 applicants (27 percent) were admitted out of 8,250 applications. 721 students enrolled. The diversity of the class is impressive: 32 percent students of color and 9 percent international. 21 percent come from New England, and 40 percent from the mid-Atlantic states. 17 percent come from the West, and 6 percent each from the South and Midwest. 57 percent graduated from public high schools, and 40 percent were admitted early decision. 6 percent are legacies, while 16 percent are first-generation to four-year colleges. 65 percent were in the top decile of their high school class (of the 51 percent who reported class rank). Median SAT scores were 700 on all three sections, and the ACT median was 31.

The Ideal Student

At a college where the students are described as liberal, progressive, activist, leftist, socially aware, politically informed and engaged, independent, and intellectual, the ideal student will likely fulfill one or more of those categories.

He or she will also be willing to get involved, take a stand, make an argument, and ask a question. The Wesleyan social environment and academic climate are not for the faint of heart. Students must be willing to work hard, pursue their curiosity, and challenge and be challenged by their faculty and classmates. They must also be open-minded about alternative lifestyles and viewpoints; sensitive to politically correct concerns; and willing to interact and learn from all manner of students.

An administrator notes, "Students develop close relationships with faculty through formal advising, informal mentoring, etc. A high percentage of classes are small which facilitates students developing relationships with faculty and with each other. Residential Life staff and the class deans weave a safety net that can identify students who are experiencing difficulty and refer them to appropriate resources. Wesleyan fosters a cooperative and collaborative learning environment—our honor code is important in this regard." Students should enjoy and be willing to take advantage of these personal interactions in and out of the classroom with professors. The honor code promotes self-awareness and responsibility in the community.

A senior notes the "positive attitudes and energy on this campus" and "the intense intellectual curiosity and proactive nature of the student body." She says that "Wesleyan encourages students to identify and follow their passion and expand their personal boundaries to strive for great achievement." The university environment is described as "Open-minded. Inspiring. Encouraging. Demanding. Proactive."

Those who do well are variously described by students as "students who are focused, hardworking, and inherently intellectual. Students that apply their time well and work very hard to achieve their passions." Another senior says, "Students who are intellectually curious. I don't think anyone has to have one particular interest or another, but students who love learning love Wesleyan because there are so many opportunities to explore a myriad of exciting new subjects." According to a classmate, "People who are passionate about creating change in the world by the way in which they choose to engage with whatever it is they are pursuing academically. Wesleyan emphasizes that students think outside of themselves and discover the role they play in the global community." Finally, a junior says, "Open, enthusiastic students who are interested in meeting new people and trying new things. People who don't mind being involved in twenty different things and sometimes having to sacrifice sleep to do everything they're interested in. Students who are willing to work hard academically and who are proactive about their education (for example, willing to approach their professors, ask questions, etc.)." She continues, "The students at Wesleyan aren't just wacky activists. The vast majority are grounded people who are serious about both their education and their extracurriculars. They are hardworking, intelligent students passionate about a wide variety of academic pursuits from molecular biology to women's studies to international relations."

Student Perspectives on Their Experience

Students at Wesleyan are generally happy and motivated by their educational experience. They are most positive about "small classes, and intimate relationships with faculty and administrators," in the words of one senior, and "its commitment to diversity, social awareness, academia, and public service." They note that "it is challenging and intensely academic, in addition to extremely aware and proactive," and value "its flexible and exciting curriculum, its size, its location, and its opportunities. I decided to attend Wesleyan after a visit to the campus because of how intellectual, committed, and inspiring the students I met here seemed to be."

Students succeed because of "the support of other students, of professors, and of administrators. Professors are incredibly supportive and always leave their doors open for students to come in and discuss the course work or to help with any difficulties. Also, programs like peer advising and writing tutors provide student-to-student formalized support at various hours of the day." Another senior notes, "I have really appreciated Residential Life and the support they provided, especially during my freshman and sophomore years. Having a residential advisor who helped me through some issues on my hall, who gave me course advice, and who was always around to point me in the right direction or check in was incredibly reassuring." A senior international student chose Wesleyan because of "the friendly and engaging dialogue that happens between on campus as seen when I visited. The passion of and giving nature of students. The lack of cutthroat competition though it is an atmosphere that strives for excellence. The prestige of the institution. The location."

According to one student, Wesleyan does these things best: "Theater, life sciences!!!!! English, psychology, film. Competitive athletics, great administration, loads of social events on campus (ranging from parties to film showings to plays, concerts and contra dance!)." "The academic environment here is very cooperative," says a junior, "so I would say that other students help students to succeed, both on an official capacity through our workshop and tutoring programs, and on a friend-to-friend, classmate-to-classmate basis. Freshman year, when half the people in your hall are taking intro biology and calculus, it's great to be able to get a big group together to study the night before a test." Another says, "Our writing tutor program was a great help to me my freshman year. The school hires and trains upperclassman English majors. A campuswide e-mail is sent to all freshmen at the beginning of their first year, inviting anyone interested to apply for a writing tutor. I took advantage of this and it was a fantastic experience. My writing tutor met with me every week of fall semester and helped me plan everything about my papers from when to start writing them, to forming my thesis and ideas, to helping me check over the final draft. It was incredibly helpful."

Students speak of their peers as involved and pursuing one or more passions with great zeal. A senior says, "Students are involved with many different activities and within many different communities on campus. There is surely overlap between student groups, and groups and communities are not segregated based on interests or activities. I think the combined aspects of diversity of thought and a supportive environment make for a wonderful atmosphere of growth and learning. Students are constantly engaged and constantly exchanging knowledge in formal and informal settings." "Wesleyan is a committed and active campus," notes a senior. "Not everyone is politically active, specifically, but everyone has a passion and is involved in that passion whether it is academically based or extracurricularly focused." A junior says, "What truly makes Wesleyan special is the enthusiasm and energy of its students. On a more specific level, I think this is demonstrated by the Wespeak section of the *Argus*. Students, faculty, alumni, and members of the Middletown community are invited to write in editorials about absolutely anything (be it a previous article in the paper or an issue on campus or in the town) and they do— typically there are between four and ten every issue, and the paper comes out twice a week."

Diversity at Wesleyan is generally seen as significant and a positive force on campus. A senior points out the "many program houses, or small residence halls for sophomores through seniors, to share in the mission toward a common cause. Many cultural clubs and annual shows happen on campus and all students are invited to join, participate in, or watch all of these events." We hear that "Wesleyan is diverse in many ways, not just in its racial, ethnic, and socioeconomic makeup but also in students' interests and how students think. Students here are passionate about such diverse topics as the molecular structures of certain plants and how those structures are affected by climate change and medieval European ballads. Politically, as well, there is a real diversity on this campus in terms of what students prioritize, what students believe, and how students vote or if they vote at all. Students also come from diverse backgrounds and the college is supportive of those identities mostly through cultural and social events celebrating difference on this campus, but also by providing supports for those students who may not agree with the majority opinion or who may not come from the privilege that many students here have."

Another student says, "Students from all walks of life attend Wesleyan and this is seen in the cultural concerts we have on campus (African, Asian, Southeast Asian, West Indian) as well residential program houses." Students suggest that "campuswide unity/solidarity events such as benefit performances, Spring Fling, and other large-scale events" help in bringing students together, as do "student involvement in policy and decision making, and campuswide forums for campus community members to voice their thoughts and concerns." Yet some feel that more could be done: "I would like to change the campus to facilitate cross-community interaction and dialogue. It is sometimes difficult for students who do not share interests or backgrounds to meet each other and sustain relationships."

Wesleyan seniors provide some great advice for prospective freshmen: "My advice would be to try new things. Wesleyan is a place where you can develop interests in new areas that you may never have heard of before. There is so much to do and learn here, and Wesleyan offers so many opportunities that are simply not available outside of the undergraduate experience, so I would encourage incoming students to close no doors and keep an open mind." Another says, "Prepare to be challenged academically, socially, and simply in the way you think!" And another insists, "Be open to everything! Sign up for a hundred different groups and figure out which ones you're truly passionate about. Take advantage of every academic opportunity you can."

What Happens after College

"One finds Wesleyan graduates in a surprisingly wide variety of work settings," says a college official, "heavily involved in service agencies, and many students continue their education through graduate programs. Because of the kind of education Wesleyan provides, a disproportionate number of our graduates end up in leadership positions in a variety of areas." Wesleyan students are well prepared for graduate studies, in fields across the arts and sciences, while psychology, English, and government are the most popular undergraduate majors. About 13 percent of graduating seniors will enroll directly in graduate school, with another more than 50 percent finding a job within six months. About sixty organizations recruit on campus.

The Career Resource Center reports, "Generally speaking, Wesleyan alumni are like most other liberal arts graduates in that their broad and rigorous academic pursuits give them the foundation skills for just about any potential career field. Employers and graduate/professional schools alike value liberal arts graduates for their ability to speak well, write effectively, think quickly on their feet, and learn continuously. Wesleyan alumni possess all of these attributes along with heightened sensitivity around issues of diversity, multiculturalism, and fairness." A survey of more than 12,000 alumni showed their involvement in these broad areas of work: business (26 percent); education (20 percent); health professions (9 percent); law (8 percent); and entertainment (6 percent). The school reported a law school admit rate for 2004–05 of 76 percent and a medical school acceptance rate for the five years 2003–07 ranging from 67 to 81 percent.

Williams College

33 Stetson Court
Williamstown, MA 01267

(413) 597-4052

williams.edu

admission@williams.edu

[
Number of undergraduates: 1,970
Total number of students: 2,018
Ratio of male to female: 47/53
Tuition and fees for 2008–2009: $37,640
% of students receiving need-based financial aid: 49
% of students graduating within six years: 96
2007 endowment: $1,892,055,000
Endowment per student: $937,589
]

Overall Features

The quintessential "small New England liberal arts college," Williams is often held up as the classic example of the type. Like its archrival, Amherst, Williams has undergone significant changes in the past few decades which demand that prospective students look well beyond its beautiful exterior in judging the amazing academic and extracurricular opportunities at hand. Founded in 1793 in the farthest northwest corner of Massachusetts, a beautiful rural outpost in the Berkshires, Williams College began its life educating ministers, doctors, and other professionals, and through the 1800s established its reputation as a college dedicated to teaching and learning in a personal, connected community. The college has maintained this focus on individualized, close instruction between faculty and students, exemplified by its tutorial program, while it has opened itself to the world socially, academically, and philosophically.

Williams moved away from organized religious observance in the early 1960s, while developing a residential house system to take the place of fraternities, thus opening up the campus to a more unifying housing and social environment. Coeducation arrived in Williamstown in 1970, making life more balanced and bearable for undergraduates spending their time 135 miles from Boston and 165 miles from New York City. Today, Williams is a fully balanced male/female environment, with a strong athletic tradition in sixteen varsity intercollegiate sports each for men and women, which attract some 40 percent of the student body. Many JV, club, and intramural sports offer additional sporting options.

Set in the great outdoors, Williams maintains an immaculate 450-acre campus and several thousand additional acres nearby. Williamstown itself is small, but the college encompasses some 100 buildings, including an observatory, a renowned art museum featuring more than 12,000 works, a theater and dance center, several libraries housing almost a million volumes as well as thousands of multimedia archives, topflight athletic facilities, and the new Paresky student center. Not all students are the outdoorsy type, but most not only tolerate but revel in the college's surroundings. Skiing on trails or in the backcountry is close by, and hiking up Mount Greylock and up and down the Berkshires is easily accessible. Nearby North Adams and Adams offer shopping and some additional social outlets. Students live in forty different residences and participate in more than 100 clubs and other organizations.

Williams employs 312 full-time faculty teaching in twenty-four departments offering thirty-three majors, as well as various concentrations and programs. The student-faculty ratio is just seven to one, and 91 percent of Williams students will graduate in four years (96 percent within six years). Williams's most popular majors include economics, English, psychology, art, political science, biology, history, and mathematics. Williams students have many exchange and study-abroad opportunities, some 150 overall, including the college's special program with Exeter College at Oxford in England, Williams in New York, and a maritime studies option at Mystic. Around 50 percent of juniors will take advantage of an off-campus study program each year.

The 27,000 or so alumni of Williams are legendary in their devotion to the college, participating at high levels in fundraising efforts, serving as ambassadors, and providing an extensive career and graduate school network. The substantial endowment Williams has amassed, one of the very few at a small college in excess of $1 billion, enables the admissions office to recruit widely and provide significant financial aid to any student with identified need. Wil-

liams is not only need-blind, but guarantees to meet 100 percent of need with no loans. With half the student body receiving aid, 60 percent of all students coming from public high schools, students of color comprising a third of the college, and 7 percent more coming from abroad, Williams is a surprisingly diverse college that is quite different than it was a generation ago.

What the College Stands For

Williams's mission statement captures the essence of a residential liberal arts college: "Williams seeks to provide the finest possible liberal arts education by nurturing in students the academic and civic virtues, and their related traits of character. Academic virtues include the capacities to explore widely and deeply, think critically, reason empirically, express clearly, and connect ideas creatively. Civic virtues include commitment to engage both the broad public realm and community life, and the skills to do so effectively. These virtues, in turn, have associated traits of character. For example, free inquiry requires open-mindedness, and commitment to community draws on concern for others." There is little doubt that Williams continues to emphasize today its core competence in connecting faculty with students, and students with one another. Teaching is at the center of the college experience, with faculty "totally dedicated to undergraduate teaching."

Within that framework, Williams seeks to inculcate in its campus community a love for learning; continual interactions between students, faculty, and administration; and a balanced approach to education. That balance includes a physically active lifestyle, whether in a competitive or recreational sport, enjoying the great outdoors, or traveling the world. The balance also involves the arts, especially visual arts. The balance means a multiplicity of students and perspectives from across the country and around the world. "Today, Williams is multicultural, ecumenical, worldly, and world-renowned," says the college. "Today, we believe the privilege is ours to welcome students from every one of the United States, many of the world's countries, and every possible background. . . . Our hope is that after four years of discovering, sharing, and discussing ideas together, students emerge ready to contribute to a world they more fully understand as a direct result of the people they met at Williams."

In its faculty, Williams seeks scholar-teachers, promoting and rewarding not only publications, but "providing each Williams student an enriching and supportive educational experience." Faculty engage in cutting-edge research, while sitting across the table from students in small group or one-on-one settings to share new knowledge and teach core subjects. Williams maintains an honor code and honor system that govern academic honesty and student interactions on campus. All this is conducted within a spirited and tradition-full setting. From Mountain Day—a surprise day in October when the president rings a bell sending students up Mount Greylock—to Ivy Planting (on the day before commencement) and Watch Dropping (from the college chapel), the college celebrates its heritage and community.

Curriculum, Academic Life, and Unique Programs of Study

The curriculum at Williams is broad, allowing for a survey of the classic arts and sciences, as well as flexible, permitting students to engage in more rigorous study in interdepartmental areas or topics of their own devising. The departments offering majors include the traditional core disciplines, but also less common programs, such as astronomy and astrophysics; classics; Asian studies, Japanese, and Chinese; geosciences; political economy; Russian; and women's and gender studies. Concentrations are available in areas as diverse as Africana studies, cognitive science, environmental studies, maritime studies, and leadership studies. Students may pursue studies in Arabic, dance, teaching, and other areas. Two small master of arts programs offer specialized graduate degrees in policy economics and art history.

Needless to say, the Williams student will not run out of course choices or academic opportunities at this small institution or on one of its study-away or specialized off-campus or on-campus programs. The central defining component of the Williams educational experience remains the tutorial program. Offered in every department, tutorials "aim to teach students how to develop and present arguments; listen carefully, and then refine their positions in the context of a challenging discussion; and respond quickly and coherently to critiques of their work." Modeled on the system common to many British colleges, tutorials enroll a group of ten or so students, who are then divided into pairs who meet with professors once a week for an hour. The students take turns preparing a paper on assigned topics and critiquing the other's work. The college offers sixty tutorials, which are quite popular in terms of student enrollment. From "Fin-de-Siecle Russia" to "Machine Learning" to "Between Art and Cinema," the topics and methods of conducting the tutorials vary widely. Says one student quoted by the admissions office, "The tutorial program at Williams . . . has been one of the defining characteristics of my Williams education. I've taken several tutorials in the sciences and humanities and no two have been alike, save that they've been extremely rewarding. While the basic format remains more or less fixed, each tutorial I've taken has exposed me to new ideas and new ways of learning."

To earn a degree at Williams, a student must complete thirty-two semester courses, four Winter Study projects, and four physical education activities. A distribution requirement comprises four components: divisional, including three courses each in languages and arts, social studies, and science and mathematics; one course in exploring diversity; writing, including two writing intensive classes; and one course in quantitative/formal reasoning. Thus all students will graduate with a fairly broad academic background. The Winter Study projects offer a wide variety of random three-week opportunities graded on an honors/pass/fail basis, both on and off campus.

To support its diversity efforts and encourage understanding and interaction on campus, Williams offers a Multicultural Center (MCC), a Jewish religious center, and space for a Queer Student Union, Black Student Union, and International Club. The MCC holds lunch forums on a regular basis, providing students, faculty, and visiting speakers the opportunity to present papers and engage in dialogue.

The Center for Environmental Studies, founded in 1967, is an anchor for Williams's extensive focus on sustainability, environmental sciences, policy, and experiential learning. Williams students have access to hundreds of acres of wilderness in their backyard, including the Hopkins Memorial Forest next to campus. There is a field station for research in the forest, and also an Environmental Analysis Laboratory in the college's recently constructed Unified Science Center. Williams has an active

Outing Club, as well as a forest garden that is maintained by members of the college community.

Whether you are interested in spending a semester studying ocean and coastal studies at Mystic Seaport in Connecticut and on several field seminars in different oceans, living at Exeter College of Oxford University for a year as a visiting student, immersing yourself for a semester in New York City's rich cultural and academic life, or spending the academic years and summers in Williamstown getting all you can out of the college environment, you will find enormous challenge and opportunities at Williams.

Major Admissions Criteria

Prospective Williams applicants face some daunting odds in the admissions process, as the college is as selective as a number of Ivy League institutions, and among the most competitive of the Hidden Ivies. 7,548 men and women applied for the class of 2012, 1,276 were admitted (17 percent), and 535 enrolled. A large proportion (42 percent) of the class was admitted via early decision, during which 223 of 600 applicants were admitted (37 percent). Eighty-eight percent of the class ranked in the top decile. Thirty-six percent had an SAT critical reading score between 750 and 800, and another 25 percent had a score between 700 and 740. The mid-Atlantic states are home to the largest proportion of students (32 percent), with 21 percent coming from New England, 16 percent from the West, 10 percent each from the South and abroad, 7 percent from the Midwest, and 4 percent from the Southwest.

According to the admissions office profile, "Each candidate's folder is read cover to cover by two or three professional admissions officers. In our reading, we focus on the quality of the candidate's academic program, particularly as it relates to the challenges available at the student's secondary school, the strength of the academic record achieved, the quality of recommendations from teachers and guidance counselors, the results of standardized testing, strength of the candidate's personal statement, and documented nonacademic achievement. In order to make an accurate assessment of a student's academic record, it is extremely helpful to the admissions committee to receive a thorough profile of courses offered by the school, and some comparative information about achievement in context (class rank, grade distribution, etc.), even for the most talented and successful students."

The Ideal Student

Given the opportunities and academic challenges available at Williams and in its special study-away programs, the most successful student is likely to be one who is outgoing, energetic, fun loving, and community oriented, and who maintains both a wide range of academic interests and abilities and a passion for learning in a particular area of knowledge. As the college notes, "The quality of campus discourse at Williams, formal and informal, is an extension of the academic quality of the students we attract and admit. Conversation here is usually challenging and thought-provoking, invariably civil and well-informed. Every Williams student is exceptionally strong academically. But nonacademic achievements—from crusading to save a local wetland to making music with a punk-funk band—make for a lively campus, too. That's why, in selecting each incoming class, we look beyond the ste-

reotypical 'well-rounded student.' Instead, we look for those who bring a mix of passions, eccentricities, and ambitions to create a well-rounded campus community. If there is a single characteristic that sets Williams students apart from other highly talented students, it is their tendency to excel in more than one way."

Williams students might be hard-charging varsity athletes during part of the day, and art history scholars during another. They might be headed to work on Wall Street or in a prestigious law firm, or to teach in the inner city or serve in public office. They create new programs, get involved, and teach each other. Students who are too shy or unwilling to engage with a diverse group of talented classmates who will ask tough questions, help each other out, and work hard during the day while playing hard at night, are unlikely to get the most of their Williams experience.

Student Perspectives on Their Experience

Most students report being happy, engaged, and challenged at Williams. Ninety-seven percent return for their second year, so the college must be doing something right! Typically, we hear that students find the academic workload demanding no matter the discipline they pursue, and that falling behind is not an option. Nevertheless, the community ethos is one of mutual support, friendship, teamwork, and self-reliance. Students respect the long-standing honor code, which promotes self-scheduled exam work and collaboration among classmates. Many feel well supported by their junior advisors (JAs), faculty, and each other, including the numerous students who work as peer tutors and in other service roles.

Students are happy with their housing and the fact that almost everyone lives together right on campus. They marvel at their physical surroundings and the special "sense of place" it engenders. Most find more than enough to do on campus, but relish the occasional opportunity to road-trip to rivals Amherst or Wesleyan, or weekend excursions in New York or Boston. They love the Winter Term, and the opportunity these special classes provide for on- and off-campus study (the term is used loosely) during the tough New England month of January.

School spirit is highly rated, and it seems impossible not to be caught up in the traditions of the college, from Mountain Day to the big games between the Ephs (Williams's nickname, a.k.a. the Purple Cows) and the Lord Jeffs of Amherst. Students are proud of their athletic prowess, as well as their institutional prestige, though many find that away from certain areas, Williams College might not ring a bell in conversation. Yet they appreciate the super-strong alumni network they will become a part of upon graduation, with so many Williams graduates holding prominent positions in law, medicine, business, the arts, and academia.

Social life revolves around dorm parties and some trips to local taverns, with alcohol readily available despite the lack of fraternities. Students report widespread partying, but not significant social pressure to partake. Options such as concerts, films, and talks provide other entertainment outlets.

For those seeking an intimate, highly personal, and bonded collegiate experience, Williams doesn't fail to please. Students will gladly boast of having turned down Harvard or another major Ivy or Hidden Ivy university for the benefits of a Williams education with few regrets.

What Happens after College

The college notes that graduates are expected, following their liberal arts education at Williams, to serve society and lead enriched lives: "From this holistic immersion students learn more than they will ever know. Such is the testimony of countless graduates—that their Williams experience has equipped them to live fuller, more effective lives. Ultimately, the college's greatest mark on the world consists of this: the contributions our alumni make in their professions, their communities, and their personal lives." According to the college, 95 percent of seniors surveyed "feel that their Williams education has contributed significantly to their ability to think critically and analytically."

More than 150 employers and graduate schools recruit on the Williams campus each year, and the college boasts these impressive numbers: a 90 percent medical school acceptance rate; a 99 percent admission rate to law, business, or science Ph.D. programs; 40 percent of recent alumni earned their medical degree from a top-25 school; 78 percent earned their business degree from a top-25 school; 55 percent earned their law degree from a top-25 school. Some 20 percent of Williams graduates enroll directly in graduate or professional school. More than 70 percent will do so after five to ten years.

FOUR

As college admissions time approaches, it is the rare parent who does not feel a wave of potentially debilitating emotions, from confusion, bewilderment, and impotence to varying degrees of panic and desperation. If these emotions strike a chord of recognition, you are ready to begin the admissions process! Let us encourage you to see the admissions search as a joyous occasion; that's right, a time for celebrating the opportunity that awaits you.

As you consider the many milestones in your child's life in which you have played a significant role already, you can feel confident that you will also meet this phase of his or her development successfully. Remember that you helped him to accomplish each of the previous milestones in good order, otherwise he would not be at this point in his journey to adulthood and independence in preparing to enter college.

Over the course of their high school junior and senior years, you will have the opportunity to communicate . . . ok, you will have to work hard to extract tentative ideas and concerns about the next phase of education from your children. They are counting on you to be available and supportive throughout the search, although at times they will resist any offer of help. This will be as critical a parental role as the early years of teaching them to take care of themselves. Imagine the opportunity you now have to support them in determining what they want their future to be, what skills and talents they can build on, and what kind of living/learning experiences they can grow from next.

Most parents who have experienced the college-search process can recount the good times they had visiting campuses around the country with their children. They remember how the kids would talk about everything while traveling in a car, even to them. Haven't you had the sense that the car was in some way a sanctuary or neutral ground where all topics small and large could be articulated? Wasn't this the setting in which you first learned of their triumphs and travails? They now may fill up all of the backseat of the car or may even be the person behind the wheel, but they still have that marvelous habit of looking straight ahead and sharing their thoughts and asking those sensitive questions. You can have a terrific time planning a tour of a group of college campuses and doing the road trip together.

With the advantage of many years of counseling parents and adolescents on their educational futures, we can share with you a number of useful tips to help you through this stressful experience. We can tell you how to support your son or daughter as he or she confronts the reality of leaving home for college. Try to imagine the emotions churning in your child's stomach as she contemplates the hurdles she faces: getting good grades in junior and senior years as she learns that her performance will define the opportunity to get into a good college; taking hours of timed tests that are seen as a measure of ability; making decisions about interests and educational choices based on little exposure and experience; and, finally, being judged by a jury of strangers who can determine her future!

What can you do to keep the family relationships amicable and your child from falling into his or her own state of panic and despair? Here are some recommendations that have helped the thousands of families whom we have counseled.

Lessons to Take to Heart

- Keep in mind that there is no one perfect institution for your child. No one college can meet the special interests and needs of every young woman or man between the ages of eighteen and twenty-two! Try to get past the rankings as your major source of

information and influence. There are dozens of outstanding colleges that can meet a good student's intellectual, academic, social, and personal interests and goals. Focus on what factors matter most for your child, taking into account the criteria we have delineated in chapter 2 in identifying worthy colleges that may not be household names to you. You will quickly learn that there is a large selection of excellent colleges that will potentially be available to your child. This can remove a good deal of the stress around getting accepted. Any of the colleges in *The Hidden Ivies* will provide the quality of education and growth you would desire for your child, as will many other institutions.

- Avoid the stampede syndrome. Groups of students, and sometimes their parents, suddenly decide with limited or dubious information what is the hot school of the decade. They treat the one or several colleges as the only place they can possibly be happy. Then they become dismayed that they will have great difficulty gaining admission. Meanwhile they may have missed the opportunity to consider carefully the host of other great colleges that would suit them well. We also refer to this as the bunching effect, which really does result in talented applicants neutralizing one another so nobody wins a coveted place.

- As the world turns ever more quickly into a knowledge-based economic community and advanced training is required for personal success and security, more specialized knowledge is required. Thus the rise in graduate school enrollments across all fields of knowledge. Undergraduate education, as we describe in chapter 1, is a step on the road to further advanced education. Those who have researched well and chosen a college that suits their abilities are the most likely to gain entrance into the select graduate schools across the nation. Not only will they gain an excellent and personally rewarding education, they will have the opportunity to explore many fields of study to decide what suits them best.

- The cost of an education, which defies Newtonian principles of gravity, will continue to rise at an annual rate far beyond the cost of living index and discretionary income for most families. Research the cost of various college options to find the best buys for your child and all potential sources of financial aid and awards to make a college diploma a reality. Cast the application net widely to ensure a good acceptance and aid catch. Keep in mind some of the criteria we have presented: The endowment will determine the scholarship generosity of any particular college, as will its founding mission to offer all or almost all deserving young men and women the opportunity to attend their institution. Accordingly, the top colleges try to maintain a policy of need-blind admissions. Also remember that the so-called tuition sticker price is not necessarily what you will be expected to pay. Academic and talent-based awards

are available at the greater number of colleges today. Keep in mind that the majority of financial aid is given on the basis of need. So go for it!

- Admissions professionals look very positively upon an extra year of high school for the student who wishes to build stronger academic credentials, and in some cases stronger muscles if they are a very competitive athlete, or gain greater maturity. In a number of cases, candidates have spent an interim year working in a responsible position or volunteering in a community service enterprise at home or abroad. We have counseled many students to act on their desire to travel abroad to perfect their language skills or work intensely with a professional art or music instructor to prepare for advanced studies later on. For anyone considering a year between high school and college, rest assured that you will meet with a favorable response from the decision makers in the colleges.

- Study the comments of the admissions officers and leaders of each college, like those presented in the individual profiles here, to help determine the rightness of the particular college for your child. A group of colleges may offer similar high standards of education and facilities and resources, but they can attract a certain type of student and deliver a unique style of education that may or may not match with your son or daughter. Note which colleges place a higher or lesser premium on test scores or other criteria in choosing their entering class each year.

- Reconsider and reconcile your dreams and ambitions for your son or daughter as you work with them through the search process. You are the adult most able to help make a realistic evaluation of your child's abilities and motivation. Over-reaching can affect their confidence for the long term, while they can lose out on many good options in the short term. This can lead to frustration and failure if they find themselves in the wrong college.

- *Visit, visit, visit*! We emphasize every day to families the importance of visiting a group of colleges of differing sizes, styles, and locales in the early stages, preferably taking the initial step onto campuses in the junior year. This is the only way a student can determine for herself what she really feels most comfortable with. High school juniors frequently start with roman-tic or exaggerated notions of what they want the college years to be like. Some typical articulations we hear: "I am tired of the weather, the people, the sameness of where I have grown up, so I want to go across the country to the other coast," or "I want to go the largest college I can identify because I am in a small high school and I know everybody all too well; my high school athletic teams are terrible so I want to go to a college with nationally ranked athletic teams that I can cheer for; I hate the structured curriculum I have had to take for four years so I want a very progressive college where I can take anything I want and have no requirements to fulfill." Well, you can rest easily if these dialogues arise at home. The great majority of students figure out in time,

especially if they have had on-site visits, that there is more to choosing a college than reacting to their high school experi-ence.

MORE ON COLLEGE RETENTION

A number of elaborate studies in recent years indicate that a sense of *social isolation* is a significant factor in students dropping out of college, and sometimes it is the best predic-tor of failure to complete a college diploma. Other studies have confirmed that emotional and social adjustment to a new environment are more accurate predictors of dropping out or successful completion than academic fac-tors. Confidence in one's abilities is another significant factor that affects success. A self-perception of being over one's head academi-cally or socially can readily lead to failure. Further, interaction both in and out of the classroom, satisfaction with the quality and content of courses undertaken, and a sense of self-confidence affect more positively the determination to finish college.

A commitment to graduating from the par-ticular college and to the institution itself translates to greater persistence to remain and receive the diploma. The obvious factors of adequate facilities, living arrangements, and financial support can affect a student's attitude toward his or her college. A percep-tion of the college as a good place that cares about its students, especially as evidenced by faculty and counselor availability and interest, will have a direct relationship to a college's retention of its students. The inability to cope with adjustment socially and academically re-sults in a greater likelihood of dropping out.

Here is further evidence of the vital importance of choosing the right college. What is
considered a good college for one student
may be entirely wrong for the next.

Practical Help a Parent Can Provide

- Help your child collect all useful resources, such as responsibly written books
  on the college admissions process, direc-
  tories of the colleges, college videos and
  Web sites, information from your child's
  high school regarding recent admissions
  results for its graduates, and articles from
  newspapers and magazines.

- Provide adequate computer capability to
  allow for browsing the Internet to collect
  up-to-date information on individual
  colleges; financial-aid resources; special
  academic, performing arts, athletic, and
  extracurricular activities of interest; and
  for downloading and completing admissions applications.

- Understand how the admissions process
  works, what is required to qualify, what
  testing is needed, and what sources of
  help for counseling and test-taking
  practice are available in your area. You
  are then able to gently guide, encourage,
  and answer questions for your child. You
  will have to dodge a number of verbal
  bullets or dark glares along the way, but
  most of your input will be taken into
  account in time.

- Put on your chauffeur's hat, top off the gas
  tank, and stand ready to get on the road to
  various college campuses. Most students

do appreciate and want their parents to
share in this adventure.

- Keep in mind that one student's opinion
  does not an entire college make! Be
  careful of the story of a friend or neighbor's child being unhappy at a college you
  and your child consider to be right on
  target. Find out the facts behind the story
  and judge the college on the basis of all
  the sources we suggest here.

Emotional Assistance a Parent Can Provide

- Share openly and honestly with your child
  your expectations for college for them. In
  most cases students believe their parents
  have unusually high ambitions and goals
  that they feel will be impossible to meet.
  This is why many students retreat from
  the process prematurely with a defeated
  attitude.

- Be realistic in helping your child evaluate
  his strengths, maturity, readiness for
  college, and the degree of competition and
  demands he can meet with success.

- Be the head of the cheerleading section.
  There are enough critics out there telling
  your child what she has not done right or
  why she will not qualify for a competitive
  college. Be as positive as you can be and
  leave the yet unmet achievements alone.

- Learn patience. Your son or daughter will
  go through a series of fits and starts,
  optimistic moments followed by crashes as
  they make their way through the steps
  toward applying. Their procrastination in

meeting deadlines, their mood swings, their failure to communicate with you at various times are all part of the proceedings. They are counting on you to go with the flow, otherwise known as possessing the patience of the saints.

- Do not make the dinner table the arena for college talk. While this is one of the few places and times when a family sits down together—or are we making this up?—you can be certain your child will disappear if he finds that the only topic of discussion is colleges and what he has done to move himself forward. Let the food table be a refuge and a comfort from the realities of this stressful time. In its place, set an appointment in another place to go through the schedule of things to be done in an appropriate order.

- Ask what you can do to ease the pressure and confusion. You might be surprised by the positive response you get when the offer to help is presented in this fashion. It will be perceived as less intrusive. For some students it will be an opportunity to express their self-doubts and worries; others may be concerned with issues of time management and executive skills to get the job done.

- Share openly your family financial situation. Will financial aid or a part-time job be necessary for your child to attend college? We are no longer surprised by the many students who have no idea if they will need to apply for scholarships, loans, or work-study programs. Give them a reasonable idea of how much from savings and income you plan to put to college

costs. This will alleviate confusion and make everyone aware of potential financial support from the beginning.

- Be thee not proud! Yell for assistance from the school guidance staff or, if not available, from professional outside counselors and test evaluators if learning issues are a possible cause of poor test and/or grade performance. A club advisor or athletic coach can often be a good source of information on college opportunities as well as a supporter of your son or daughter when they apply.

- Take every opportunity to get a good night's sleep. You will need it as you get farther along in the admissions steps. And you will be better able to remember the recommendations you have just read.

- Live in a state of suspended disbelief through the process! Have faith in your child and his ability to be accepted at a college that truly reflects the hard work he has put into his high school studies and activities. Have confidence that there is a logical, sequential set of steps that will increase his chances of getting into his several top choices. Explore college options beyond the "usual suspects" or familiar names.

FROM VETERAN PARENTS: WHAT WE WISH WE HAD KNOWN WHEN WE BEGAN THE ADMISSIONS PROCESS

1. We thought our son understood that we only wanted him to attend a college where he would be happy and success-

ful, rather than a name college where he might be unhappy while we basked in the glow of its reputation.

2. We thought he understood how much we could afford to pay for his college education; we did not realize it would help to share our family financial situation with him so that he could take advantage of the financial-aid opportunities available today.

3. We did not understand that the great majority of private colleges provide larger amounts of financial aid than do public institutions, to make up the great difference in tuition costs. We considered allowing him to apply only to state colleges.

4. We had no idea that college admissions officers welcomed parents to participate in the visits to campus for tours and information sessions.

5. We did not realize that the selective colleges are actively seeking disadvantaged and minority students, that our child could have a potential advantage in gaining admission rather than being discriminated against. All of the recent publicity about affirmative action being done away with is not true for the private colleges.

6. We wish we had known earlier that 80 percent of American colleges and universities offer scholarship grants to talented students regardless of the parents' financial circumstances. We thought only the very needy student could receive aid.

7. We needed someone to explain the differences among all those early application plans and if any of them would help get our child get into a good college. We now understand what early-decision and early-action and rolling-admission plans are and the advantages each can have in admission decisions.

8. We thought that a student could get accepted into a graduate school only if she attended the most prestigious college. We should have known that what really counts is doing very well in good courses in a respected college, that a high grade point average gives her the real advantage.

9. We did not know that major colleges and universities actively encourage students who have demonstrated academic excellence to apply for transfer admission, that a student could start at a college at a comfortable and realistic level for him, do well, and then transfer.

10. We eliminated a number of college choices that our daughter was really interested in because her test scores did not match the average score the guidebooks listed. We now understand that strong grades in top courses and special talents far outweigh the importance of the admissions tests like the SAT.

11. We thought a student had to participate in many different school activities and play a sport. The colleges, we have learned, far prefer the student who has shown major dedication to one or two activities and assumed a leadership role in them. We also saw that only a few terrific athletes were helped to get into college because of their sports involvement.

12. We pushed our child too hard and controlled too much of the admissions

process, taking too much of the decision making and responsibility from him.

Advice for Students

- Be conscious of the fact that nothing remains the same in life. What you consider to be your future major interests and goals right now are likely to change as you encounter new ideas and fields of study through your college years. The career you may have in mind today may not even exist by the time you graduate, while new jobs and careers are being created at an exponential rate. In the 1930s, Alfred North Whitehead, one of the great educator/philosophers of the twentieth century, wrote this about change: What differentiated the present from the recent past was a rate of progress that would call upon an individual to face new situations at a pace which found no parallel in history. "The fixed person for the fixed duties who, in older societies, was such a godsend, in the future will be a public danger." What he was saying is that a person trained to do one thing well will quickly become obsolete. His words have more relevance today than ever before. You need to remain flexible and keep an open mind regarding what you study in college and what your future will, in fact, turn out to be. Thus, do not choose a college solely or primarily because it offers a particular academic discipline.

- Be sure the colleges you are investigating provide a broad range of fields of study and extracurricular activities. Trust your instincts and listen to your inner voice rather than that of your peers, even your friends. You can ask for their thoughts, but ultimately it is up to you to determine what is the most comfortable and best place for you. Be careful about choosing a college because a boy/girlfriend is going there or will be nearby.

- It is ok to feel anxious as you launch yourself into the admissions game. We would be more concerned if you were not uptight as you join the competition for a place at the selective colleges. The best way to manage the tension is to turn this energy force into action. Get going and do what is necessary, step by step, and watch how much better you will feel.

- Do not look over your shoulder at what your classmates are planning to do for college or compare their grades and test scores to yours unfavorably. A college of major interest to you can decide to admit you and students with very different profiles if you are all qualified. Decide what you want to strive for and what type of college will let you reach your full potential and simply encourage your friends and peers to do the same for themselves, thank you very much.

- Start a list of your special talents, accomplishments, and the contributions you feel you can make to a college community. Keep adding to this list as you enter your senior year. You should see that you have more to offer than you probably realized. As you read the college materials pay attention to what characteristics and skills they look for in reviewing applicants and determine which of the colleges match up strongly with your list of strengths and interests.

- Avoid a focus on names and rumors about individual colleges. Suspend your disbelief, just as your parents should, and instead explore many college choices; make your own preference known, and above all, look for a place where you will find the programs you want to pursue. You can "survive" and attain a good education in many institutions, but discover the one where you will be most comfortable, challenged, stimulated, and successful.

- Can you afford these colleges? We hope that it has been clear that you should never say never. A clear theme throughout these pages has been the enormous amount of need-based (and in many cases also merit-based) financial aid available at the Hidden Ivies and their great interest in attracting a diverse student body. Apply first. If you are accepted, these schools will try to make it work for you.

How to Make It into a Hidden Ivy: Use the Five Ps

How do admissions officers at the nation's top colleges and universities decide which students to admit from the large pool of outstanding applicants? As you have learned from the comments of the admissions directors and college officials quoted throughout chapter 3, there are certain factors that count most in identifying students they would like to have in their colleges. One way to understand the admissions decision process and to determine how you can

improve your odds of being accepted is to concentrate on what we call the Five Ps.

Program

If you are determined to be accepted into a competitive college you should take the most rigorous course load you can handle, starting freshman year of high school. You should tailor a demanding curriculum with a focus on your particular strengths and interests. If you got off to a slower start in the first two years of high school, you can make up for this by taking as many enriched courses in the junior and senior years as you can manage comfortably. You can even use the summers to enhance your academic profile and explore potential future interests. The Hidden Ivies make it clear that they look for students who have taken advantage of academic opportunities in and beyond high school and who have well-developed intellectual interests and pursuits.

Example:

A prospective English major should take Advanced Placement English, literature, and possibly history, but not necessarily Advanced Placement physics or chemistry if these are areas of weakness.

Performance

You have now heard from every admissions director at the Hidden Ivies that grades matter greatly as they choose applicants from the mass of candidates. Outstanding performance in a high-level curriculum remains the greatest single factor in a student's chances for admission to an elite college. Serious students should aim for an overall grade point average of 3.5 or higher. Keep in mind that a rigorous curriculum will always show better than a higher average from less challenging courses. It is not necessary to take every honors or advanced placement course your school offers. Concentrate on those that are of great interest to you and that will allow you to excel. Finding a balance is the overriding theme here.

Preparation

By junior year students should know what tests they need to take in order to be considered by their potential colleges of choice, and they should begin making plans to take them at the most advantageous time. Follow this rule: Virtually all selective colleges require the SAT and two or three SAT tests. The alternative testing program that most will accept is the ACT. The Math level I or II is the two most preferred subject test. The second and third subject should be your choice and should be those which will give you your best results. You may take additional SAT subject tests to demonstrate your enriched academic background. You should also check with individual colleges of interest to know if there are more specific test requirements. AP tests are another way for students to show course mastery, and those with a language other than English as a first language will need to take the TOEFL. You should also know if the college programs in which you are interested require particular high school subjects for consideration. If standardized tests like the SAT are not your cup of tea, remember that some of the Hidden Ivies and other colleges have made them optional.

Example:

The senior applicant who is short of the required tests is not likely to be considered, as this re-flects a lack of planning or lack of qualifications to be taken seriously as a candidate. If in the application you express a serious interest in specific fields of study, you must demonstrate this in your choice of courses in the last two years of high school. An intended engineering or science major should have advanced math, chemistry, and physics; a prelaw major a strong English, history, and economics background; an international relations major a strong language(s) and social studies foundation. And students should have standardized test scores in these areas to help support their applications.

Passion

The elite colleges are in the enviable position of selecting only highly intelligent candidates to fill their classes. But they also want young men and women who pursue nonacademic interests with an energy and a passion that make them special. Students who are talented musicians, artists, athletes, community caregivers, school leaders, writers, and so forth make a college campus a rich and exciting environment. This is why students want to attend one of these colleges in the first place.

Parents should be sensitive in the early stages of middle school and high school to those activities in which their child shows special interest or talent and help them to cultivate these. They should help them find outlets whether within their school community or outside it and encourage them to pursue these interests with energy and delight. The smartest course is to drop those activities of minor interest and to pursue those of major interest and enjoyment. The goal is to demonstrate talent, leadership, and commitment. Many students make the mistake of joining a wide range of activities and accomplishing little of note in any of them.

Summertime presents a great window of opportunity to strengthen a skill or interest. Here is where research into exciting and relevant programs can be of immeasurable help.

For students, the rule of thumb should always be: Do what you enjoy, and do a lot of it. Yes, you may do some activities because they are required, or are "the right thing to do." But try to avoid involving yourself in something because you think it will look good to a college. And when you find one or two areas that you truly enjoy, pursue them with commitment, dedication, and fun.

Presentation

The personal interview is for many colleges a thing of the past. Few top colleges require a personal meeting with an admissions officer on campus today. They do, however, encourage a campus visit that includes a student-led tour, a group informational session with an admissions representative, and possibly a meeting with a coach or professor in the student's special interest area. Check with the particular colleges to learn if they do interview on campus, and also consider alumni interviews, which are typically offered once a student has submitted an application.

What has become of indispensable importance to the admissions committee is the personal application. Every top college requires applicants to write several personal statements and include an activity résumé in order to provide insight into their personal qualities and strengths. Most admissions officers place a great deal of emphasis on the personal application. Their feeling is that the essays are very revealing about a student's character, interests, experiences, special circumstances, and, of equal importance, their ability to communicate in writing. We can attest to the fact that many qualified applicants have been rejected by their dream college because of the content or poor quality of their presentation.

You should write about topics that are most meaningful to you, rather than those you think will impress the admissions committee. You should have a trusted teacher or counselor or mature friend read your essays and full application for critical feedback. What is most important is to convey who you are, what the ingredients are that make you the whole person that you know you are and what you will become. Grammar, syntax, and spelling matter greatly, as well, but it is the essence of who you are and what you have accomplished that will make the difference among equally qualified candidates.

It can also be invaluable to send with your application supplemental materials that demonstrate your passions and talents. The admissions committee members are not put off by this. Quite the contrary, they are helped in comparing candidates who profess to have certain talents by evidence of their work and the positive addition this can bring to their campus.

Example:

A résumé of theater productions and training, a musical résumé and tape of a performance, a portfolio of art or photography, a bound set of writing pieces, or an athletic résumé and film of competitive play can have a significant impact on the committee. Letters of recommendation from appropriate teachers, advisors, employers, or coaches are also of value.

Your Charge

The great Irish poet W. B. Yeats described education not as the filling of a pail with information and facts, but as the igniting of a flame.

We hope we have made the point throughout this book that the happiest and most fulfilled students are those who experience a faculty passionate about teaching and mentoring, administrators and advisors who care about supporting students in their personal needs, and fellow students who have come to learn in the broadest meaning of this term. These are the key factors that can ignite the love of learning and of exploring new ideas and interests with other talented men and women who inhabit the best of the colleges. How do you gauge if the choice of college is a correct one? Here is what a parent will want to hear from a son or daughter once he or she is enrolled in college; it is also what a student might expect to be thinking. Parents, don't be fussy regarding the means of communication: Count your blessings if you hear from your child by telephone, e-mail, or even via the back of a napkin from the local fast food restaurant:

1. The work is really hard here, but I like my teachers, especially Professor X in the department I think I am going to major in, because he knows his subject and is enthusiastic and energetic in the classroom.

2. Three out of four of my classes are pretty small so I can ask a lot of questions and also learn what the other kids are thinking. The teachers are cool; the first day they gave us their e-mail addresses and a couple even gave out their home telephone numbers in case we have a question or a problem.

3. I can always get to my professor when I have a problem with an assignment. He has already been really helpful. I know I can always get in touch with all my teachers when I really need to and they will be willing to meet with me.

4. I was invited to assist my science (read any subject here) instructor in the research she is carrying out on a grant. I will have the opportunity to work with her on the experiments she is doing and to help write up the findings.

5. I had a great meeting with my advisor to help me plan my courses for the next couple of terms. This way I can keep a number of options for my major open and meet the requirements for those I am considering.

6. I followed up on my promise and met with the director of financial aid today. He's really friendly and sympathetic to my concern about the number of hours I have to work to pay for all those books I had to buy this term. He is going to look into my aid package and see if there are any extra dollars in their budget to allow me to cut back on my job.

7. I'm impressed by how hardworking and smart the kids are. It makes me nervous, to be honest. But I know I can handle the work and I find it exciting to talk about a book or topic we just covered in class in the student center or the dorm. I'm glad that I took those AP courses last year so I can keep up with the reading and writing they require here.

8. You know what? The admissions representative who came to the high school last year was on target. It's cool on this campus to be smart and study a lot and have a good social life at the same time.

9. I really like the fact that there are so many different kinds of people here. It is so different from high school where most

of my classmates seemed the same to me. I am meeting lots of new people with different ideas and opinions, so we can talk a lot about them but no one tries to jam their opinions down my throat.

10. Hey, Mom, the food here sure isn't like yours, but it will do. I have a bunch of options for eating around campus so I get to meet more people and we laugh together about the food. It's the kids here who lift your spirits every day.

11. Well, my roommate is definitely different. Count on that when you visit on Parents' Weekend. But you know, she has a lot of interesting ideas that are nothing like what my friends at home would think. I am learning a lot about her family and their culture. I think we will become good friends over time. We have our differences about things, like what music to play or when to have quiet study time, but we seem to work it out.

12. I'll tell you what is really amazing here. There are so many clubs and groups I can't decide which ones I want to join. I knew varsity sports would take up a ton of time so I am glad I can play on the intramural teams and still be able to write for the paper.

13. You guys will be proud of me. I went to the Career Counseling Office yesterday to find out what I need to do if I want to get into medical school. They had all this great information in a booklet. It tells me what courses I have to take and what exams I'd need to qualify. I also got information on how to pay for grad school and talked with a career and graduate admissions advisor about my options.

I think I am going to give it a try now that I know what I have to do.

14. Hey, I hope you won't be upset when you get my final grades for the term. This is a tough place and the professors push you to do your best if you want a top grade. I have learned an incredible amount so far this year in just about every course. I already know what I can do next year to improve. You know, I don't care too much about my grades even though I got mostly A's in high school because I like this place so much and I have made so many friends. Can't wait to see you and thanks for letting me choose to come here.

Now about that car we talked about at spring break . . .

A Concluding Note

We hope that the information we have presented throughout *The Hidden Ivies* has been helpful to students, parents, and others interested in college admissions and in learning about some of the wonderful academic environments offering an outstanding education in the liberal arts today. We would like to encourage students, in particular, to listen to their forebears as they describe their colleges. Look for themes, for the personalities of the schools, and the consistently reoccurring messages. These include calls for your active involvement, participation, going the extra mile, speaking out, taking chances, making connections, and making a difference. The Hidden Ivies are all unique, but they represent top quality liberal arts institutions, each in it's own way spurring students to think critically, to challenge themselves and others, to be leaders. These are communities, families in a way, that invite you in to join them, to contribute, and to take from them knowledge and the means to keep learning throughout your lifetime. These are not places where you can hide, shrink from tough academic work, or values-based decisions. They are not places to float through, where it will be easy to get lost in the crowd or avoid taking responsibility. If you give them a chance and put your heart into any one of these places, you will come out not only a better student, but a better person. Oh, and remember, it's all about the professors, your access to them, and their concern for and ability to inspire you.

APPENDIX I

The Hidden Ivies

Amherst College

Barnard College

Bates College

Boston College

Bowdoin College

Bryn Mawr College

Bucknell University

Carleton College

University of Chicago

Claremont McKenna College

Colby College

Colgate University

Colorado College

Davidson College

Duke University

Emory University

Georgetown University

Grinnell College

Hamilton College

Haverford College

The Johns Hopkins University

Kenyon College

Lafayette College

Lehigh University

Macalester College

Middlebury College

Mount Holyoke College

Northwestern University

University of Notre Dame

Oberlin College

Pomona College

Reed College

Rice University

University of Richmond

University of Rochester

Smith College

University of Southern California

Stanford University

Swarthmore College

Trinity College

Tufts University

Tulane University

Vanderbilt University

Vassar College

Wake Forest University

Washington and Lee University

Washington University in St. Louis

Wellesley College

Wesleyan University

Williams College

Other Colleges and Universities of Excellence

Brandeis University

Case Western Reserve University

Connecticut College

Denison University

DePauw University

Dickinson College

Franklin and Marshall College

Furman University

George Washington University

College of the Holy Cross

Lawrence University

St. Olaf College

Scripps College

Skidmore College

University of the South (Sewanee)

Southern Methodist University

Trinity University (Texas)

Union College

Wheaton College (Massachusetts)

Whitman College

Predominately Scientific and Technical Universities

California Institute of Technology

Carnegie Mellon University

Harvey Mudd College

Massachusetts Institute of Technology

Rensselaer Polytechnic Institute

Rochester Institute of Technology

Institutional Leaders Questionnaire

We sent this set of questions to the president, provost, chancellor, dean of admissions, director of admissions, and those officials responsible for academic affairs, alumni relations, community services, first-year students, minority students, placement and career services, public relations, residential life, and undergraduate studies, as appropriate, at each institution.

Name _____

Position _____

Years with YOUR COLLEGE?

Graduate of YOUR COLLEGE?
YES _____   NO _____

1. What are the educational goals of YOUR COLLEGE for your students?

2. How do you know when you have succeeded in accomplishing these goals with your students?

3. How has YOUR COLLEGE changed over time?

4. What has caused YOUR COLLEGE to change?

5. What helps students to succeed at YOUR COLLEGE?

6. How do you expect YOUR COLLEGE to change in the future? Where would you like to see the institution go?

7. Where do your students go after graduating from YOUR COLLEGE? What are they doing?

8. What programs, policies, or practices have been particularly important to YOUR COLLEGE? Why?

9. What qualities are you looking for in your students at YOUR COLLEGE, and who does best there?

10. What are the three most important criteria in the selection of applicants at YOUR COLLEGE?

11. What kind of cultural, ethnic, racial, socioeconomic, or other diversity exists on your campus, and how does YOUR COLLEGE make itself more accessible and friendly to students with different backgrounds and experiences?

12. What makes YOUR COLLEGE unique in the world of higher education?

13. What have we not asked that we should have, and what particular traits would you like families, students, and educators to know about YOUR COLLEGE?

Student Questionnaire

We sent this set of questions to about a dozen students on each campus, sometimes directly through the offices of admissions or public relations.

A cover letter to the students read, in part, "Our goal in surveying students at the colleges and universities we are writing about is to gain an accurate sampling of perspectives from a diverse set of students. We hope to hear from about a dozen students on each campus, representing current opinions from those who know their college well and who can speak to its unique attributes, strengths, and weaknesses. We are seeking sincere and thoughtful responses that will help prospective students and parents understand the 'personality' of each environment. Some of those we hope to hear from include: student govern-ment leaders, fraternity/sorority and residence hall leaders, international students, multicultural students, athletes, newspaper editors, scholarship students, and so forth. We hope to hear from first-year students through seniors, men and women (as appropriate), and campus leaders."

Name _____

College or university attending _____

Year _____

Membership and/or leadership positions in college or class organizations?

_____

_____

High school graduated from _____

Year _____

Please respond to the following questions as openly and thoroughly as possible, and use any additional space as necessary. Again, you may return the survey to us by e-mail, fax, or regular mail, as works best for you. We will use the information we gather to discuss your school in depth, and with the benefit of your knowledge and insights, we hope to capture its unique qualities. We will report responses in our book anonymously. We appreciate your time and your candor, and we value your contribution to this discussion of excellent institutions of higher education.

1   Why did you choose your college?

2.   What does the school do best?

3.   What does it do worst?

4.   Who does best at your school?

5.   What helps students to succeed there?

6.   What would you like to change about the college?

7.   What are your educational goals?

8.   What programs, policies, or practices have been particularly important to you at the college? Why?

9.   What do you plan to do after graduation?

10.   What kind of cultural, ethnic, racial, socioeconomic, or other diversity exists on your campus, and how does the college make itself more accessible and friendly to students with different backgrounds and experiences?

11.   What makes your college special?

12.   What haven't we asked that we should have, and what would you like families, students, and educators really to know about the school?

INDEX

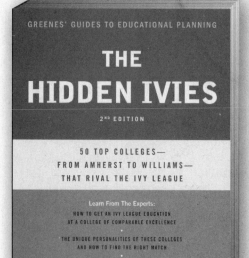